ROSS and WILSON

Foundations of
Anatomy and Physiology

Janet S. Ross

RGN   RFN   RNT

Sister Tutor Certificate (University of Edinburgh)

Formerly: Director of Nurse Training, North
Edinburgh School of Nursing; Principal Tutor,
Western General Hospital, Edinburgh; Examiner to
the General Nursing Council for Scotland;
Departmental Sister, Orthopaedic Department, Royal
Infirmary, Glasgow; Nursing Sister and Sister Tutor,
Colonial Nursing Service, Gold Coast.

Kathleen J. W. Wilson

OBE   BSc   PhD   RGN   SCM   RNT

Examiner, Degree courses for nurses, Leeds
Polytechnic and Chelsea College, University of
London; Member, Regional Nurse Training
Committee, West Midlands; Editor, *International
Journal of Nursing Studies*; Honorary Senior
Research Fellow, University of Birmingham.
Formerly: Senior Lecturer, Department of Nursing
Studies, University of Edinburgh; Visiting Professor,
Katholieke Universiteit, Leuven, Belgium (WHO);
Nursing Research Liaison Officer, West Midlands
Region; Principal Tutor, Royal Infirmary, Edinburgh.

ROSS and WILSON

Foundations of Anatomy and Physiology

Fifth Edition

Susan Schooling.

Revised by
Kathleen J. W. Wilson

CHURCHILL LIVINGSTONE
EDINBURGH LONDON MELBOURNE AND NEW YORK 1981

CHURCHILL LIVINGSTONE
Medical Division of Longman Group Limited

Distributed in the United States of America by
Churchill Livingstone Inc., 1560 Broadway,
New York, N.Y. 10036, and by associated companies,
branches and representatives throughout the world.

First Edition 1963
Second Edition 1966
Third Edition 1968
ELBS Edition first published 1971
Fourth Edition 1973
ELBS Edition of Fourth Edition 1973
Fifth Edition 1981
ELBS Edition of Fifth Edition 1981
Fifth Edition reprinted 1982 (twice)
ELBS Edition reprinted 1983
Reprinted 1983

ISBN 0 443 01681 X

British Library Cataloguing in Publication Data
  Ross, Janet Smith
    Foundations of anatomy and physiology. – 5th ed.
    1. Human physiology
    2. Anatomy, Human
    I. Title   II. Wilson, Kathleen Jean Wallace
    612′.002′4613   QP34.5   79–41706

Printed in Hong Kong by
C & C Joint Printing Co., (H.K.) LTD.

Preface to the Fifth Edition

In this edition quite fundamental changes have been made in the text, in the illustrations, and in the order of presentation of the material. A new first chapter has been written, using an approach to the subject which is different from that of previous editions and of the subsequent chapters of this edition. Its purpose is to outline the functions of the body from the point of view of the essential physiological principles which apply to all mammals, and to try to explain the need for specialisation of function in multicellular organisms. The illustrations of these functions are diagrammatic, and are not intended to be anatomically accurate.

A brief description of the skeletal system and some illustrations have been included at the end of Chapter 1 because so many of the 'landmarks' used in association with other systems are expressed in terms of bones. The more detailed description of bones and joints, and of the muscles which move them, has been transferred to the end of the book because it was considered more important to bring forward the chapters on the cardiovascular and lymphatic systems, i.e. the *internal transportation systems* as they are described in Chapter 1.

The opening page of each chapter contains an outline of its theoretical material. These outlines represent ways of thinking about the content and are not always arranged in the same order as the chapter itself.

All the illustrations from the Fourth Edition which have been retained have been redrawn with modifications appropriate to the revised text. Most of the new illustrations have been devised to explain physiological phenomena.

I am grateful to readers of the previous edition who have made helpful suggestions, and I have been pleased to note that most of these comments have been made by nurses in training.

I would like to express my thanks to Mrs June Clain, the artist, who illustrated this edition, and to Mrs Audrey Young, who typed large parts of the revised text, with both of whom it has been a pleasure to work; to Dr Jennifer Boore SRN and Mr Ian Kirkland FRCS for reading the text and making helpful suggestions; and to the staff of Churchill Livingstone for their constant encouragement and practical help throughout the preparation of this major revision.

Dilwyn, 1981   Kathleen J. W. Wilson

Contents

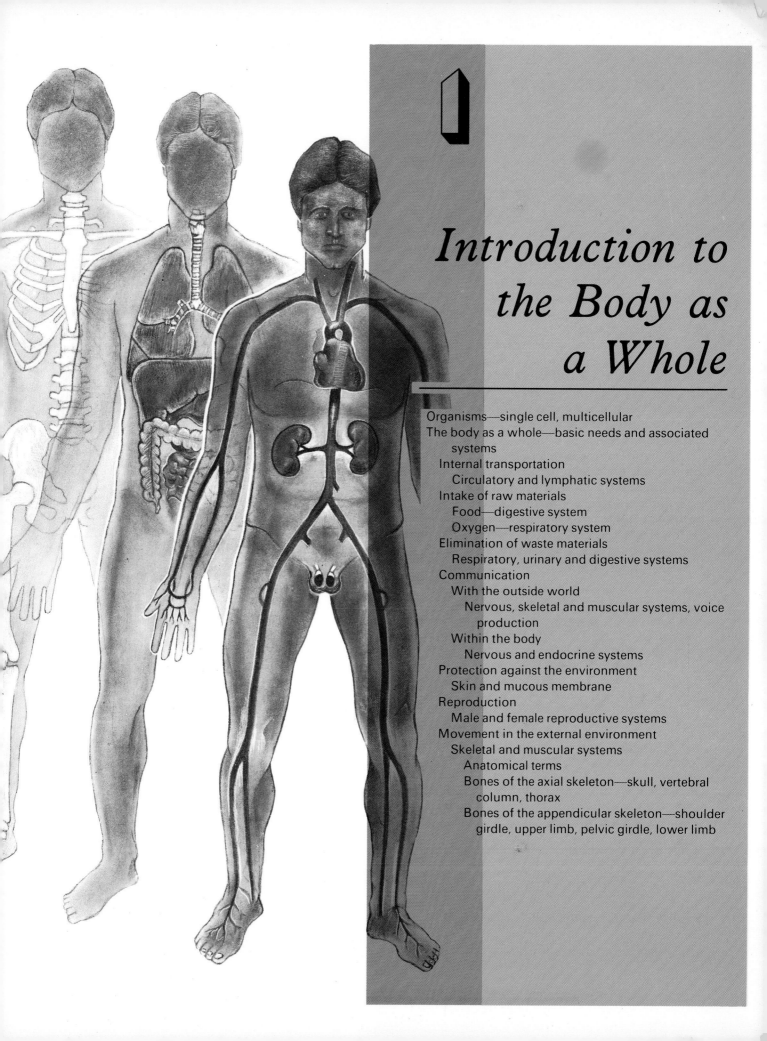

Introduction to the Body as a Whole

1. Introduction to the Body as a Whole

The human animal is a very complex multicellular organism in which the maintenance of life depends upon a vast number of physiological and biochemical activities. The sum of these activities enables the human being to live in and utilise his environment and to maintain the species by reproducing.

There is a considerable variation in the complexity of organisms. At one end of the continuum there is the *single cell organism* such as the amoeba, and at the other, the highly complex *multicellular human animal*. A cell is the smallest functional unit of an organism, thus a single cell organism is the simplest kind of organism that can exist independently.

In order to survive, each species, simple and complex, must be able to perform certain functions. A single cell organism can carry out all these functions because all the parts of the cell have easy access to its external environment. The inside of the organism is separated from its environment by only one porous membrane, which is described as *semipermeable* because small particles can pass through it while large ones cannot (Fig. 1:1). It is not

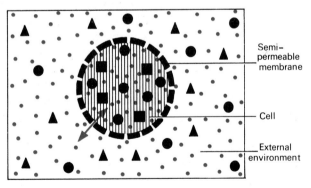

=Small particles able to pass through pores in the cell membrane

▲ =Large particles—outside—cannot pass into the cell

■ = Large particles—inside—cannot pass out of the cell

● = Large particles—inside and outside the cell which cannot pass through its membrane

Figure 1:1 Diagram of a single cell with a semipermeable membrane.

possible for all the cells of the human animal to be in close contact with the environment so, in order to survive, specialisation of cells has evolved and functional specialisa-

tion has taken place in parallel with structural specialisation. Figure 1:2 shows a multicellular structure.

The cells of the body are too small to be seen with the naked eye, but when magnified by using a microscope different types of cells can be distinguished by their size, shape and the types of dye by which they can be stained in the laboratory.

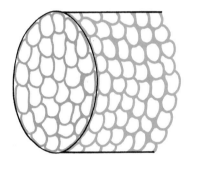

◯ =Cells in contact with internal environment

◯ =Cells not in contact with external environment

Figure 1:2 Diagram of a multicellular structure.

Groups of cells which have the same physical characteristics tend to have similar specialised functions, and a large number of cells which perform the same functions are described as *tissues*. *Organs* are made up of a number of different types of tissue, and *systems* consist of a number of organs and tissues. Each system contributes to the carrying out of one or more of the vital functions of the body. However, because of specialisation of cells, none of the systems can exist in isolation.

Basic Needs of the Body and Associated Systems

It will be noted from the list on page 3 that some of the systems contribute to a number of different basic needs. The contribution that each system makes to each need is described briefly in this chapter. Internal transportation is described first, because all the systems depend upon the circulation of blood. The description of the systems involved in meeting the other needs assumes that an effective circulatory system is operating.

A more detailed description of the structure and functions of each system is provided in later chapters.

Needs	System(s)
1. Internal transportation	Circulatory (Ch. 5)
	Lymphatic (Ch. 6)
2. Intake of raw materials	
A. Food	Digestive (Ch. 9)
B. Oxygen	Respiratory (Ch. 7)
3. Elimination of waste materials	Urinary (Ch. 10)
	Respiratory
	Digestive
4. Communication	
A. With the outside world	Nervous (Ch. 12)
	Respiratory—voice production
	Skeletal (Ch. 16)
	Muscular (Ch. 18)
B. Within the body	Nervous
	Endocrine (Ch. 14)
5. Protection against the external environment	Skin (Ch. 11)
	Membranes lining passages which open on to the surface of the body
6. Reproduction	Male and female reproductive (Ch. 15)
	Endocrine
7. Movement within the external environment	Skeletal
	Muscular

1. Internal Transportation

The need for functional specialisation among the cells of the body was mentioned earlier. Because of this specialisation, a sophisticated transport system is needed to ensure that all cells have access to the *external* and *internal environments* of the body.

1. The external environment surrounds the body and provides the oxygen and nutritional materials required by all the cells of the body. Waste products of cell activity are excreted into the external environment.
2. The internal environment provides chemical substances produced by specialised cells. All living cells in the body are bathed in fluid (interstitial fluid). Oxygen, nutritional materials, chemicals produced by the body, and waste products, pass through the interstitial fluid between the cells and the internal transportation system.

The systems involved are the *circulatory* and *lymphatic*.

THE CIRCULATORY SYSTEM (Chs. 3, 5 and 14. Figs. 1:3 and 1:4)

This system consists of the *blood*, the *blood vessels* and the *heart*.

THE BLOOD

This consists of two parts—a sticky fluid called *plasma* and *cells* which float in the plasma.

The plasma

This consists of water and chemical substances dissolved or suspended in it. These are:
1. Nutrient materials absorbed from the intestine
2. Oxygen absorbed from the lungs
3. Chemical substances synthesised by body cells
4. Waste materials produced by body cells to be eliminated from the body by excretion.

The blood cells (Fig. 1:3)

There are three distinct groups, which are classified according to their functions.

1. *Erythrocytes* (red blood corpuscles) are mainly concerned with transporting oxygen from the lungs to the body cells. They contain a substance called *haemoglobin* which combines with oxygen in the lungs and gives it up to the cells in all parts of the body. The supply of oxygen available to cells varies according to how active they are. When the activity of a group of cells increases, they need and obtain increased amounts of oxygen.

There are about 5×10^{12} erythrocytes in each litre of blood and the adult body contains between 5 and 6 litres of blood.

2. *Leucocytes* (white blood cells) are mainly concerned with protecting the body against the successful invasion by micro-organisms, some of which may cause disease. There are several different types of leucocyte which carry out their protective functions in different ways.

These cells are larger than erythrocytes and are less numerous: 4×10^9 to 11×10^9 in each litre of blood.

3. *Thrombocytes* (platelets) are tiny cells which play an essential part in the very complex process of blood clotting. A blood clot is a 'plug' consisting of blood cells and fibrous material which forms in the cut or torn ends of a blood vessel. It prevents excessive loss of blood.

There are 100×10^9 to 500×10^9 thrombocytes in every litre of blood.

THE BLOOD VESSELS

There are three types:
1. *The arteries*, which convey blood away from the heart.
2. *The veins*, which return blood to the heart.
3. *The capillaries*, which link the arteries and veins. These are tiny blood vessels with very thin walls consisting of only one layer of cells. Between these cells there are very small openings or pores, which allow some of the constituents of blood to pass through, such as oxygen, nutritional materials, some chemical substances synthesised in the body and waste products from cells. Larger sized substances, such as erythrocytes and large molecule proteins,

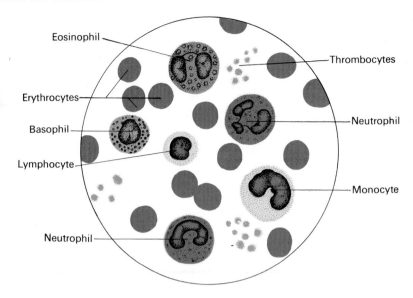

Figure 1:3   Blood cells after staining in the laboratory.

cannot pass through the *semipermeable* capillary walls (Fig. 1:1).

THE HEART

The heart is a muscular sac. It pumps the blood into the blood vessels which transport it:

1. To the lungs (*pulmonary circulation*) where oxygen is absorbed from the air in the lungs and at the same time carbon dioxide is excreted from the blood into the air

2. To the cells in all parts of the body (*general circulation*)

The layer of muscle in the heart wall is not under the control of the will. At rest, the heart contracts between 65 and 75 times per minute, but the rate may be greatly increased during physical exercise when the oxygen and nutritional needs of the muscles which move the limbs are increased.

The rate at which the heart beats can be counted by taking the *pulse*. This is a wave of expansion of artery walls which occurs when the heart contracts and pushes blood into the *aorta*, the first artery of the general circulation. The pulse can be felt most easily where an artery lies close to the surface of the body and can be pressed gently against a bone. The wrist and the temple are the sites most commonly used for this purpose.

THE LYMPHATIC SYSTEM   (Ch. 6. Fig 1:5)

This system is a subsidiary of the circulatory system. It consists of a number of lymph vessels which begin as blind end tubes in the area containing tissue fluid

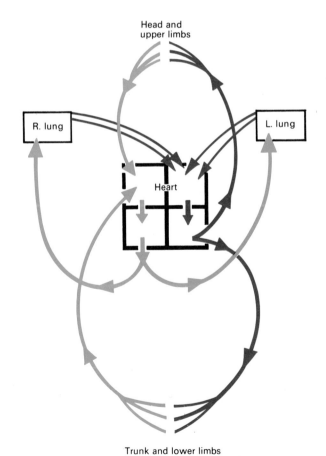

Figure 1:4   Diagram representing the circulatory system.

between the blood capillaries and the cells (Fig. 1:5). Structurally they are similar to veins and blood capillaries but the pores in the walls of the lymph capillaries are larger than those of the blood capillaries. This means that water containing large molecule substances, fragments of damaged cells and foreign matter such as micro-organisms drain away from the interstitial spaces in lymph vessels. The two largest lymph vessels empty lymph into large veins near the heart.

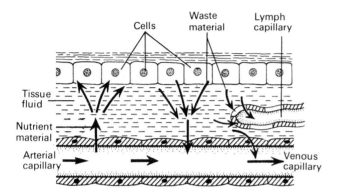

Figure 1:5   Diagram showing the beginning of a lymph capillary in the interstitial space.

There are collections of tissue called *lymph nodes* situated at various points along the length of the lymph vessels. Lymph from the interstitial spaces is filtered as it passes through the lymph nodes, and micro-organisms and other noxious substances are removed.

2. Intake of Raw Materials

A. FOOD (Chs. 8 and 9)
Food is one of the sources of raw materials which cells must obtain from the external environment, but it is not always in a form which cells can use. A specialised system has developed to modify or *digest* food to make it usable.

THE DIGESTIVE SYSTEM (Ch. 9. Figs. 1:6 and 1:7)
This system consists of:
1. *The alimentary canal* which begins at the mouth and continues through the pharynx, the oesophagus, the stomach, the small and the large intestines, the rectum and the anus.
2. *Glands** which form *digestive enzymes*† and discharge them into the canal through ducts or little tubes. Many glands are in the walls of the alimentary canal but those situated outside the canal with ducts

*Glands are aggregates of secretory cells enclosed within a sheet, or membrane, of connective tissue.
†*Enzymes* are chemical compounds which cause, or speed up, chemical changes in substances with which they are in contact. They are organic catalysts.

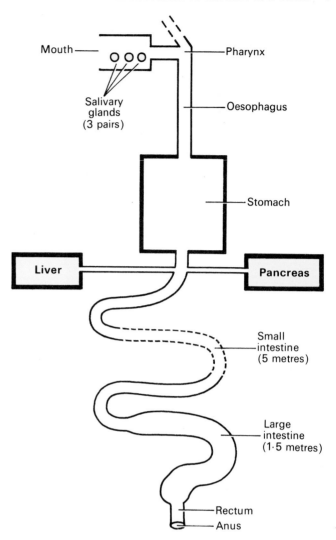

Figure 1:6   Diagram representing the digestive system.

leading into it are the *salivary glands*, the *pancreas* and the *liver*.

Food, which is chemically complex, is taken in at the mouth and simplified by *physical* and *chemical* means to forms in which it can pass through the walls of the alimentary canal into the blood to be transported around the body (Fig. 1:7).

The substances absorbed from the alimentary tract are:

Water
Carbohydrates in the form of monosaccharides
Fats in the form of fatty acids and glycerol
Proteins in the form of amino acids
Mineral salts
Vitamins

These substances provide the materials required by the body for:

Energy production—	carbohydrates and fats
Cell building and repair—	proteins

Chemical synthesis—   mineral salts, proteins,
                      fats, carbohydrates

Medium for all chemical
    activity—   water

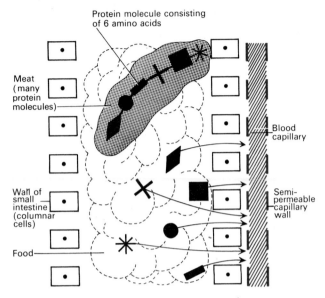

Figure 1:7   Diagram representing the digestion and absorption of food.

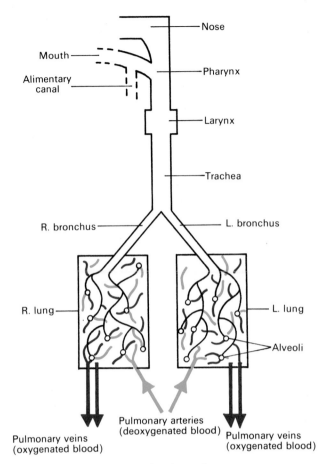

Figure 1:8   Diagram representing the respiratory system.

B. OXYGEN (Ch. 7)

Oxygen is a gas which makes up about 21 per cent of the atmospheric air. It is essential for human life because most of the chemical activities which take place in the cells can only occur in its presence. Oxygen is involved in the series of chemical changes which result in the release of the energy from nutrient materials. This energy is essential for all the cellular activities.

Metabolism

This is the sum total of the chemical activity in the body. It consists of two groups of processes:

1. *Anabolism*, which is the process of building or synthesising new products
2. *Catabolism*, which is the breaking down of substances, either for excretion as waste or to provide raw materials for anabolism

The release of energy from nutritional materials, such as carbohydrates and fats, is a catabolic process.

THE RESPIRATORY SYSTEM (Ch. 7. Fig. 1:8)

This is the system through which oxygen is taken into the body from the external environment and carbon dioxide, a waste product of cell metabolism, is excreted. It consists of a series of tubes or *respiratory passages* which carry air from the nose to the *lungs* and a network of *blood capillaries* in the lungs. The respiratory passages, which are continuous with each other, are the

nose, the *pharynx* (also part of the alimentary canal), the *larynx* (voice box), the *trachea*, two *bronchi* (one bronchus to each lung), a large number of *bronchial tubes* which subdivide and lead to millions of tiny air sacs called *alveoli*. Air may also enter the pharynx through the mouth.

The lungs are two in number and are situated one on each side of the heart in the thoracic cavity. They consist of bronchial tubes, alveoli, blood vessels and nerves all of which are supported by connective tissue. Each alveolus is surrounded by a dense network of blood capillaries.

Respiration

Air containing oxygen and carbon dioxide is breathed into the lungs, filling the alveoli, and it is separated from the blood in the capillaries by two semipermeable membranes. These are the walls of the alveoli and the capillary walls, each of which is only one cell thick. Oxygen is in higher concentration in the alveoli so it passes from the alveoli to the blood; carbon dioxide is in higher concentration in blood so it passes in the opposite direction, from the blood to the alveoli (Fig. 1:9). Oxygen is carried in the blood in *solution in the blood water* and in *chemical combination with haemoglobin in the red blood corpuscles*. The cells throughout the body obtain oxygen and get rid of carbon dioxide by the reverse process. Oxygen is in higher concentration in the blood than in the cells and the interstitial fluid around the cells, so it diffuses down the concentration gradient from the capillaries to the cells. Carbon dioxide, which is a waste product of cell metabolism, is in higher concentration outside the capillaries and diffuses down the concentration gradient from the cells to the blood.

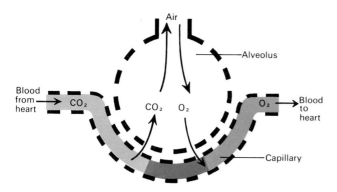

Figure 1:9   Diagram showing the interchange of gases between air in an alveolus and the blood in a capillary.

Breathing is the regular inflation and deflation of the lungs which maintains a steady concentration of atmospheric gases in the alveoli, that is, the constant intake of oxygen and output of carbon dioxide.

Nitrogen, which is present in atmospheric air, is breathed in and out with oxygen and carbon dioxide but, in this gaseous form, it cannot be used by the body. The nitrogen which is needed by the body is present in protein foods, for example in meat and fish.

3. Elimination of Waste Materials

Substances in the body are regarded as waste materials if they cannot be used by the cells and if their accumulation will upset the fine balance which must be maintained between chemical substances in the internal environment. One of the most important of these is the acidity/alkalinity balance for which the *scale of measurement* is pH. The scale extends from 0 to 14 with 7 the *neutral point*. From 7 to 0 represents *increasing acidity* and from 7 to 14 *increasing alkalinity* (Fig. 1:10).

Varying quantities of acids and alkalis are produced by metabolism. It is an important function of the systems involved in elimination to control excretion of these substances in order to maintain the optimum blood pH, which is in the region of 7·4. In addition, other waste materials which do not affect the pH are also excreted. The respiratory, urinary and digestive systems are those involved in elimination.

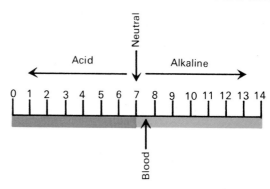

Figure 1:10   The pH scale.

THE RESPIRATORY SYSTEM

This system excretes carbon dioxide, as described above. Carbon dioxide dissolved in water forms an acid which must be excreted in appropriate amounts to maintain the optimum pH of the blood at about 7·4.

THE URINARY SYSTEM   (Ch. 10. Fig. 1:11)

This system is involved in:

1. Maintaining the appropriate balance between water and substances dissolved in it, and between acids and alkalis
2. The elimination of the waste products resulting from the catabolism of cell protein (urea and uric acid)

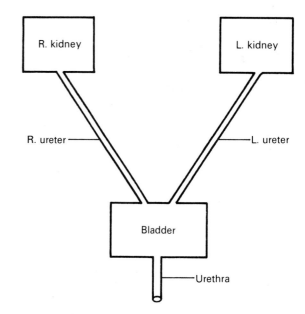

Figure 1:11   Diagram representing the urinary system.

The system consists of:

2 kidneys, situated one on each side of the vertebral column on the posterior wall of the abdominal cavity
2 ureters, one of which extends from each kidney to the bladder

The *urinary bladder*, situated in the pelvis

The *urethra*, which extends from the bladder to the exterior

THE DIGESTIVE SYSTEM

The large intestine excretes *faeces* which contain:

1. Food residue which remains in the alimentary canal because it cannot be digested and absorbed
2. Bile from the liver, which contains the waste products resulting from the catabolism of erythrocytes

---

4. Communication

In order to live in and be able to adapt to the external environment, the individual must be able to communicate with it. Similarly, communications are necessary for the stimulation, regulation and co-ordination of activities within the body. In both cases communication involves a cycle of receiving, collating and giving information.

A. WITH THE OUTSIDE WORLD

THE NERVOUS SYSTEM (Ch. 12. Fig. 1:12)

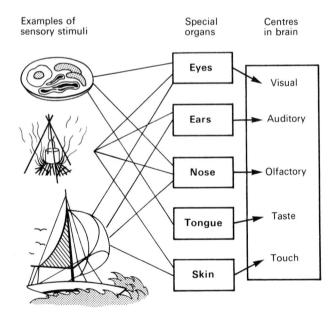

Figure 1:12 Diagram representing the special senses.

The brain receives communications from outside the body through the five senses (Fig. 1:13). These senses and the special organs which are involved are:

Sight— the eyes

Hearing— the ears

Smell— the nose

Taste— the tongue

Touch— the skin

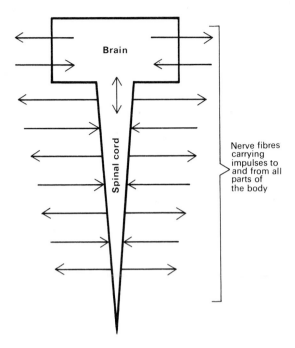

Figure 1:13 Diagram representing the nervous system.

Although the senses are considered different and separate from each other, one sense is rarely used on its own. For example, when the smell of smoke is perceived, other senses such as sight and hearing are likely to be used to try to locate the source of the smoke. Similarly, taste and smell are closely associated in the enjoyment, or otherwise, of food.

Nerve endings are stimulated by phenomena outside the body and the resultant nerve impulses are transmitted to the brain by nerve fibres for 'interpretation' or *perception*. The brain collates this material with information obtained from the memory, and the result is co-ordinated and regulated communication with the outside world. The nervous system consists of:

1. *The brain*, situated inside the skull
2. *The spinal cord*, which extends from the base of the brain to the lumbar region and is protected from injury by the bones of the vertebral column
3. *Nerve fibres*, of two types: *sensory* or *afferent nerves*, which provide the brain with 'input' from organs and tissues, and *motor* or *efferent nerves*, which convey nerve impulses from the brain to organs and tissues

VOICE PRODUCTION

This mechanism provides a means of communication with the outside world. Sound is produced in the larynx as a result of blowing air through the *vocal cords* during expiration. What is known as speech is the *manipulation of sound* by the contraction of the muscles of the throat and the cheeks and the movements of the tongue and lower jaw.

THE SKELETAL AND MUSCULAR SYSTEMS
(Chs. 16, 17 and 18)

These systems combined produce the posture and movements associated with non-verbal communication with the outside world. The skeletal system provides the bony framework of the body, and movement takes place at joints between bones. The muscles which move the bones lie between them and the skin, and they are stimulated by the part of the nervous system which is under the control of the will. Some non-verbal communication, for example changes in facial expression, may not involve the movement of bones.

B. WITHIN THE BODY

THE NERVOUS SYSTEM

The brain receives information from inside the body by means of:

1. Nerve impulses transmitted by nerve fibres
2. Chemical substances circulating in the blood which supplies the brain

The brain collates this information and responds by initiating efferent nerve impulses and secreting chemical substances. These pass from the brain to other parts of the body to stimulate, regulate and co-ordinate the activities of other organs and systems. Some of this activity is under the control of the will but much of it goes on without the individual being conscious of it. For example, the individual is not usually aware of minor changes in the heart rate; emptying of the stomach; dilatation of the pupils of the eyes.

THE ENDOCRINE SYSTEM (Ch. 14)

This system consists of a number of *endocrine glands* which are situated in different parts of the body. They communicate with each other, and with other organs and tissues of the body, by means of chemical substances called *hormones* which they synthesise. The glands are 'informed' about conditions within the body by variations in the concentration of chemical substances in the blood which supplies them and by nerve impulses. Changes in the amount of a hormone secreted result in stimulation or depression of activity in a target organ.

5. Protection Against the External Environment

All the living cells in the body are in a watery environment and the substances which enter and leave the cells are either dissolved or suspended in water.

Although the *skin* and the *mucous membrane* lining the passages which open to the surface of the body are not normally classified as systems, they provide a barrier between the 'dry' external environment of the body and the watery environment of the body cells.

THE SKIN (Ch. 11. Fig. 1:14)

The skin is described in two parts, the *epidermis* and the *dermis*, each of which consists of a number of layers of cells.

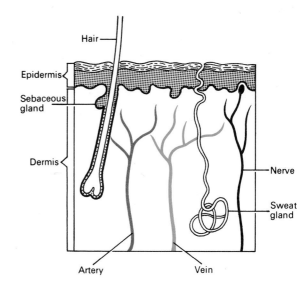

Figure 1:14   Diagram of the skin.

The epidermis lies superficially and is composed of cells which grow from the dermis towards the surface. The cells of the epidermis are *dead* cells and the surface layer is rubbed off by clothes, etc. and is replaced from below. The epidermis constitutes the barrier between the moist environment of the living cells of the body and the dry atmosphere of the external environment.

The dermis consists of *living* cells which gradually grow towards the surface. It contains tiny *sweat glands* with little canals or ducts which lead to the surface. When sweat from these glands reaches the surface it evaporates, cooling the body and thus playing a major part in the maintenance of a temperature level inside the body which is consistent with the life of the cells.

Hair, like the epidermis, is not living material. Hairs grow from *follicles* in the dermis which have groups of specialised cells called *sebaceous glands*. These glands secrete *sebum* into the follicles and so on to the skin surface. The oily materials present in sebum keep the skin soft, pliable and to a large extent waterproof.

Sebum and tears contain chemical substances called *lysozymes* which kill micro-organisms. In this way sebum enhances the protective function of the skin and tears protect the eyes.

Sensory nerve endings present in the dermis are stimulated by pain, temperature and touch. If the finger touches a very hot plate it is removed immediately. This cycle of events is called a *reflex action*, that is, a very rapid motor response (contraction of muscles) to a sensory stimulus (stimulation of sensory nerve endings in the skin). This type of reflex action is an important protective mechanism.

MUCOUS MEMBRANE (Ch. 2)

This type of membrane consists of living cells which keep themselves moist on their free surface by secreting a thick sticky fluid called *mucus*. Mucus protects the mouth and oesophagus from mechanical injury by food and the lungs and respiratory passages from dust and micro-organisms which may be inhaled.

6. Reproduction (Ch. 15. Fig. 1:15)

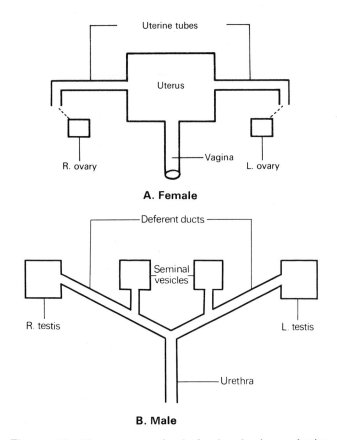

A. Female

B. Male

Figure 1:15 Diagram representing the female and male reproductive systems.

Successful reproduction is essential in order to ensure the continuation of a species from one generation to the next. *Bisexual reproduction* results from the fertilisation of the female egg cell or *ovum* by the male sperm cell or *spermatozoon*. Ova are produced by two *ovaries* situated in the female pelvis. Usually only one ovum is released at a time and it travels towards the *uterus* in the *uterine tube*. The spermatozoa are produced in large numbers by the two *testes* which are situated in the *scrotum*. From each testis spermatazoa pass through a duct called the *deferent duct* to the *seminal vesicles* then to the urethra. During coitus the spermatozoa are deposited in the female *vagina*, then they pass upwards through the uterus and fertilise the ovum in the uterine tube. The fertilised ovum then passes into the uterus, embeds itself in the uterine wall and grows to maturity in about 40 weeks. When a baby is born it is entirely dependent on other people for the food and protection provided before birth by its mother's body.

One ovum is produced about every 28 days during the child-bearing years, which begin at *puberty*, at about 13 years of age, and end at the *menopause*, about 35 years later. When the ovum is not fertilised it passes out of the uterus accompanied by some bleeding called *menstruation*. The cycle, called the *menstrual cycle* in the female, has recognisable phases which are associated with changes in the concentration of hormones. Although there is no similar cycle in the male, the same hormones are involved in the production and development to maturity of the spermatozoa.

7. Movement within the External Environment

The essential purposes of the physical movement of the whole, or parts, of the body within the environment are:
   To obtain food and water
   To avoid injury
   To reproduce

Most of the body movement is under the control of the will and is initiated consciously. The exceptions are some protective movements which are carried out before the individual is aware of them, for example, the reflex action of removing the finger from a very hot surface.

The systems involved in movement are the *skeletal* and the *muscular*.

Before going on to discuss the individual bones which make up this framework it is necessary to be familiar with certain anatomical terms and their meaning.

The anatomical position. This is the position assumed in all anatomical descriptions. The body is in the upright position with the head facing forward, the arms at the sides with the palms of the hands facing forward and the feet together.

Median plane. When the body, in the anatomical position, is divided *longitudinally* into two equal parts it has been divided in the median plane. Any structure which is described as being *medial* to another is, therefore, nearer the midline and any structure which is *lateral* to another is farther from the midline or at the side of the body.

Proximal and distal. These terms are used when describing the bones of the limbs. The proximal end of a bone is that which is nearest the point of attachment of the limb, and the distal end that which is farthest away from the point of attachment of the limb.

Anterior or ventral. This indicates that the part being described is nearer the front of the body.

Posterior or dorsal. This means that the part being described is nearer the back of the body.

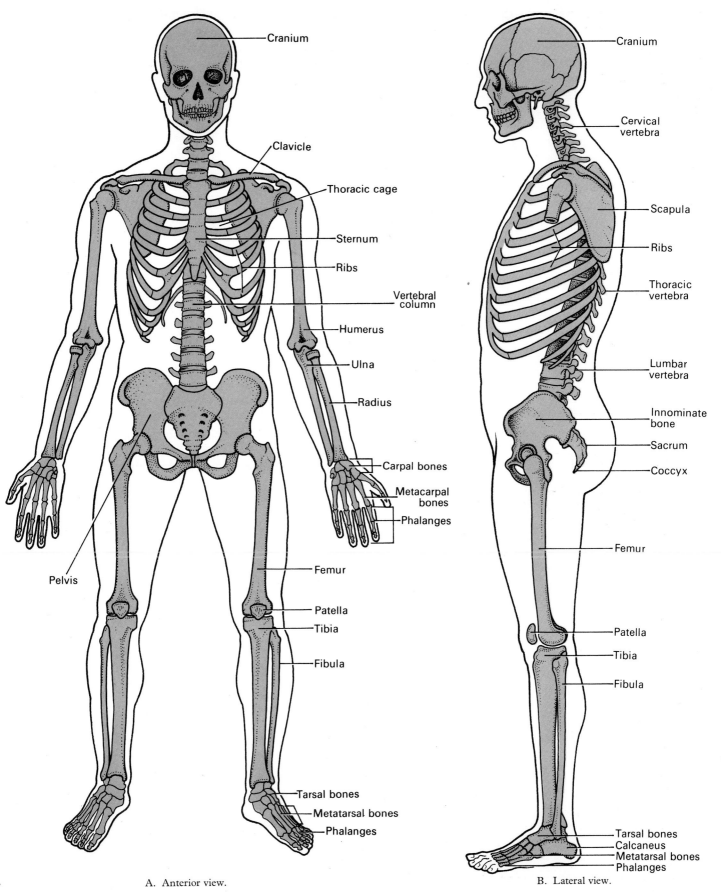

Cranium

Clavicle

Thoracic cage

Sternum

Ribs

Vertebral column

Humerus

Ulna

Radius

Carpal bones

Metacarpal bones

Phalanges

Femur

Patella

Tibia

Fibula

Pelvis

Tarsal bones

Metatarsal bones

Phalanges

A. Anterior view.

Cranium

Cervical vertebra

Scapula

Ribs

Thoracic vertebra

Lumbar vertebra

Innominate bone

Sacrum

Coccyx

Femur

Patella

Tibia

Fibula

Tarsal bones

Calcaneus

Metatarsal bones

Phalanges

B. Lateral view.

Figure 1:16 The bony skeleton.

Superior. This indicates a structure nearer the head.

Inferior. This indicates a structure farther away from the head.

Border. This is a ridge of bone which separates two surfaces.

Spine, spinous process or crest. This is a sharp ridge of bone.

Trochanter, tuberosity or tubercle. These are roughened bony projections usually for the attachment of muscles or ligaments. The different names are used according to the size of the projection. Trochanters are the largest and tubercles the smallest.

Styloid process. This is a sharp downward projection of bone which gives attachment to muscles and ligaments.

Fossa (plural fossae). This is a hollow or depression.

Foramen (plural foramina). This is a hole in a structure.

Sinus. This is a hollow cavity within a bone.

Meatus. This indicates a tube-shaped cavity within a bone.

Articulation. This is a joint between two or more bones.

Suture. This is the name given to an immovable joint, e.g., between the bones of the skull.

Articulating surface. This is the part of the bone which enters into the formation of a joint.

Facet. This is a small, generally rather flat, articulating surface.

Condyle. This is a smooth rounded projection of bone which takes part in a joint.

Septum. This is a partition separating two cavities.

Fissure or cleft. This indicates a narrow slit.

THE SKELETAL AND MUSCULAR SYSTEMS

The bones of the skeleton provide the framework for the body, and movement is achieved when the muscles which lie between the skin and the bones contract and move them at the joints.

The central part of the skeleton is described as the *axial skeleton*. The bones of the limbs and those which attach them to the axis make up the *appendicular skeleton*. (See Figs. 1:16A and B.)

The axial skeleton consists of:

The skull
The vertebral column
The sternum or breast bone
The ribs

The appendicular skeleton consists of:

The bones of the upper limbs, the two clavicles and the two scapulae
The bones of the lower limbs and the innominate bones of the pelvis

THE AXIAL SKELETON (Figs. 1:17, 1:18, 1:19)

THE SKULL

The skull is described in two parts, the *cranium* which contains the brain, and the *face*. It consists of a number of bones which develop separately but fuse together as they mature. The only movable bone is the mandible or lower jaw. The names and position of the individual bones of the skull can be seen in Figure 1:17.

Functions

The various parts of the skull have specific and different functions:

1. The cranium protects the delicate tissues of the brain.
2. The bony eye sockets provide the eyes with some protection against injury and give attachment to the muscles which move the eyes.
3. The temporal bone protects the delicate structures of the ear.
4. Some bones of the face and the base of the skull give resonance to the voice because they contain cavities

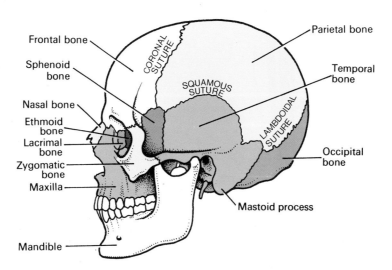

Figure 1:17   The bones of the skull and face—lateral view.

called *sinuses* which are filled with air. The sinuses have tiny openings into the nose.

5. The bones of the face form the walls of the posterior part of the nasal cavities. They keep the air passage open during breathing.

6. The maxilla and the mandible provide alveolar parts in which the teeth are embedded.

7. The mandible is the only movable bone of the skull and chewing food is the result of raising and lowering the mandible by contracting and relaxing some muscles of the face, the muscles of mastication.

THE VERTEBRAL COLUMN

This consists of 24 separate movable bones (vertebrae) plus the sacrum and the coccyx. The vertebral column is described in 5 parts and the bones are numbered from the skull downwards (Fig. 1:18):

    7 cervical
12 thoracic
  5 lumbar
  1 sacrum (5 fused bones)
  1 coccyx (4 fused bones)

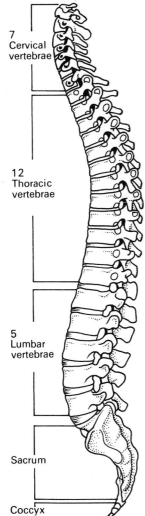

7
Cervical
vertebrae

12
Thoracic
vertebrae

5
Lumbar
vertebrae

Sacrum

Coccyx

Figure 1:18　The vertebral column—lateral view.

The first cervical vertebra, called *the atlas*, articulates with the skull. Thereafter each vertebra forms a joint with the vertebrae immediately above and below. In the cervical and lumbar regions more movement is possible than in the thoracic region.

The sacrum consists of five vertebrae fused into one bone which articulates (forms joints) above with the 5th lumbar vertebra and an innominate (pelvic) bone at each side.

The coccyx consists of the four terminal vertebrae fused into a small triangular bone which articulates with the sacrum above.

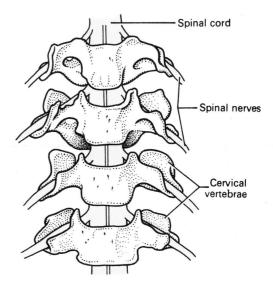

Spinal cord

Spinal nerves

Cervical
vertebrae

Figure 1:19　The cervical vertebrae separated to show the spinal cord and the spinal nerves.

Functions

The vertebral column has several important functions:

1. It protects the spinal cord. In each bone there is a hole or *foramen* and when the vertebrae are arranged one above the other, as shown in Figure 1:18, the foramina form a canal. The spinal cord, which is an extension of nerve tissue from the brain, lies in this canal (Fig. 1:19).

2. Adjacent vertebrae form openings for the passage of the spinal nerves. There is a small foramen on each side of the vertebral column through which nerves from the spinal cord pass to all parts of the body (Fig. 1:19).

3. In the thoracic region the ribs articulate with the vertebrae forming joints which move during respiration.

THE THORACIC CAGE

The thoracic cage is formed by:

12 thoracic vertebrae
12 pairs of ribs
 1 sternum or breast bone

The arrangement of the bones can be seen in Figure 1:20.

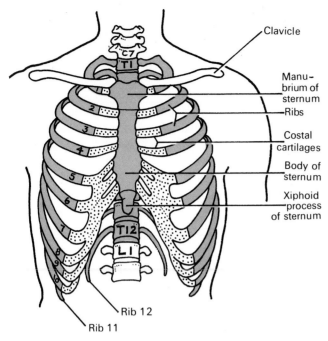

Figure 1:20   The structures forming the walls of the thoracic cavity.

Functions

The functions of the thoracic cage are:

1. To protect the thoracic organs. The bony framework protects the heart, the lungs, the large blood vessels and other structures present in the thoracic cavity.
2. To form joints between the upper limbs and the axial skeleton. The upper part of the sternum, the *manubrium*, articulates with the clavicles forming the only joints between the upper limbs and the axial skeleton.

3. To give attachment to the muscles of respiration:
   (a) Intercostal muscles occupy the spaces between the ribs and when they contract the ribs move upwards and outwards and inspiration (breathing in) occurs.
   (b) The diaphragm is a dome-shaped muscle which separates the thoracic and abdominal cavities. It is attached to the bones of the thorax and when it contracts it assists with inspiration. Structures which extend from one cavity to the other must pass through the diaphragm.

THE APPENDICULAR SKELETON

The appendages are:

1. The upper limbs and the shoulder girdle
2. The lower limbs and the innominate bones of the pelvis.

The names of the bones involved, their position and their relationship to other bones are shown in Figure 1:21.

Functions

The appendicular skeleton has two functions:

1. Voluntary movement. The bones, muscles and joints of the limbs are involved in voluntary movement, which may range from the very fine movements of the fingers associated with writing, to the co-ordinated movement of all the limbs when the whole body moves in activities such as running and jumping.
2. Protection of delicate structures. Structures such as blood vessels and nerves lie along the length of bones of the limbs and are protected from injury by the muscles and skin. These structures are most vulnerable where they cross joints and where bones can be felt near the skin.

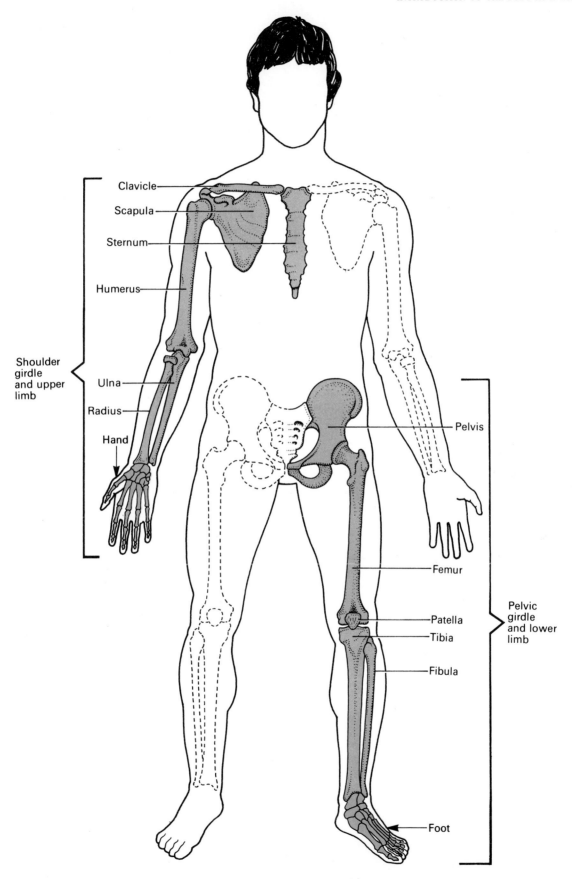

Figure 1:21   The bones of the upper and lower limbs and their relationships to the axial skeleton.

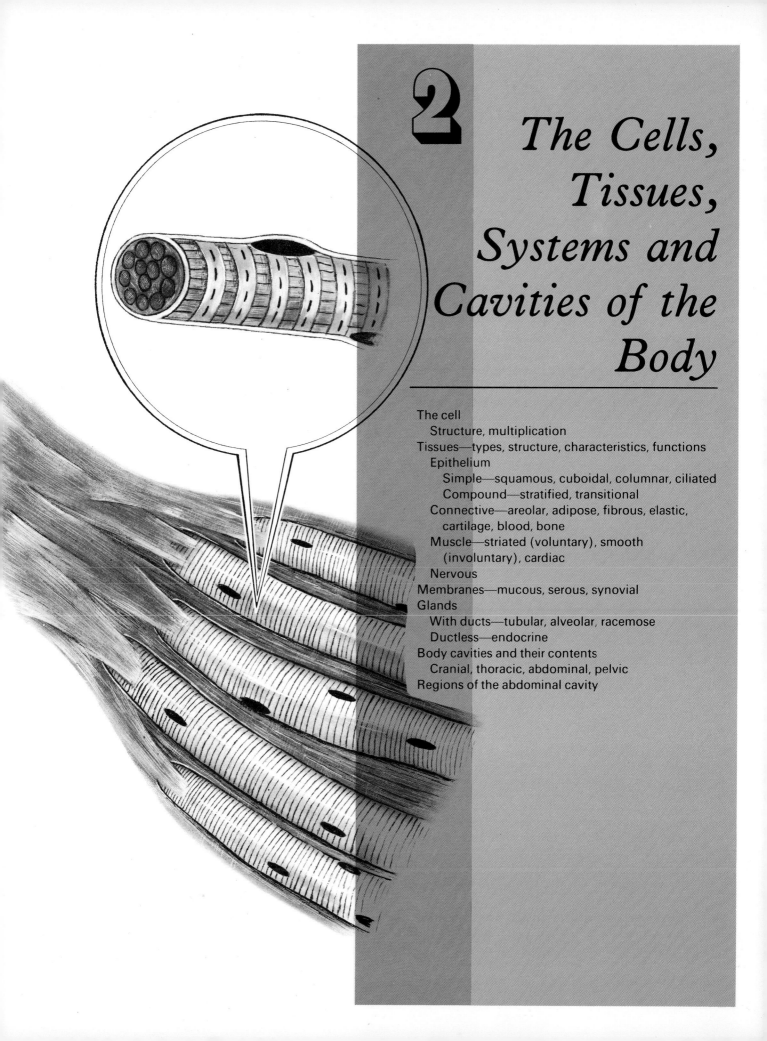

2 The Cells, Tissues, Systems and Cavities of the Body

2. The Cells, Tissues, Systems and Cavities of the Body

The Cell

The human body develops from a single cell called the *zygote* which results from the fusion of the ovum (female egg cell) and the spermatazoon (male germ cell). All the specialised cells which form the tissues, organs and systems originate from the zygote, and groups of specialised cells which have the same functions have the same appearance. Tissues, such as the lining of the mouth, can be seen with the naked eye, but the millions of cells which make up the tissues are so small that they can only be seen with the aid of a microscope. When they are stained in the laboratory and viewed through a microscope the cells can be recognised by their characteristic size and shape and by the dyes which they absorb.

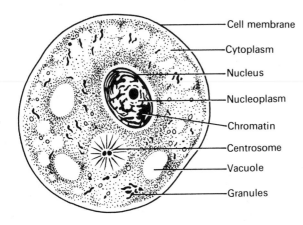

— Cell membrane

— Cytoplasm

— Nucleus

— Nucleoplasm

— Chromatin

— Centrosome

— Vacuole

— Granules

Figure 2:1 The simple cell.

STRUCTURE (Fig. 2:1)
All living cells are made up of a substance called *protoplasm*—a slightly opaque, colourless, soft jelly-like substance consisting of water and the following substances in solution or suspension:

Organic and inorganic salts
Carbohydrates
Lipids (fatty substances)
Nitrogenous substances (amino acids obtained from proteins)
Compounds of the above substances

THE CELL MEMBRANE
The protoplasm of the cell is contained within a very fine membrane consisting of protein threads and lipids. Substances pass into and out of the cell across this membrane in a variety of ways. Only three main methods are described here.

1. *By diffusion.* The cell membrane has tiny holes or pores between the protein threads and lipids through which very small molecules of chemical substances such as oxygen and carbon dioxide can pass. When diffusion takes place substances always pass from a higher concentration on one side of the membrane to a lower concentration on the other (see Fig. 4:6).

2. *By dissolving in the lipids in the cell membrane.* Fat-soluble substances which cannot diffuse through the pores of the membrane pass through by dissolving in the lipid part.

3. *By active transport.* This system applies to substances which are too large to diffuse through the cell membrane and are not soluble in fat. The substance being transported links with a carrier substance in the cell wall. It is then 'carried' across and released on the other side. Carrier substances are specific. Glucose and amino acids are transported in this way but the glucose carrier cannot transport amino acids and vice versa.

THE CYTOPLASM
The cytoplasm is the protoplasm inside the cell but outside the nucleus (Fig. 2:1). It contains molecules known as *ribonucleic acids* (RNA) and granular structures called *mitochondria*. Mitochondria are involved in oxidative reactions which result in the controlled release of energy from the nutrient materials and the formation of *adenosine triphosphate* (ATP), the main energy carrier within the cell. Within the cytoplasm there are clear circular spaces called *vacuoles* which may contain waste materials or secretions formed by the cell.

THE NUCLEUS
Within the cell there is a central mass called the *nucleus* which is surrounded by a *nuclear membrane*. The protoplasm within the nucleus is *nucleoplasm*. Near the nucleus there is a small spherical body called the *centrosome* surrounded by a radiating thread-like structure. The

centrosome contains two minute, dark, circular bodies called *centrioles* which participate in the early stages of cell division.

CELL DIVISION (MULTIPLICATION)

Within the nucleus there are *chromatin* threads which carry the *genes*, composed of *deoxyribonucleic acid* (DNA). The genes are the inherited material: they ensure that when cells divide the 'daughter cells' are identical to the 'parent cell'. This type of cell division or multiplication, called *mitosis*, goes on throughout the life of the individual. A number of genes are linked together to form thread-like structures called *chromosomes*. Each cell has 46 chromosomes arranged in 23 pairs.

The ovum and the spermatazoon each have 23 chromosomes so that when they combine, the zygote which is formed has the full complement of 46 chromosomes. Thus the new individual has a mixture of DNA, half of which is obtained from the mother and half from the father. This type of cell division or multiplication is called *mieosis*.

Mitosis

The multiplication of cells by mitosis occurs throughout the life of the individual. It occurs at a more rapid rate until growth is complete and thereafter new cells are formed to replace those which have died. Nerve cells are a notable exception: when they die they are not replaced.

There are several fairly well-defined stages in the process of cell division by mitosis of which four are described here (Fig. 2:2).

1. *The prophase.* The centrosome divides into two parts which migrate to either pole of the cell, and the two parts remain attached by fine thread-like spindles. The chromatin becomes concentrated and more definite in shape and forms dark, rod-shaped structures known as the *chromosomes*. There are 23 pairs of chromosomes and they contain the genes (DNA) which determine the inherited traits and characteristics of the individual. Each chromosome divides into two and the newly formed pair is held together at the middle by a *centromere*.

2. *The metaphase.* The nuclear membrane disappears and the chromosomes arrange themselves at the centre of the cell and are attached to the spindles of the centrosomes, which are now at opposite poles of the cell.

3. *The anaphase.* The centromere divides and the two identical chromosomes move apart, still attached to the spindles. They arrange themselves at opposite poles of the cell. At the end of this stage the spindles break.

4. *The telophase.* The nuclear membrane reappears, the spindles completely disappear and a constriction develops round the middle of the cell body.

The constriction of the cytoplasm increases until the cell divides, the chromosomes disappear and the thread-like structure known as chromatin reappears. The cell has now completed its division and two identical *daughter cells* have been formed.

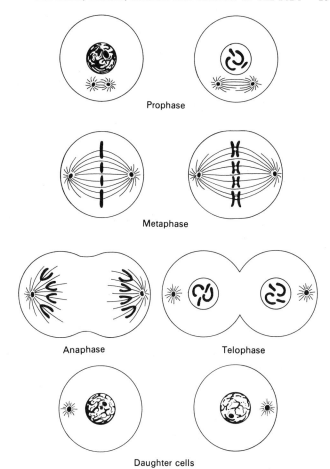

Figure 2:2 Some stages in cell multiplication by mitosis.

CHROMOSOMES

As indicated above, the nucleus of every human cell contains 46 chromosomes. The chromosomes are arranged in pairs, one member of each pair being derived from the male parent and the other from the female parent. That is, 23 chromosomes in each cell are *inherited* from the mother and 23 from the father.

The chromosomes contain the basic hereditary substances which determine the individual's characteristics and traits. Hereditary characteristics include colour of hair and eyes, structure of bones and teeth, the height of an individual and the ability of the cells of the body to produce the enzymes necessary for its metabolic processes.

Determination of sex depends on the *sex chromosomes*, that is, one pair in the male and one pair in the female. In the female the sex chromosomes are the same in size and shape and are called X chromosomes. In the male they are slightly different, one is an X chromosome and the other is a slightly smaller Y chromosome. Therefore the female sex chromosomes are XX and the male XY.

In the ovum and in the spermatozoon the number of chromosomes is reduced to 23 single chromosomes. This means that all ova have an X chromosome and half the spermatozoa have an X chromosome and half a Y chromosome.

In conception, if an X-bearing spermatozoon fertilises an ovum the offspring will be female and if a Y-bearing spermatozoon fertilises an ovum the offspring will be male.

Sperm X + ovum X = offspring XX = female child
Sperm Y + ovum X = offspring XY = male child

Within a pair of genes, one may exert a stronger influence than the other. The gene exerting the stronger influence is termed *dominant* and the gene which is less effective is described as *recessive*. The characteristics of the offspring such as height, colour of eyes and hair and other familial traits depend upon the dominance of the parents' genes.

The Elementary or Fundamental Tissues

The tissues of the body consist of large numbers of cells and they are classified according to the size, shape and functions of these cells. There are four main types of tissue, each of which has subdivisions. They are:

Epithelial tissue or epithelium
Connective tissue
Muscle tissue
Nervous tissue

EPITHELIAL TISSUE
The cells forming this tissue are very closely packed together and the intercellular substance, called the *matrix*, is minimal. The cells usually lie on a *basement membrane* which is an inert connective tissue.

Epithelial tissue is divided into two types: *simple* and *compound*.

SIMPLE EPITHELIUM
Simple epithelium consists of a single layer of cells and is divided into four types. The types are named according to the shape of the cells, which differ according to the functions they perform.

Squamous or pavement epithelium
This is composed of a single layer of flattened cells (Fig. 2:3). The cells fit closely together like flat stones and they form a very smooth membrane.

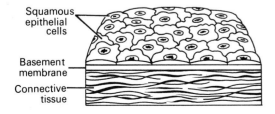

Figure 2:3 Squamous epithelium.

The function of squamous epithelium is to provide a thin, smooth, inactive lining for the following structures:
The heart
The blood vessels
The alveoli of the lungs
The lymph vessels

Cuboidal epithelium
This consists of cube-shaped cells fitting closely together lying on a basement membrane (Fig. 2:4). It forms the tubules of the kidneys and is found in some glands. Cuboidal epithelium is actively involved in secretion and absorption.

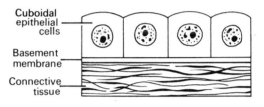

Figure 2:4 Cuboidal epithelium.

Columnar epithelium
This is formed by a single layer of cells, shaped like elongated cubes, on a basement membrane (Fig. 2:5). It is found lining the organs of the alimentary tract and consists of a mixture of cells, some of which absorb the products of digestion and others secrete *mucus*. Mucus is a thick sticky substance secreted by special columnar cells called *goblet cells*.

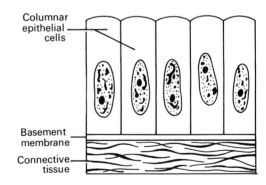

Figure 2:5 Columnar epithelium.

Ciliated epithelium (Fig. 2:6)
This is formed by columnar cells which have fine, hair-like processes called *cilia* on their free surface. The wave-like movement of many cilia propels the contents of the tubes that they line in one direction only.

Ciliated epithelium is found lining most of the respiratory passages and the uterine tubes. In the respiratory passages it propels mucus towards the throat (Ch. 7) and in the uterine tubes it propels the ova towards the uterus (Ch. 15).

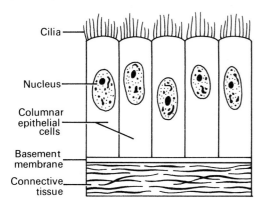

Figure 2:6   Ciliated columnar epithelium.

COMPOUND EPITHELIUM

Compound epithelium consists of several layers of cells. The superficial layers grow up from below and basement membranes are usually absent. The main function of compound epithelium is to protect underlying structures. There are two main types: stratified and transitional.

Stratified epithelium (Fig. 2:7)

This is composed of a number of layers of cells of different shapes. In the deepest layers the cells are mainly columnar in shape and as they grow towards the surface they become flattened.

Nonkeratinised stratified epithelium is found on wet surfaces that may be subjected to wear and tear, such as the conjunctiva of the eyes and the linings of the mouth, the pharynx and the oesophagus.

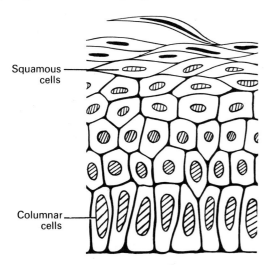

Figure 2:7   Stratified epithelium.

Keratinised stratified epithelium is found on dry surfaces, that is skin, hair and nails. The surface layers of keratinised cells are dead cells. They give protection to

and prevent drying of the cells in the deeper layers from which they develop. The surface layer of skin is rubbed off by the clothes and is replaced from below.

Transitional epithelium (Fig. 2:8)

This is composed of several layers of pear-shaped cells and is found lining:

    The ureters
    The urinary bladder

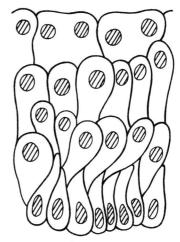

Figure 2:8   Transitional epithelium.

CONNECTIVE TISSUE

The cells forming the connective tissues are more widely separated from each other than those forming the epithelium and the intercellular substance is present in considerably larger amounts. There may or may not be fibres present in the matrix, which may be of a semisolid jelly-like consistency or dense and rigid, depending upon the position and function of the tissue.

The connective tissues are sometimes described as the supporting tissues of the body because their functions are mainly mechanical, connecting other more active tissues.

AREOLAR TISSUE (Fig. 2:9)

This is the most generalised of all connective tissue. The matrix is described as semisolid with the cells called

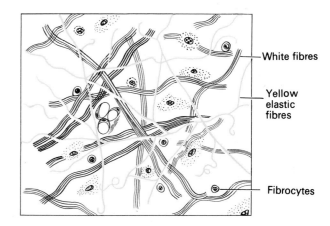

Figure 2:9   Areolar tissue.

fibrocytes widely separated by *yellow elastic* and *white fibres*. It is found in almost every part of the body connecting and supporting other tissues, for example:

Under the skin

Between muscles

Supporting blood vessels and nerves

In the alimentary canal

In glands supporting secretory cells

ADIPOSE TISSUE (Fig. 2:10)

Adipose tissue consists of a collection of *fat cells* containing fat globules. These cells are present in varying numbers in the matrix of areolar tissue.

Fatty or adipose tissue is found supporting organs such as the kidneys and the eyes. It is also found between bundles of muscle fibres and, with areolar tissue, under the skin giving the body a smooth continuous outline.

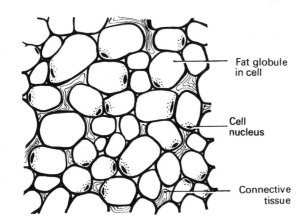

Figure 2:10 Adipose tissue.

WHITE FIBROUS TISSUE (Fig. 2:11)

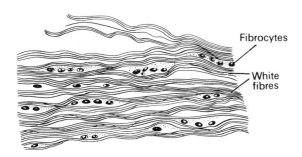

Figure 2:11 White fibrous tissue.

This is a strong connecting tissue made up mainly of closely packed bundles of *white fibres* with very little matrix. There are very few cells and these lie in rows between the bundles of fibres. Fibrous tissue is found:

Forming the ligaments which bind bones together

As an outer protective covering for bone, known as the *periosteum*

As an outer protective covering of some organs, for example, the kidneys, lymph nodes and the brain

Forming muscle sheaths called *muscle fascia*, which extends beyond the muscle to become the tendon that binds the muscle to bone

YELLOW ELASTIC TISSUE (Fig. 2:12)

Yellow elastic tissue is capable of considerable extension and recoil. There are few cells and the matrix consists mainly of masses of *elastic fibres*. It is found in organs where alteration of shape is required, for example:

In the arteries, particularly the large arteries

In the trachea and bronchi

In the lungs

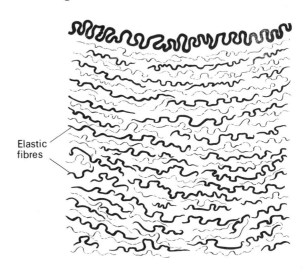

Figure 2:12 Yellow elastic tissue.

LYMPHOID TISSUE (Fig. 2:13)

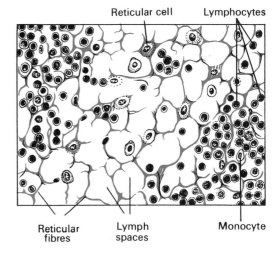

Figure 2:13 Lymphoid tissue.

This tissue has a semisolid matrix with fine branching

fibres. The cells are specialised and are known as *lymphocytes*. Lymphoid tissue is found:

    In the lymph nodes

    In the spleen

    In the palatine and pharyngeal tonsils

    In the vermiform appendix

    Forming the solitary and aggregated glands in the small intestine

    In the wall of the large intestine

CARTILAGE

Cartilage is a much firmer tissue than any of the other connective tissues; the matrix is quite solid. For descriptive purposes cartilage is divided into three types:

    Hyaline

    White fibrocartilage

    Yellow or elastic fibrocartilage

Hyaline cartilage (Fig. 2:14)

Hyaline cartilage appears as a smooth bluish-white tissue. Under the microscope the cells appear in groups of two or more and where they come in contact with one another their edges appear to be flattened. The matrix is solid and smooth. Hyaline cartilage is found:

    On the surface of the parts of the bones which form joints

    Forming the costal cartilages which attach the ribs to the sternum

    Forming part of the larynx, trachea and bronchi

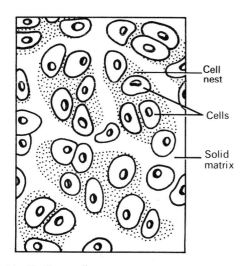

Figure 2:14 Hyaline cartilage.

White fibrocartilage (Fig. 2:15)

This consists of dense masses of white fibres in a solid matrix with the cells widely dispersed. It is a tough, slightly flexible tissue. White fibrocartilage is found:

    As pads between the bodies of the vertebrae called the intervertebral discs

Between the articulating surfaces of the bones of the knee joint known as the semilunar cartilages

Surrounding the rim of the bony sockets of the hip and shoulder joints deepening the cavities

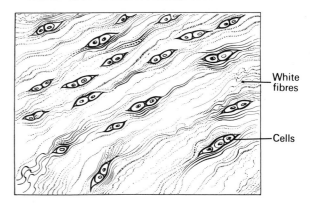

Figure 2:15 White fibrocartilage.

Yellow elastic fibrocartilage (Fig. 2:16)

This consists of yellow elastic fibres running through the solid matrix. The cells lie between the fibres. It forms the pinna or lobe of the ear and the epiglottis.

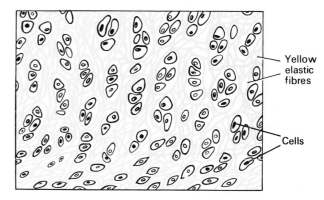

Figure 2:16 Yellow elastic fibrocartilage.

BLOOD

This is a fluid connective tissue and will be described in detail in Chapter 3.

BONE

Bone is one of the hardest connective tissues in the body and when fully developed it is composed of:

    Water (25 per cent)

    Organic material (30 per cent)

    Inorganic salts (45 per cent)

There are two types of bone tissue, *compact* and *cancellous*.

Compact bone

To the naked eye compact bone appears to be a solid structure, but when examined under a microscope large numbers of units with the same structure can be identified.

These are called *haversian systems* (Fig. 2:17) and they have several well-defined characteristics:

1. A central *haversian canal* runs longitudinally and contains blood and lymph capillaries and nerves.
2. Surrounding this central canal there are concentric plates of bone known as the *lamellae*.
3. Between the lamellae are spaces or *lacunae* containing lymph and bone cells called *osteocytes*.
4. Running between the lacunae and the haversian canal are fine channels called the *canaliculi*. Lymph carrying nourishment to the bone cells flows through the canaliculi.
5. In the spaces between the haversian system there are interstitial lamellae.

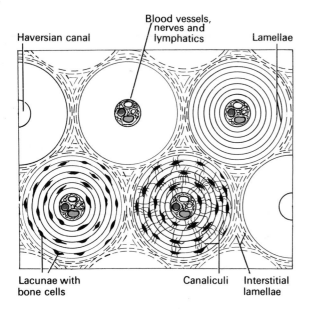

A. Cross section.

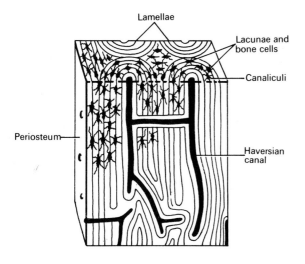

B. Longitudinal section.

Figure 2:17   Microscopic structure of bone.

Cancellous bone

To the naked eye, cancellous bone looks like a sponge. Microscopically, the haversian canals are much larger than in compact bone and there are fewer lamellae, giving the appearance of a honeycomb. *Red bone marrow* is always present within cancellous bone.

Red bone marrow (Fig. 2:18) is the tissue in which erythrocytes, leucocytes and thrombocytes are formed and grow to maturity before entering the blood (Ch. 3).

Yellow bone marrow (Fig. 2:18) is a fatty substance found in the hollow shafts of long bones in adults. Before birth and in early childhood all bone marrow is red and forms blood cells, but by the time adulthood is reached red bone marrow is found only in cancellous bone.

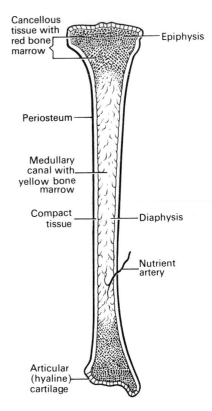

Figure 2:18   A long bone—longitudinal section.

The outer surface of bone is covered by a vascular fibrous membrane called *periosteum*, which has several functions:

1. It forms an outer protective covering for the bone.
2. It gives attachment to the muscle tendons.
3. It gives attachment to ligaments.
4. In its deeper layer there are many bone forming cells called *osteoblasts* which are responsible for the deposition of new bone tissue.

Periosteum covers bones except for those surfaces which take part in the formation of freely movable joints, where it is replaced by *hyaline cartilage*.

The skeleton

The skeleton or bony framework of the body is made up of a variety of bones which are classified as follows:

Long
Irregular
Flat
Sesamoid

These bones are composed of both *compact* and *cancellous* bone tissue.

Long bones. When long bones are examined by the naked eye they are described as having a *diaphysis* or shaft and *two epiphyses* or extremities. *The diaphysis* is composed mainly of compact bone with a central canal known as the *medullary canal*. In adult life this canal is filled with *yellow bone marrow. The epiphyses* are composed of a thin layer of compact bone with cancellous bone tissue and red bone marrow inside. Where an epiphysis forms a joint with another bone its surface is covered by hyaline cartilage.

Irregular, flat and sesamoid bones. These are composed of a thin layer of compact bone surrounding an inner mass of cancellous bone and red bone marrow and, like long bones, they are encased in periosteum except on their articular surfaces where they are covered by hyaline cartilage.

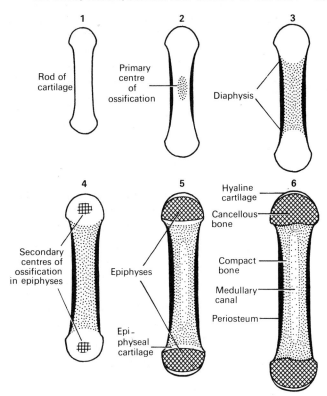

Figure 2:20   Diagram of the stages of development of long bones.

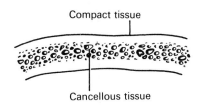

A. Flat Bone

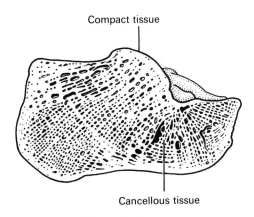

B. Irregular bone

Figure 2:19   Flat and irregular bones.

The development of bones (Fig. 2:20)

The development of bones begins before birth and is not complete until about the 25th year of life. Long, short and irregular bones develop from cartilage models, flat bones from membrane and sesamoid bones from tendon. These original tissues are gradually *replaced* by bone.

The cells which form bone are called *osteoblasts* and after they are established in their lacunae, they are called *osteocytes.*

The other cells found in bone are the *osteoclasts*, which are responsible for resorption of bone. The functions of osteoclasts are:

To maintain the shape of growing bone
To form the medullary canal in long and short bones
To form the sinuses (air-filled cavities) in some of the bones of the face and skull

Development of long bones. Long bones develop by a complicated series of changes which result in the removal of cartilage and its replacement by bone tissue. About the eighth week of fetal life a group of bone-forming cells called a *primary centre of ossification* appears in the middle of the rod of cartilage. By the action of enzymes on the bone cells (osteoblasts) of the primary centre, bone formation takes place together with the deposition of mineral salts, mainly calcium phosphate.

From this primary centre of ossification the *diaphysis* of the bone is formed and ossification spreads from the middle towards each end. The thickening of the shaft occurs when bone and inorganic salts are deposited by the osteoblasts present in the periosteum. *Secondary centres of ossification* appear in the parts of the cartilage which will eventually be the *epiphyses* of the bone. Some epiphyses have more than one secondary centre from which ossification spreads until the replacement of the cartilage model is complete. Most secondary centres of ossification appear after birth.

Bone grows in length at the *epiphyseal cartilage,* which is situated between each epiphysis and the diaphysis. The growth of the bone is not complete until the rate of ossification overtakes the rate of growth of the epiphyseal cartilage. Until ossification is complete a thin clear 'epiphyseal line' shows on X-ray plates.

Flat, irregular and sesamoid bones develop from one or more centres of ossification.

Functions
The bones of the skeleton have a number of functions:
1. To form a supporting framework for the body
2. To form the boundaries for the cranial, thoracic and pelvic cavities
3. To give protection to delicate organs
4. To form the joints which are essential for the movement of the body
5. To provide attachment for the voluntary muscles which move the joints
6. To form blood cells in red bone marrow in cancellous bone
7. To provide a store of calcium salts

MUSCLE TISSUE
Muscle tissue is composed of:
Water (75 per cent)
Protein (20 per cent)
Mineral salts, glycogen, glucose and fat (5 per cent)

There are three main types of muscle tissue:

1. Striated or voluntary muscle
2. Smooth or involuntary muscle
3. Cardiac muscle

Striated or voluntary muscle (Fig. 2:21)
This may be described as *skeletal, striated, striped* or *voluntary* muscle. It is known as voluntary because it is under the control of the will.

When voluntary muscle is examined microscopically the cells are found to be roughly cylindrical in shape and their length varies from 10 to 40 millimetres. Each fibre has several nuclei situated just under the *sarcolemma,* which is a fine sheath surrounding each muscle fibre. The muscle fibres lie parallel to one another and, when viewed under the microscope, they show well marked transverse dark and light bands, hence the name striated or striped muscle.

A muscle consists of a large number of muscle fibres. In addition to the sarcolemma mentioned previously, each fibre is enclosed in and attached to fibrous tissue

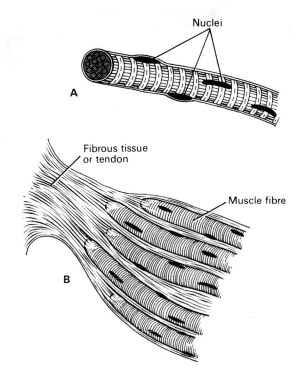

Figure 2:21 A. A striated muscle fibre. B. A bundle of striated muscle fibres and their connective tissue.

called *endomycium.* Small bundles of fibres are enclosed in *perimycium* and the whole muscle in *epimycium.* The fibrous tissue enclosing the fibres, the bundles and the whole muscle extends beyond the muscle fibres to become the *tendon* which attaches the muscle to bone or skin.

Smooth or involuntary muscle (Fig. 2:22)

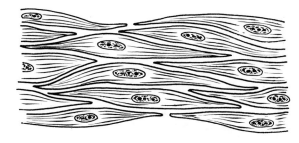

Figure 2:22   Smooth muscle fibres.

Smooth muscle may be described as *involuntary, plain* or *visceral* muscle. It is not under the control of the will. It is found in the walls of blood and lymph vessels, the alimentary tract, the respiratory tract, the urinary bladder and the uterus.

When examined under a microscope the cells are seen to be spindle-shaped with only one central nucleus. There is no distinct sarcolemma but a very fine membrane surrounds each fibre. Bundles of fibres form sheets of muscle which form parts of the walls of the above structures.

Cardiac muscle (Fig. 2:23)

This type of muscle tissue is found exclusively in the wall of the heart. It is not under the control of the will but, when viewed under a microscope, cross stripes, which are characteristic of voluntary muscle, can be seen. Each fibre (cell) has a nucleus and one or more branches. The ends of the cells and their branches are in very close contact with the ends and branches of adjacent cells. Microscopically these 'joints' or *intercalated discs* can be seen as lines which are thicker and darker than the ordinary cross stripes. This arrangement gives cardiac muscle the appearance of a sheet of muscle rather than a very large number of individual fibres. The end to end contiguity of cardiac muscle cells has significance in relation to the way the heart contracts. Each fibre does not need to be stimulated as the impulse of contraction spreads from cell to cell across the intercalated discs.

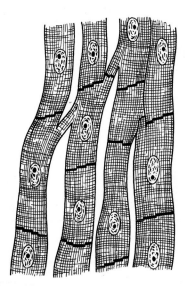

Figure 2:23   Cardiac muscle.

Functions

The function of muscle is to *contract* and when it contracts it *pulls*. A muscle fibre contracts when it is stimulated by one of the following means:

Electrical
Mechanical
Chemical
Thermal

In the human body the necessary stimulus (chemical) is supplied by *nerves* which pass to the muscle and break up into minute nerve endings, each one of which stimulates a single muscle fibre (Ch. 12). When the muscle fibre contracts it follows the *all or none law*, which means that it contracts to its full capacity or it does not contract at all. When they contract, muscle fibres become shorter and thicker.

In order to contract when it is stimulated a muscle fibre must have an adequate blood supply to provide sufficient oxygen and nutritional materials and to remove waste products.

Muscle tone

This is a state of partial contraction of muscles. It is achieved by the contraction of a few muscle fibres at a time. Muscle tone in relation to striated muscle is associated with the maintenance of posture in the sitting and standing positions. The muscle is stimulated to contract through a system of spinal reflexes. Stretching the muscle or its tendon stimulates the reflex action (Ch. 12). A degree of muscle tone is also maintained by smooth and cardiac muscle.

Force of muscle contraction

The force of the contraction of the muscle depends upon the *number of fibres* which contract at the same time. The more fibres involved the stronger is the force of contraction, e.g. when lifting a heavy as distinct from a light object.

Effect of temperature on muscle tissue

If muscle tissue is warm it responds to stimulation more rapidly, that is, there is less delay between stimulation and response.

Muscle fatigue

If a muscle is stimulated to contract at very frequent intervals, it gradually becomes depressed and will no longer respond. This is called muscle fatigue and is usually due to an inadequate blood supply. Fatigue is prevented during sustained muscular effort because the fibres usually contract in series. All the fibres of a muscle rarely contract at the same time and if they do so the contraction can be maintained for only a short time.

Energy source for muscle contraction

The energy which muscles require is derived from the breakdown (catabolism) of carbohydrate and fat molecules inside the fibres. Each molecule undergoes a series of changes and, with each change, small quantities of energy are released. For the complete breakdown of these molecules and the release of all the available energy, an adequate supply of oxygen is required. If the individual undertakes excessive exercise the oxygen supply may be insufficient to meet the metabolic needs of the muscle fibres, which may result in the accumulation of intermediate metabolic products, such as lactic acid. Where the breakdown process and the release of energy are complete, the waste products are carbon dioxide and water (Ch. 9).

Further features of striated muscle

The voluntary muscles are those which produce the movements of the body. Each muscle consists of a *fleshy* part made up of striped fibres and *tendinous* parts consisting of white fibrous tissue usually at both ends of the fleshy part. The muscle is attached to bone or skin by these tendons. When the tendinous attachment of a muscle is broad and flat it is called an *aponeurosis*.

To be able to produce movement at a joint a muscle or its tendon *must stretch across the joint*. When a muscle contracts, its fibres *shorten and it pulls* one bone towards another, for example, bending the elbow.

The muscles of the skeleton are arranged in groups, some of which are *antagonistic* to each other. To produce movement at a joint one muscle or group of muscles contracts while the antagonists relax; for example, to bend the knee the muscles at the back of the thigh contract and those on the front relax. The constant adjustment of the contraction and relaxation of antagonistic groups of muscles is well demonstrated in the maintenance of balance and posture when sitting and standing.

Individual muscles and groups of muscles have been given names that reflect certain characteristics, for example:

1. *The shape* of the muscle—the trapezius is shaped like a trapezium.
2. *The direction* in which the fibres run—the oblique muscles of the abdominal wall.
3. *The position* of the muscle—the tibialis is associated with the tibia.
4. *The movement* produced by contraction of the muscle—flexors, extensors, adductors.
5. *The number of points of attachment* of a muscle—the biceps muscle has two tendons at one end.
6. *The names of the bones to which the muscle is attached*—the carpi radialis muscles are attached to the carpal bones in the wrist and to the radius in the forearm.

NERVOUS TISSUE
The structure of nervous tissue is described in Chapter 12.

MEMBRANES
Some of the tissues which line or cover organs are described as *membranes*. These are composed of types of cells which have already been described. When they are called membranes the cells form *sheets* which cover or line structures. The more important membranes can be classified as follows:

Mucous
Serous
Synovial

Mucous membrane
This is the name given to the lining of the alimentary tract, respiratory tract and genito-urinary tracts. The cells forming the membrane produce a secretion known as *mucus*. This is a slimy tenacious fluid containing a protein material known as *mucin*. This secretion is formed within the cytoplasm of the cells and as it accumulates the cells become distended and finally burst, discharging the mucus on to the free surface. As the cells fill up with mucus they have the appearance of a goblet or flask and are known as the *goblet cells* (Fig. 2:24). Thus the organs lined by mucous membrane have a moist slippery surface. Mucus protects the lining membrane from mechanical and chemical injury and in the respiratory tract it traps foreign particles, preventing them from entering the lungs.

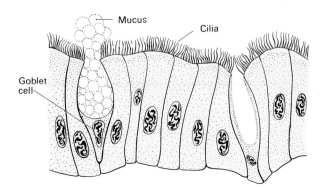

Figure 2:24 Ciliated columnar epithelium with goblet cell.

Serous membrane
Serous membranes consist of a double layer of tissue. The *visceral* layer closely invests the organ and the *parietal* layer lines the space in which the organ lies. The visceral layer is reflected off the organ to become the parietal layer and the two layers are separated by a watery or *serous fluid* secreted by the membrane.

There are three sites in which serous membranes are found. In the thoracic cavity the *pleura* surround the lungs and the *pericardium* surrounds the heart, and in the abdominal cavity the *peritoneum* surrounds the abdominal organs.

The presence of serous fluid between the visceral and parietal layers of serous membrane enables an organ to move without being damaged by friction between it and adjacent organs. For example, the heart changes its shape and size during each beat and friction damage is prevented by the arrangement of pericardium and its serous fluid.

Synovial membrane
This membrane is found lining the joint cavities and surrounding tendons which could be injured by rubbing against bones. It is made up of a layer of fine, flattened cells on a layer of delicate, connective tissue.

Synovial membrane secretes a clear, sticky, oily fluid known as *synovial fluid* which acts as a lubricant to the joints and helps to maintain their stability. Synovial fluid also helps to nourish the hyaline cartilage which covers the articular surfaces of the bones, and synovial cells remove damaging material from within the joint cavity.

GLANDS

Secreting epithelial cells line many of the organs of the body. The number of secretory cells is usually greatly increased by the invagination of the lining to form *glands*.

The glands vary in shape and complexity. The more complex glands require *canals* or *ducts* to carry their secretion to the surface of the organ. The names of the glands vary depending upon their complexity (Fig. 2:25).

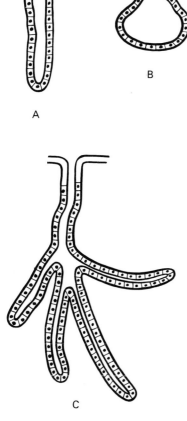

Figure 2:25   A. Simple tubular gland. B. Alveolar gland.
C. Branched tubular gland.

Simple tubular glands

The cells that make up these glands form a single tube which opens directly on to the free surface of the organ.

Branched tubular glands

Here the deep part of the tube becomes branched, having a more complex appearance and providing a larger number of secretory cells.

Alveolar glands

The cells of these glands enclose little spherical cavities called *alveoli*.

Compound alveolar or racemose glands (Fig. 2:26)

The cells that make up these very complex glands form many alveoli. Ducts carry away the secretion from each alveolus joining up with other ducts and finally empty the contents into one large duct which leads to the surface. The salivary glands are classical examples of *compound racemose glands*.

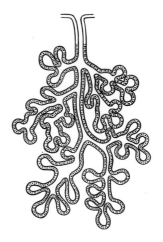

Figure 2:26   Compound racemose gland.

The secretions of the glands described above are carried away by the ducts. There are other glands found in the human body which secrete directly into the blood stream. These are known as the *ductless* or *endocrine glands* and are described in Chapter 14.

The Systems and Cavities of the Body

A *system* is a group of structures or organs which together carry out essential and related functions. The main systems of the body are:

Circulatory
Respiratory
Digestive
Urinary
Nervous
Endocrine
Reproductive
Osseus or skeletal
Muscular

The body as a whole is built round the bony framework or skeleton and consists of a number of different parts:

The *head* and *neck*
The *trunk*, which can be divided into the *chest* or *thorax*, *abdomen* and *pelvis*
The *limbs*, both upper and lower

For descriptive purposes the human body is divided into *cavities* and the main organs of the body are contained in these cavities. There are four main cavities:

Cranial
Thoracic

Abdominal
Pelvic

THE CRANIAL CAVITY

The cranial cavity contains the *brain*, and its *boundaries* are formed by the bones of the skull (Fig. 2:27).

Anteriorly—	1 frontal bone
Laterally—	2 temporal bones
Posteriorly—	1 occipital bone
Superiorly—	2 parietal bones
Inferiorly—	1 sphenoid and 1 ethmoid bone and parts of the frontal, temporal and occipital bones

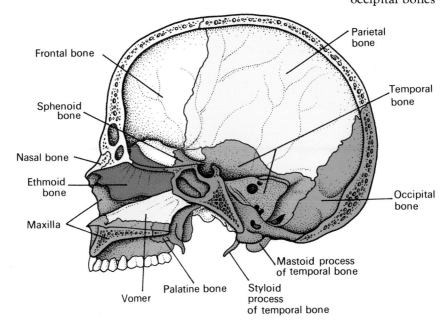

Figure 2:27   Bones forming the right half of the cranial cavity and the face—viewed from the left.

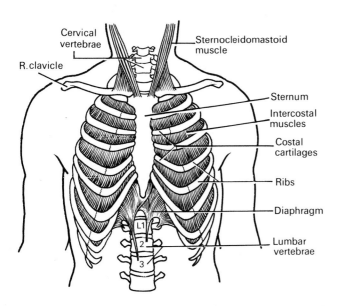

Figure 2:28   Structures forming the walls of the thoracic cavity.

THE THORACIC CAVITY

This cavity is situated in the upper part of the trunk. Its *boundaries* are formed by a bony framework and supporting muscles (Fig. 2:28).

Anteriorly—	the sternum and costal cartilages of the ribs
Laterally—	12 pairs of ribs and the intercostal muscles
Posteriorly—	the thoracic vertebrae and the intervertebral discs which lie between the bodies of the vertebrae
Superiorly—	the structures forming the root of the neck
Inferiorly—	the diaphragm, a dome-shaped muscle

Contents

The main organs and structures contained in the thoracic cavity are (Fig. 2:29):

The lungs
The heart
The trachea
The bronchi
Nerves
Lymph nodes
The oesophagus
The aorta

The superior and inferior venae cavae
Other blood vessels
Lymph vessels
Glands

The *mediastinum* is the name given to the space between the lungs.

THE ABDOMINAL CAVITY

This is the largest cavity in the body and is oval in shape (Figs. 2:30 and 2:31). It is situated in the main part of the trunk and its *boundaries* are:

Superiorly— the diaphragm, which separates it from the thoracic cavity

Anteriorly— the muscles forming the anterior abdominal wall

Posteriorly— the lumbar vertebrae and muscles forming the posterior abdominal wall

Laterally— the lower ribs and parts of the muscles of the abdominal wall

Inferiorly— the pelvic cavity with which it is continuous

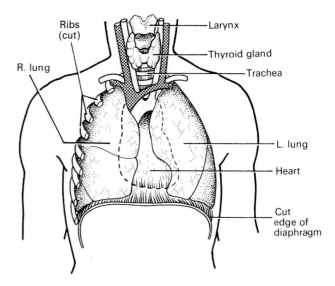

Figure 2:29   Main structures in the thoracic cavity and the root of the neck. Dotted line—heart.

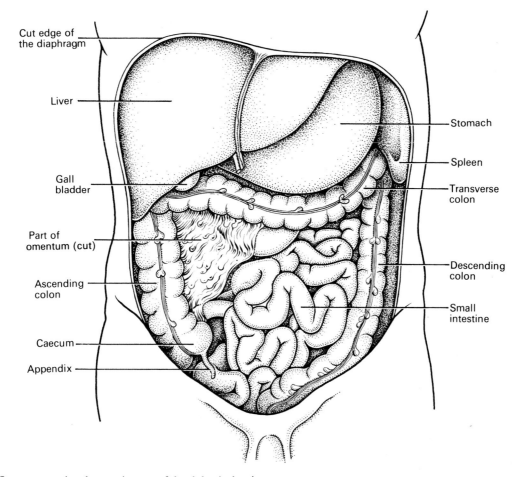

Figure 2:30   Organs occupying the anterior part of the abdominal cavity.

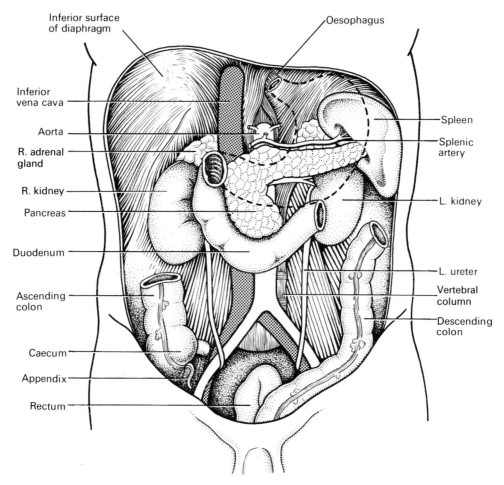

Figure 2:31   Organs occupying the posterior part of the abdominal cavity.

The abdominal cavity is divided into nine regions for the purpose of identifying the positions of the abdominal organs. These are named in Figure 2:32.

Contents

Most of the space in the abdominal cavity is occupied by the organs and glands involved in the digestion and absorption of food. These are:

    The stomach
    The small intestine
    Most of the large intestine
    The liver
    The gall bladder
    The bile ducts
    The pancreas

Other structures include:

    The spleen
    The kidneys and the upper part of the ureters
    The adrenal or suprarenal glands
    Numerous blood vessels, lymph vessels, nerves
    Lymph nodes

THE PELVIC CAVITY

The pelvic cavity is roughly funnel-shaped and extends from the lower end of the abdominal cavity (Figs. 2:33 and 2:34). The *boundaries* of the pelvic cavity are:

Superiorly—	it is continuous with the abdominal cavity
Anteriorly—	the pubic bones
Posteriorly—	the sacrum and coccyx
Laterally—	the innominate bones
Inferiorly—	the muscles of the pelvic floor

Contents

The pelvic cavity contains the following structures:

    The lower part of the large intestine, the rectum and the anus
    Some loops of the small intestine
    The urinary bladder
    The lower parts of the ureters

The urethra

In the female, the organs of the reproductive system—the uterus, uterine tubes, ovaries and vagina (Fig. 2:33)

In the male, some of the organs of the reproductive system—the prostate gland, seminal vesicles, spermatic cords, deferent ducts and ejaculatory ducts (Fig. 2:34)

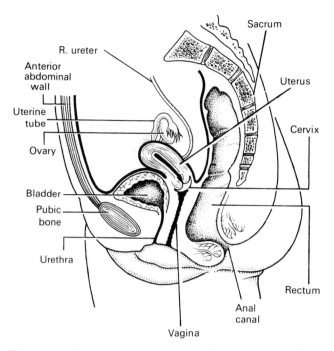

Figure 2:33 Organs in the pelvic cavity in the female.

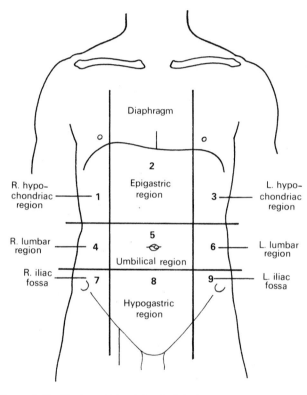

Figure 2:32 Regions of the abdominal cavity.

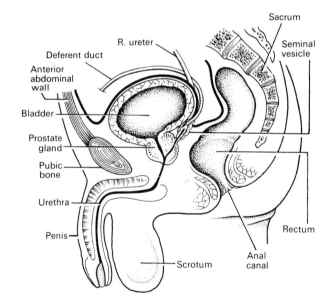

Figure 2:34 Organs in the pelvic cavity in the male.

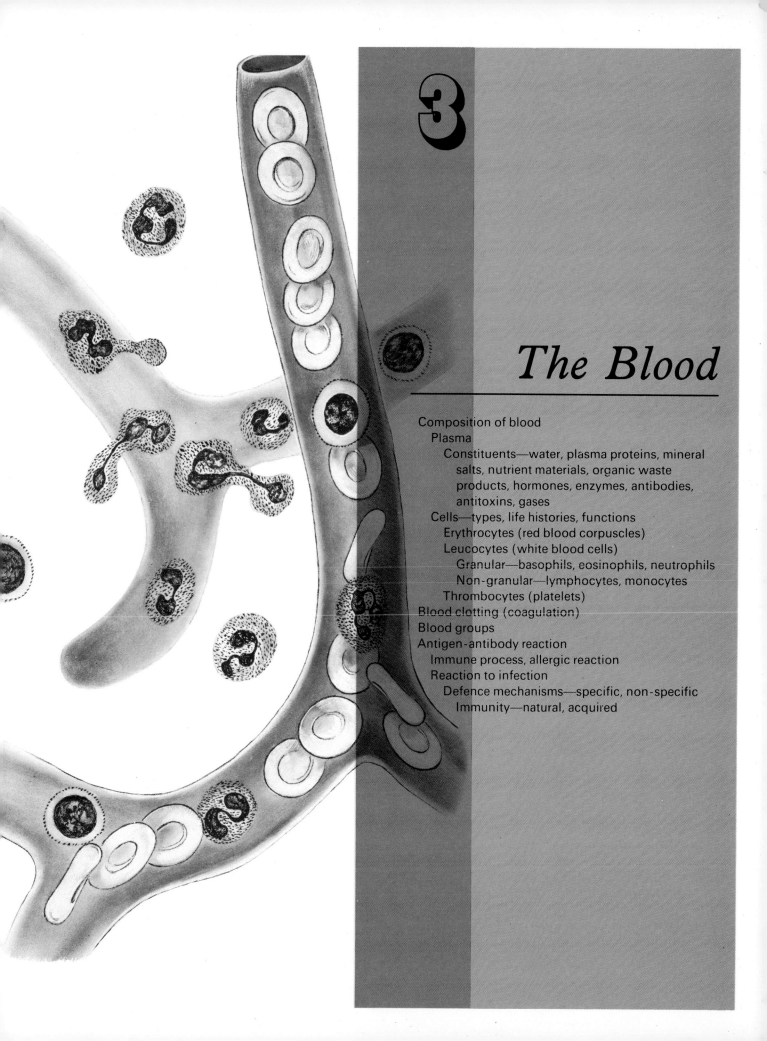

3

The Blood

3. The Blood

Blood is commonly described as a connective tissue. It provides one of the methods of communication between the cells of different parts of the body and indirectly with the external environment. It carries oxygen from the lungs to the tissues; nutrient materials from the alimentary tract to the tissues; hormones manufactured in the endocrine glands to their target organs; waste products to excretory organs such as the kidneys and the lungs; heat produced in a very active organ to less active areas; protective substances such as antibodies to areas of infection.

The blood in the blood vessels is always in motion. Cells take up raw materials to meet their needs and add waste products and newly formed substances required in other parts of the body. When an organ is highly active it takes up extra supplies to meet its needs and adds increased amounts of waste products. However, the flow of blood is such that a fairly stable environment is maintained around the cells and any changes in the composition of the blood as a whole remain within fairly narrow limits.

The blood constitutes approximately 8 per cent of the body weight or 5·6 litres in a 70 kg man.

Composition of Blood (Fig. 3:1)

The blood is composed of a faintly yellow transparent fluid known as the *plasma* and floating in this fluid are different types of *cells* and *corpuscles*.

The plasma constitutes approximately 55 per cent and the cells and corpuscles approximately 45 per cent of the blood volume.

THE PLASMA
The plasma is composed of the following constituents:

Water—	90 to 92 per cent
Plasma proteins—	60 to 80 g/l
albumin—	35 to 50 g/l
globulin—	20 to 37 g/l
fibrinogen—	2 to 4 g/l
prothrombin—	100 to 150 mg/l

Mineral salts (inorganic salts)
   sodium chloride and sodium bicarbonate. Also small amounts of potassium, magnesium, phosphorus, calcium, iron, copper, and iodine.

Nutrient materials (from digested foods)
   amino acids
   monosaccharides, mainly glucose
   fatty acids and glycerol
   vitamins

Organic waste products
   urea
   uric acid
   creatinine

Hormones

Enzymes

Antibodies and antitoxins

Gases
   oxygen, carbon dioxide and nitrogen

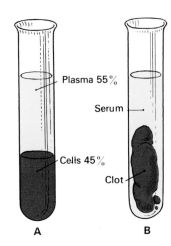

Figure 3:1   Blood in a test tube. A. Shows the percentage composition of plasma and cells. B. Shows the presence of a blood clot in serum.

PLASMA PROTEINS
The plasma proteins have several important functions to perform.
   1. They exert an osmotic force of about 25 mmHg (3·3 kPa)★ across the capillary wall (*oncotic pressure*).
   2. They give viscosity to the blood, thus assisting to some extent in the maintenance of the *blood pressure*.

★ 1 kilopascal (kPa) = 7·5 millimetres of mercury (mmHg)
   1 mmHg = 133·3 Pa = 0·1333 kPa

3. Some substances are transported around the body bound to plasma proteins.

4. One form of globulin is a carrier of antibodies (see p. 43).

Albumin is formed in the liver.

Fibrinogen is produced in the liver and is necessary for the clotting (coagulation) of blood. Plasma from which fibrinogen has been removed due to clotting is known as *serum*.

Globulin is divided into different types. γ-globulin, which is associated with the immune response, is formed by white blood cells called lymphocytes. The other types are formed by the liver.

Prothrombin is formed in the liver and is an essential substance in the mechanism of blood coagulation (see p. 41). Vitamin K, a fat-soluble vitamin found in green vegetables, is essential for the formation of prothrombin. Because of this it is sometimes referred to as the *anti-haemorrhagic* vitamin.

MINERAL SALTS

These salts are necessary for the formation of cells, the contraction of muscles, the transmission of nerve impulses and the maintenance of balance between acids and alkalis in the body. In health the blood is *slightly alkaline* in reaction. This is expressed by the symbol pH, which represents the hydrogen ion concentration in the blood (Fig. 3:2). The pH of the blood is maintained at about 7·4 by a complicated series of chemical changes occurring within the body.

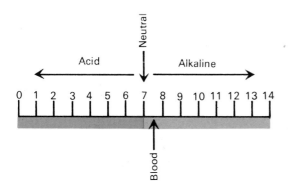

Figure 3:2   Illustration of the pH scale.

NUTRIENT MATERIALS

Amino acids, glucose, fatty acids, glycerol and vitamins are absorbed into the blood from the alimentary tract. These are the nutrient materials derived from the digestion of carbohydrates, proteins and fats which are eaten as food and are required to maintain the functioning of the body cells.

ORGANIC WASTE PRODUCTS

Urea, uric acid and creatinine are waste products of protein metabolism. They are formed in the liver and are conveyed by the blood to the kidneys to be excreted.

HORMONES

Hormones are chemical substances which are formed by certain glands. They are passed directly into the blood, which transports them to other organs (target organs) where they influence activity.

ENZYMES

Enzymes are chemical substances which can produce or speed up chemical changes in other substances without themselves being changed.

ANTIBODIES AND ANTITOXINS

These are protective substances consisting of complex proteins which are produced by *plasma cells* found mainly in lymph glands and in the spleen. Foreign proteins, such as micro-organisms or other large molecule substances, when introduced into the body may act as antigens and stimulate the production of specific antibodies. Antibodies which provide immunity against one antigen are not effective against another (see p. 43).

GASES

Oxygen, carbon dioxide and nitrogen are dissolved in and transported by the plasma water. Oxygen and carbon dioxide are also transported by other means which are described later (p. 39).

CELLULAR CONTENT OF BLOOD

There are three varieties of cells or corpuscles present in blood.

Erythrocytes or red blood corpuscles
Leucocytes or white blood cells
Thrombocytes or platelets

ERYTHROCYTES OR RED BLOOD CORPUSCLES (Fig. 3:3)

These are described as circular bi-concave non-nucleated

Figure 3:3   Red blood corpuscles.

discs, about 7 micrometres* (μm) in diameter. The central part of the corpuscle is thinner than the circumference, thus the term bi-concave.

The normal erythrocyte count is usually slightly lower in women than in men:

Women: $4\cdot5 \times 10^{12}/l$ to $5 \times 10^{12}/l$ of blood ($4\cdot5$ to 5 million per c.mm.)

Men: $5 \times 10^{12}/l$ to $5\cdot5 \times 10^{12}/l$ of blood (5 to $5\cdot5$ million per c.mm.)

It is the erythrocytes which give the blood its colour.

THE DEVELOPMENT AND LIFE HISTORY OF ERYTHROCYTES

The erythrocytes are formed in the *red bone marrow*. Red bone marrow is present in the cancellous bone which is found in the extremities of long bones and between layers of compact bone in flat and irregular bones, such as the sternum, skull, ribs and vertebrae.

*1 micrometre $= 10^{-6}$ metre.

The erythrocytes pass through several stages of development in the red bone marrow before reaching maturity. When mature they pass from the bone marrow into the circulating blood where they survive for about 120 days (Fig. 3:4).

Formation

There are two main lines of development to the stage of mature erythrocyte. One is the maturation of the erythrocyte itself and the other the formation of the substance called *haemoglobin* capable of transporting oxygen. Erythrocytes entering the blood are not normal unless they develop satisfactorily along *both* lines.

The normal maturation of the cell itself requires the presence of a number of different chemical substances, one of the most important of which is vitamin B_{12} (cyanocobalamin). This substance is present in food and

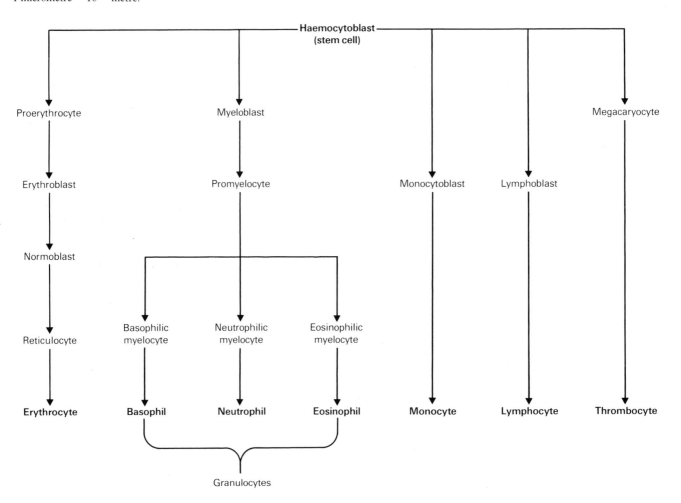

Figure 3:4 Stages in the development of blood cells in the bone marrow.

its absorption from the small intestine is dependent upon a glycoprotein secreted by the stomach called the *intrinsic factor*. When there is not sufficient vitamin B_{12} circulating to the red bone marrow large immature cells enter the blood stream and are the cause of a disease called *pernicious anaemia*.

Haemoglobin is a complex protein which is synthesised inside the immature erythrocyte in the red bone marrow. *Iron* is one of the chemicals necessary for its formation. The normal haemoglobin levels in the blood are slightly different in men and women:

Men: 14–17 g/dl (grams per 100 ml)
Women: 12–16 g/dl

Amounts of iron and vitamin B_{12} absorbed from the alimentary tract which are in excess of immediate requirements are stored in the liver.

The haemoglobin in the erythrocytes combines with oxygen to form *oxyhaemoglobin*. In this way the bulk of the oxygen breathed in is transported to the cells of the body. It is also involved in transporting the waste product, carbon dioxide, to the lungs for excretion.

Erythropoiesis, or the development of red blood cells, appears to be controlled by what is known as a feed-back mechanism. This means that the body produces erythrocytes at the same rate as they are destroyed. If there is an insufficient supply of oxygen to the body cells, for example following haemorrhage or because an individual lives at a high altitude where the oxygen pressure in the atmospheric air is reduced, the bone marrow is stimulated to increase its production of erythrocytes. Deficiency of oxygen (hypoxia) as such, and a hormone called *erythropoietin*, produced mainly by the kidneys, are the two main stimulants. After haemorrhage it is normal to find that the number of immature cells, reticulocytes, in the blood is increased.

Figure 3:4 shows the stages of development of erythrocytes, granulocytes (granular leucocytes) and platelets which originate from the same parent cell in the red bone marrow, the myeloblast.

Destruction
The life span of erythrocytes is believed to be about 120 days. The breakdown of these cells is called *haemolysis*, which is carried out by the cells of the *reticulo-endothelial system*. These cells are described as *phagocytic*, i.e. capable of ingesting and destroying foreign particles such as micro-organisms and worn out erythrocytes.

Reticulo-endothelial cells are to be found in the spleen, the liver and in the walls of blood vessels. During haemolysis iron is extracted and re-used for haemoglobin synthesis in the red bone marrow. The protein part of the erythrocyte is converted to a substance called *biliverdin*, which is almost completely changed to *bilirubin*, a yellow pigment: normal plasma level 6 to 12 μmol/l (up to 0·7 mg/100 ml). While circulating in the blood bilirubin is bound to the plasma protein, albumin. In the liver it is again changed to what is described as a *conjugated form* and excreted in bile as bile pigment. When tests are carried out to find the level of bilirubin in the blood the *unconjugated bilirubin* is described as *indirect-reacting* and the *conjugated bilirubin* as *direct-reacting*. When there is excess bilirubin in the blood the skin, the whites of the eyes and the mucous membrane become yellow. This is known as *jaundice* and the cause may be identified by measuring the levels of direct- and indirect-reacting bilirubin.

LEUCOCYTES OR WHITE BLOOD CELLS
The leucocytes differ from the erythrocytes in that they possess a nucleus, are fewer in number and larger in size. They vary in size from about 8 to 15 micrometres (μm) and there are 6×10^9 to 11×10^9 per litre of blood (6000 to 11 000 per ml).

The leucocytes are divided into two main varieties:

Granular or polymorphonuclear leucocytes
Nongranular or mononuclear leucocytes

GRANULAR LEUCOCYTES (GRANULOCYTES)
(Fig. 3:5)
These cells grow in the red bone marrow and pass through several stages of development before entering the blood system. They originate from the same parent cells as the erythrocytes, i.e. myeloblasts (Fig. 3:4). As the cells develop granules form in the cytoplasm.

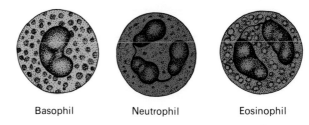

Basophil   Neutrophil   Eosinophil

Figure 3:5 The granular leucocytes.

The granulocytes constitute about 75 per cent (6.5×10^9/l) of all the white blood cells; three varieties are described:
Neutrophils—approximately 70 per cent (6×10^9/l)
Eosinophils—approximately 4 per cent (0.4×10^9/l)
Basophils—approximately 1 per cent (0.1×10^9/l)

These names are derived from the fact that the *granules* of the different varieties when stained in the laboratory absorb different dyes. The eosinophils take up a red acid dye called eosin, the basophils take up methylene blue which is alkaline in reaction and the neutrophils absorb both dyes and are therefore purple in colour.

Functions

Neutrophils. The neutrophils are responsible for providing the body with a defence against invading micro-organisms. They are attracted in large numbers to any area of the body which has been invaded by micro-organisms. The attraction comes from chemical substances liberated by the infected tissue. This process is called *chemotaxis*.

Neutrophils leave the blood by insinuating themselves through the walls of the capillaries in the infected area, a process known as *diapedesis* (Fig. 3:6). Thereafter they ingest and kill the organisms by digesting them, a process

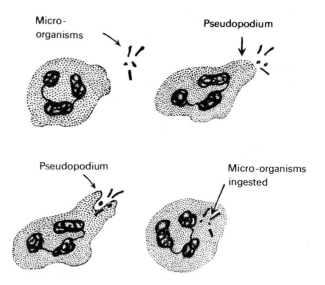

Figure 3:7   Diagram of phagocytic action of neutrophils.

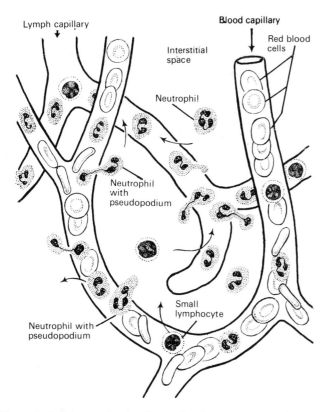

Figure 3:6   Diagram showing diapedesis of leucocytes.

known as *phagocytosis* (Fig. 3:7). The pus which may exude from an infected area consists of destroyed tissue, live micro-organisms and dead neutrophils which have ingested more micro-organisms than they could digest.

The life-span of neutrophils which remain in the blood vessels is believed to be about 30 hours, and those which leave the blood do not return.

Eosinophils and basophils. Their functions are not completely understood. They are less actively motile than the neutrophils. It is thought that the eosinophils phagocytose the particles which are formed when antigens and antibodies react (see p. 43).

The numbers of eosinophils and basophils are reduced when there is an increase in the amount of the hormone

hydrocortisone, produced by the cortex of the adrenal gland, in the blood (see p. 222).

Basophils and other similar cells when found in the tissues are frequently called *mast* cells. *Histamine* and the anticoagulant *heparin* are produced by mast cells in the tissues and probably by basophils in the blood.

NONGRANULAR LEUCOCYTES (Fig. 3:8)

There are two varieties of nongranular leucocytes and they comprise about 25 per cent of all leucocytes:

Lymphocytes—about 23 per cent (1.7×10^9/l)
Monocytes—about 2 per cent (0.1×10^9/l)

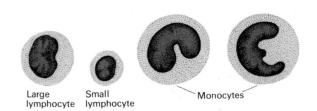

Figure 3:8   The nongranular leucocytes.

Lymphocytes

Lymphocytes are formed in the bone marrow and are found in all tissues of the body except in the brain and spinal cord. The greatest concentrations are in the lymph glands, the spleen and the walls of the large intestine.

They are classified according to their size, large or small; their life-span, 2 to 3 days or about 200 days; their functions.

After their formation in the bone marrow lymphocytes develop into two types, T and B types, each of which has a different function.

T-lymphocytes migrate to the *thymus gland* just before

and very soon after birth where they develop the characteristics associated with the rejection of tissues which are 'foreign' to the individual, such as a kidney transplant from a donor. After this 'processing' the T-lymphocytes move from the thymus gland via the blood and lymph vessels to the lymph glands and other aggregates of lymphoid tissue. In these sites they reproduce and retain the specific characteristics which could result in the rejection of a particular type of transplanted tissue.

B-lymphocytes have been given this name because they were first discovered in birds in the bursa of Fabricius which is not present in man. The site of sensitisation of these cells in man may be aggregates of lymphoid tissue, such as the tonsils. The B-lymphocytes are largely responsible for the development of specific immunity against foreign substances, or antigens, such as micro-organisms. When B-lymphocytes are stimulated by an antigen they enlarge to become *plasma cells* and form into groups. These groups or *clones* produce *specific antibodies* which act against the antigen (see p. 43).

Monocytes

These are large cells with a large nucleus; they are even bigger than the large lymphocytes and are relatively few in number. They are thought to originate in red bone marrow and in *reticuloendothelial* tissue found in organs such as the liver and spleen and in the walls of the blood vessels.

Function

The function of the monocytes closely resembles that of the neutrophils in that they are actively motile, phagocytic in action and will leave the blood stream to ingest micro-organisms and other foreign material which may be introduced into the tissues.

THROMBOCYTES OR PLATELETS

These are very small cells, smaller than the erythrocytes. They do not possess a nucleus, but have granules in their protoplasm. They have their origin in the red bone marrow and are derived from myeloblasts (see Fig. 3:4).

There are approximately $300 \times 10^9/l$ of blood (300 000 per ml).

Function

Thrombocytes play a very important part in the clotting of blood.

The Clotting or Coagulation of Blood

When any blood vessel is damaged blood will escape. To prevent blood loss the body reacts by a mechanism which is termed the *coagulation* or *clotting of blood*. This is a very complex process and is described here in its simplest terms.

There are some substances which must be present before clotting occurs:

    Prothrombin
    Calcium
    Fibrinogen
    Thromboplastin

Prothrombin, calcium and fibrinogen are all normal constituents of the blood. *Thromboplastin*, on the other hand, is released only when a blood vessel and tissue cells are damaged, for example by a cut, when it is liberated from the injured tissue cells and from the damaged thrombocytes. The release of thrombo-

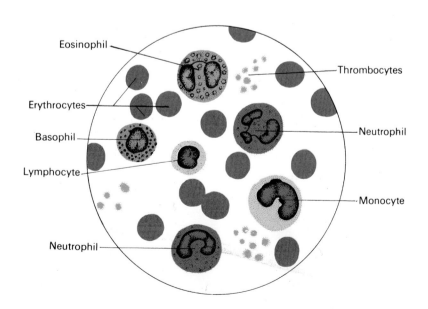

Figure 3:9 Normal blood cells after staining in the laboratory.

plastin starts off a series of events which will eventually produce a jelly-like mass known as a blood clot.

Prothrombin as such is inactive but when acted upon by *thromboplastin* in the presence of *calcium* it is converted into an active substance known as *thrombin*.

Thrombin in turn acts upon *fibrinogen* which is converted into an insoluble thread-like structure known as *fibrin*.

The *fibrin threads* enmesh the cellular content of the blood, that is, the erythrocytes and the leucocytes. This combination of fibrin and blood cells occludes the opening in the wall of the blood vessel and stops the bleeding.

After a time the clot shrinks and a clear sticky fluid is released. This is *serum*, i.e. plasma from which fibrinogen has been removed.

The mechanism of clotting can be expressed simply by the following formula:

prothrombin + calcium + thromboplastin = thrombin (active)
(inactive)   (from damaged
                     tissue cells
                     and platelets)

thrombin + fibrinogen   = fibrin (fine threads)
          (inactive)

fibrin + blood cells   = clot

The following is a list of the factors known to be involved in the coagulation of blood. They are sometimes referred to by number instead of by name.

Factor
I   Fibrinogen
II   Prothrombin
III   Thromboplastin
IV   Calcium
V   Accelerator globin or labile factor (AcG)
VII Stable factor (SPCA)
VIII Antihaemophilic factor (AHF)
IX   Plasma thromboplastic component. Christmas factor (PTC)
X   Stuart-Prower factor
XI   Plasma thromboplastin antecedent (PTA)
XII Hageman factor
XIII Fibrin-stabilising factor
(There is no factor VI.)

FACTORS AFFECTING BLOOD CLOTTING

Vitamin K

Vitamin K is necessary for the satisfactory formation of prothrombin which is made in the liver. Because of this, vitamin K is sometimes termed the *anti-haemorrhagic vitamin*. It is obtained from green foods such as cabbage, spinach and lettuce. Some vitamin K is formed by bacterial action in the large intestine but it is doubtful if significant amounts of this are absorbed.

Heparin

This is a substance which prevents the coagulation of blood and is therefore known as an *anti-coagulant*. Heparin is found in circulating basophils and in *mast cells*, which are wandering cells found in most tissues and in profusion in loose connective tissue. There is a substantial amount of heparin secreted into loose connective tissue *surrounding* the capillaries, which prevents blood coagulation. Heparin prevents the conversion of prothrombin to thrombin.

Blood Groups

Blood transfusions are frequently carried out in hospital wards. However, blood may not be taken indiscriminately from one person and transfused into another.

The membranes of erythrocytes contain antigens called *agglutinogens*, and some people have natural antibodies called *agglutinins* in their plasma which reacts against specific agglutinogens.

Individuals are divided into four main blood groups, A, B, AB and O. Group A blood always contains agglutinins which cause clumping or *agglutination* of Group B blood. Similarly Group B blood causes agglutination of Group A blood. Individuals with Group AB blood have no agglutinins and those with Group O blood have agglutinins against Group A and Group B blood. Figure 3:10 summarises the compatibility of the different blood groups.

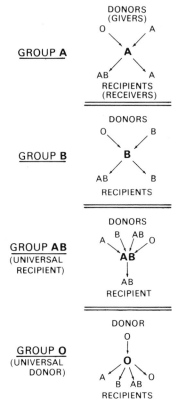

Figure 3:10 Blood groups showing their compatibility.

If incompatible blood is transfused the agglutinated erythrocytes block the small blood vessels and eventually they are haemolysed.

The frequency of occurrence in Caucasians of the *four blood groups* is as follows:

Group A— approximately 42 per cent
Group B— approximately 8 per cent
Group AB—approximately 4 per cent
Group O— approximately 46 per cent

THE RHESUS FACTOR

Since the above blood groups were described further agglutinogens have been discovered. One of the most important of these is the Rhesus (Rh) factor, named after the Rhesus monkey used in the original investigations.

In about 85 per cent of Caucasians the Rh factor is present (Rh positive) and in the remaining 15 per cent it is absent (Rh negative). In Africans, Chinese and Japanese 99 to 100 per cent are Rh positive.

TRANSFUSION OF RH+VE BLOOD

If the blood of a person who is Rh + ve is transfused into a person who is Rh − ve the recipient's body slowly produces an antibody to the agglutinogens. This antibody is termed the *anti-Rh factor*. There may be no indication of this incompatibility following the first transfusion, but if a second similar transfusion is given 10 days or even years later a serious, often fatal, reaction may occur because the anti-Rh factor causes severe damage to the transfused erythrocytes.

If, in an emergency, there is not time to test the donor's and recipient's blood for compatibility it is customary for the patient to be given Group O, Rh − ve blood.

CONCEPTION OF RH+VE CHILD BY RH−VE MOTHER

If a mother is Rh − ve and a father is Rh + ve the child may inherit the Rh factor from its father. The danger is that the erythrocytes of the fetus containing the Rh factor may escape into the mother's circulation stimulating the production of the *anti-Rh factor* by the mother. When the anti-Rh factor formed by the mother crosses the placenta into the fetal blood it causes haemolysis of the fetal erythrocytes. This factor is slow to develop and no serious effects will be noticed in the first child, but in subsequent pregnancies anti-Rh factor is very likely to be formed in quantities large enough to destroy the erythrocytes of a Rh + ve baby.

Antigen-Antibody Reaction

THE IMMUNE PROCESS

The immune response is the effect on the body of contact with antigens or substances of various types which are 'foreign' to the individual. Antibodies which are specific to the individual antigens, and will neutralise them, are produced by the lymphocytes and lymphoid tissue in the lymph glands and in the spleen (p. 37). Some of the antigens which stimulate such reactions in the body are:

Micro-organisms
Protein molecules from animals
Pollen from plants and flowers
Polysaccharides
Foreign tissues, for example, transplanted organs
Drugs, for example, large molecule drugs such as penicillin

After the first contact with an antigen there is an interval of about two weeks before antibodies are found in the blood. During this period there is intense activity within the lymphoid tissue. The antibody concentration in the blood which results from this *primary response* does not reach a high level and does not persist unless a second dose of antigen is encountered. As a result of this *secondary response* there is a marked and rapid increase in the antibody level in the blood which is sustained for a considerable period of time (Fig. 3:11).

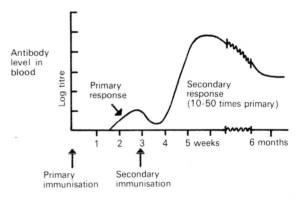

Figure 3:11 The antibody response.

Individuals are always exposed to antigens in the air, for example pollen from plants. In this way over a period of years the human body builds up a variety of antibodies to many naturally occurring antigens.

ALLERGIC REACTIONS OR HYPERSENSITIVITY

An allergic reaction is an undesirable response by the body to the invasion of a particular antigen which has previously been encountered.

The initial number of antibodies formed in response to an antigen may be small. Some of these antibodies may remain in the circulation and some may be absorbed by the mast cells.

If an individual is exposed to the same antigen again there may be sufficient antibodies in the circulation to neutralise the antigen, but if large amounts of the antigen are inhaled, eaten or come in contact with the skin there may not be sufficient *circulating* antibodies to neutralise them. The antigens may then combine with the antibodies *within the cells*. When this antigen-antibody reaction takes place within the cells a so-called *allergic reaction* occurs.

INFECTION

Infection occurs when micro-organisms which invade the body, grow, multiply by reproduction and produce toxins.

The human body is continuously exposed to contact with a wide variety of micro-organisms, such as, bacteria, fungi, protozoa and viruses. The micro-organisms which cause disease in man are described as *pathogenic*. Not all micro-organisms are pathogenic, in fact many of the *non-pathogens* are beneficial to man. *Commensal* organisms are non-pathogenic in their normal habitat but become pathogenic if they gain access to another part of the body.

The blood and tissue fluids of the body contain sufficient nutrient material for the survival, growth and reproduction of pathogenic micro-organisms. However, the normal healthy body possesses a variety of defence mechanisms which enable it to withstand the successful invasion of micro-organisms.

REACTION TO INFECTION: DEFENCE MECHANISMS

The individual is in contact with many pathogenic micro-organisms which may not gain entry to the body or which may be killed soon after entry and before they can multiply and cause disease. There is a number of non-specific and specific protections against invading organisms.

NON-SPECIFIC DEFENCE MECHANISMS

Skin and mucous membrane

When skin and mucous membrane are healthy and intact they provide a physical barrier to invading micro-organisms. The outer horny layer of the skin can be penetrated by only a few micro-organisms and the mucus secreted by the mucous membrane traps organisms on its sticky surface. Sebum and sweat, which are secreted on to the surface of the skin, contain bactericidal and fungicidal substances.

Bactericidal substances in body secretions

Hydrochloric acid, present in high concentrations in gastric juice (pH 1 to 2), kills the majority of ingested organisms.

Lysozyme is a small molecule protein with bactericidal properties present in tears, in granulocytes and in most body secretions. It is not present in sweat, urine and cerebrospinal fluid.

Saliva is secreted into the mouth and washes away food debris which could serve as culture media for micro-organisms. Its slightly acid reaction (pH about 6.7) inhibits the growth of some organisms.

Phagocytosis. When micro-organisms invade the body they immediately stimulate activity in neutrophils, monocytes and large cells described as *macrophages* found in the reticuloendothelial cells of the liver, spleen, bone marrow and connective tissue. The process of phagocytosis was described on page 40.

SPECIFIC DEFENCE MECHANISMS

Specific defence mechanisms depend upon the production of antibodies by lymphocytes in response to the presence of specific antigens. If the virus of measles invades the body antibodies are formed to destroy this particular pathogen. If the diphtheria bacillus invades the body a different antibody is produced to destroy it. In this way the body produces a multiplicity of antibodies which provide resistance to infection.

IMMUNITY

Immunity is said to occur when there are sufficient antibodies within the body to prevent the successful invasion of a particular type of pathogenic micro-organism.

Immunity can be acquired *naturally* or *artificially* and both forms may be acquired *actively* or *passively*. When immunity is acquired actively the individual has responded to the antigen and has produced his own antibodies. Passive immunity occurs when the individual has been given specific antibodies which have been produced by someone else.

Active naturally acquired immunity

This occurs when the body is involved *actively* in producing antibodies. This type of immunity can be acquired in two ways.

1. *By actually having the disease*. The micro-organisms invade the body successfully where they grow and reproduce in sufficient numbers to produce the clinical features of the disease. During the course of the illness antibodies are produced in sufficient numbers to overcome the micro-organisms and the person recovers. This sensitisation remains as a future protection against this particular pathogen.
2. *By having a subclinical or subliminal infection*. In this instance the body is exposed to minute numbers of micro-organisms which are insufficient to give rise to recognisable disease but are sufficient to stimulate the production of antibodies. Again sensitisation remains to provide future protection.

Passive naturally acquired immunity

In this instance the antibodies are passed from the mother to the fetus through the placenta before the baby is born. This type of immunity is thought to be very short lived.

Active artificially acquired immunity

This type of immunity develops in response to the

administration of a suspension of killed or attenuated micro-organisms or detoxicated toxins. *Attenuation* means that the organisms retain their antigenic identity but cannot cause the disease. This is achieved by heat treatment, exposure to chemicals such as formalin or by growing successive generations over a long period of time on artificial media in the laboratory.

The suspensions of micro-organisms which are dead or attenuated are called *vaccines* and the detoxicated toxins are called *toxoids*.

Vaccines and toxoids are prepared so that they will not cause the disease but are sufficiently powerful to stimulate the production of antibodies and thus build up an active immunity.

Many diseases can be prevented by artificial active immunisation. Some of these are:

Anthrax
Diphtheria
Measles
Poliomyelitis
Smallpox
Tetanus
Tuberculosis
Typhoid fever
Whooping cough
Yellow fever

Passive artificially acquired immunity

In passive immunity the individual plays no active part in the production of antibodies.

The ready-made antibodies obtained from human or animal serum are injected into the individual. Human serum contains the *ready-made antibodies* if the individual has recovered from an infectious disease, for example, measles. Animal serum contains the antibodies following its active immunisation.

Animal anti-sera may be used both *prophylactically* (to prevent infection) or *therapeutically* (to treat an infection).

A summary of the types of immunity is given below:

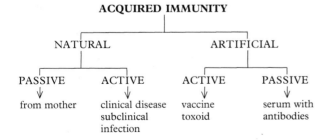

Summary of the Functions of the Blood

1. The blood transports oxygen as oxyhaemoglobin from the lungs to the body cells, and returns carbon dioxide from the cells to the lungs for excretion.

2. The blood is the means whereby all nourishment is transported to the cells. This nutritive material is absorbed into the blood stream from the small intestine in the form of monosaccharides, amino acids, fatty acids, glycerol, vitamins, mineral salts and water.

3. The blood removes all waste products from the tissues and cells. These waste materials are transported to the appropriate organs for excretion or to the liver to be prepared for excretion. After preparation by the liver the waste products of protein metabolism are in the form of urea, uric acid and creatinine and as such, they are transported in the blood to the kidneys for excretion.

4. The blood transports hormones and enzymes from their cells of origin to their target organs and tissues.

5. The blood aids in the defence of the body against the invasion of micro-organisms and their toxins due to:

(a) the phagocytic action of neutrophils and monocytes;

(b) the presence of antibodies and antitoxins.

6. By the mechanism of clotting, loss of body fluid and loss of blood cells is prevented.

7. The blood helps to maintain the body temperature. Due to the chemical activity in the cells and tissues heat is produced and the circulating blood is warmed. If too much heat is produced the blood vessels near the surface of the body dilate and heat is lost by radiation, conduction, convection and the evaporation of sweat. If the temperature of the outside atmosphere is low the superficial blood vessels constrict and heat loss is prevented.

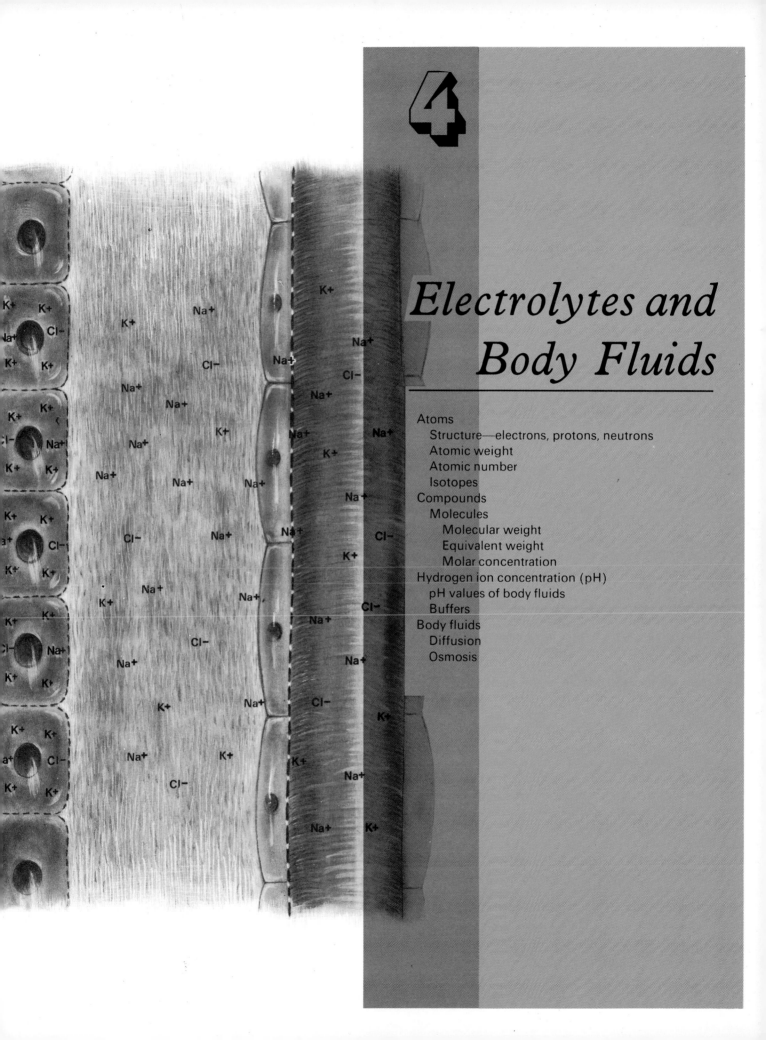

4

Electrolytes and Body Fluids

4. Electrolytes and Body Fluids

Before going on to consider the importance and nature of body fluids it is necessary to understand the meaning of such terms as electrolytes, isotopes, pH and buffer substances. If these terms are to be understood it is essential to survey some of the principles of the physics and chemistry of atoms, elements and compounds. Compounds which contain carbon are classified as organic—all others, as inorganic. The body fluids contain both.

Before discussing the compounds found in solution in the water of the body, the atom and its main constituent parts have to be considered.

The Atom

The atom is described as the smallest particle of an element which can take part in a chemical change. All atoms of an element are identical, however, the atoms of any one element are different from those of all other elements. This will be more clearly understood when the structure of the atom has been described.

Structure

Atoms are made up of a considerable number of particles, but only three are considered here.

Protons are particles which are present in the nucleus or central part of the atom. Each proton is described as having one unit of positive electrical charge and one unit of mass.

Neutrons are also found in the nucleus of the atom. They have no electrical charge and one unit of mass.

Electrons are particles which revolve in orbit around the nucleus of the atom at a distance from it, as the planets revolve round the sun. Each electron carries one unit of negative electrical charge and its mass is so small that it can be disregarded when considering the weight of an atom as a whole.

Table 4:1 summarises the characteristics of these subatomic particles.

In all atoms the number of positively charged protons in the nucleus is *equal* to the number of negatively charged electrons in orbit around the nucleus.

The difference which exists between elements is to be found in the *numbers* of these essential particles which

Table 4:1

Particle	Mass	Electric charge
Proton	1 unit	1 positive
Neutron	1 unit	neutral
Electron	negligible	1 negative

make up their atoms. The planetary electrons revolve in rings or shells around the nucleus and there is an optimum number of electrons in each shell (Fig. 4:1).

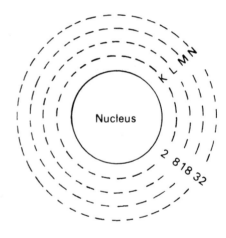

Figure 4:1 Diagram of the atom showing the nucleus and four electron shells.

Atomic number

The atomic number of an element is the number of protons, or electrons in orbit around the nuclei of the atoms of that element. For example, the atomic number of oxygen is 8 because it has 8 protons and 8 planetary electrons, hydrogen is 1 and sodium 12 (Fig. 4:2).

Formation of compounds

It was mentioned earlier that the atoms of each element have a specific number of electrons in orbit around the nucleus. When the number of electrons in the outer shell of an element is the optimum number (Fig. 4:1), the element is described as inert or chemically unreactive. It will not easily form compounds by combining

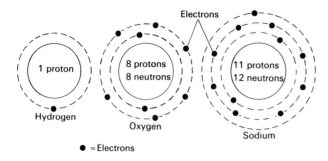

Figure 4:2 Diagram of the structure of atoms of different elements showing protons, neutrons and orbiting electrons.

with other elements. The inert gases—neon, argon, krypton, xenon and radon—come into this category.

Elements which have incomplete outer shells of electrons are reactive and will combine with other elements which also have incomplete outer electron shells. In the formation of *electrovalent* or *ionic compounds* electrons are transferred from one element to another. For example, when sodium (Na) combines with chlorine (Cl) to form sodium chloride (NaCl) there is the transfer of one electron, that is, the only electron in the outer shell of the sodium atom is transferred to the outer shell of the chlorine atom. This makes the outer shell (the M shell) of the chlorine up to its full capacity of 8 electrons and leaves the sodium part of the compound with a complete L shell (Fig. 4:3).

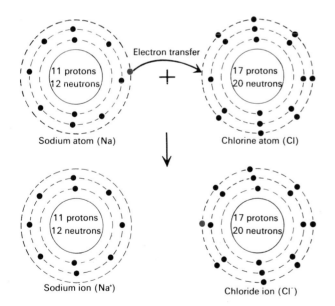

Figure 4:3 Diagram showing the formation of the ionic compound sodium chloride.

The number of electrons is the only change which occurs in the atoms in this type of reaction. There is no change in the number of protons or neutrons in the nuclei of the atoms. The chlorine atom has gained one

electron, therefore it now has *18 electrons*, each with one negative electrical charge, and *17 protons*, each with one positive electrical charge. When sodium chloride is dissolved in water this imbalance of protons and electrons results in the formation of a negatively charged particle, that is an *anion* or *negative ion*, which is written thus, Cl^-. The sodium ion, on the other hand, has lost one electron, leaving *10 electrons*, each with one negative electrical charge in orbit round a nucleus containing *11 protons*, each with one positive electrical charge. The imbalance of electrical charges in this case results in a positively charged particle called a *cation* or *positive ion*, which is written Na^+. The number of electrons which transfer in the formation of ionic compounds is indicated by the number of superscript plus or minus signs, for example, the magnesium ion is written Mg^{++} and the sulphide ion S^{--} because two electrons have transferred.

The nature of electrical charge, which is used daily in lighting, heating and so on, is closely related to the structure of the atom. Power initiated by water falling from a height or by steam under pressure is used to drive a dynamo to generate electricity, which is the movement of electrons. The electric current is then led away along a metal wire called a *conductor*. A substance which prevents the flow of electricity, or electrons, is a *non-conductor* or *insulator*.

Faraday, *c.* 1830, discovered that some substances when in solution conducted electricity while others did not. It was later discovered that the solutions which conduct electricity consist of ionic compounds. These solutions are called *electrolytes* and those which do not are *non-electrolytes*.

Sodium chloride is an inorganic compound and when dissolved in water it is said to *ionise* or *dissociate* into a number of anions (Cl^-) and an equal number of cations (Na^+) and is therefore an electrolyte. Sodium chloride is present in the human body in solution in water.

In this discussion, sodium chloride has been used as the example of the formation of an ionic compound and to illustrate electrolyte activity. There are, however, many other electrolytes within the human body which, though in relatively small quantities, are equally important. Although these substances may enter the body in the form of compounds, such as sodium bicarbonate, they are usually discussed in the ionic form, that is as sodium ions (written Na^+) and bicarbonate ions (written HCO_3^-).

The bicarbonate part of sodium bicarbonate is derived from carbonic acid (H_2CO_3). All inorganic acids contain hydrogen combined with another element or with a group of elements called *radicals*, which act like a single element. Hydrogen combines with chlorine to form hydrochloric acid (HCl) and with the *phosphate radical* to form phosphoric acid (H_3PO_4). When these two acids ionise they do so thus:

$$HCl \longrightarrow H^+Cl^-$$
$$H_3PO_4 \longrightarrow 3H^+PO_4^{---}$$

In the second example three atoms of hydrogen have each lost one electron all of which have been taken up by one unit, the phosphate radical, to make a phosphate ion with three negative charges.

A large number of compounds present in the body are not ionic and therefore have no electrical properties when dissolved in water. Some organic compounds, such as carbohydrates, are not ionic.

Isotopes

In an over simplification it was stated earlier that all atoms of an element are identical. This is true as far as the number of protons and electrons are concerned, but some atoms of the same element have a *different number of neutrons in the nucleus*. This does not affect the electrical activity of these atoms but it does affect their weight. For example, there are three forms of the hydrogen atom. The most common form has one proton in the nucleus and one orbiting electron. Another form has one proton and *one neutron* in the nucleus. A third form has one proton and *two neutrons* in the nucleus and one orbiting electron. These three forms of hydrogen are called *isotopes* (Fig. 4:4).

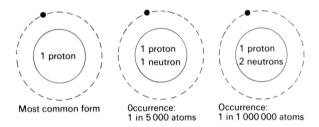

Most common form | Occurrence: 1 in 5 000 atoms | Occurrence: 1 in 1 000 000 atoms

Figure 4:4 Diagram of the isotopes of hydrogen.

Atomic weight

The atomic weight of an element is the sum of the protons and neutrons in the nucleus of the atoms of the element. Taking into account the isotopes of hydrogen and the proportions in which they occur, the atomic weight of hydrogen is 1·008, although for many practical purposes it can be taken as 1.

Oxygen has an atomic weight of 16. This means that a specific number of atoms of oxygen weigh 16 times more than the same number of atoms of hydrogen. Any standard scale of weight may be used for this purpose, for example, milligrams, grams, pounds or ounces.

Chlorine has an atomic weight of 35·5, because it exists in two isotopic forms, one form has 18 neutrons in the nucleus, and the other 20. Because the proportion of these two forms are not equal, the *average atomic weight* which emerges is 35·5.

Molecules and molecular weight

Molecules are the units of substances consisting of two or more atoms. For example, a molecule of sodium chloride (NaCl) consists of one atom of chlorine and one atom of sodium; and a molecule of atmospheric oxygen (O_2) consists of two atoms of oxygen.

The molecular weight of a compound is the sum of the atomic weights of the elements which form the molecules of the compound. For example:

A. Water (H.OH)
   2 hydrogen atoms (atomic weight =   1)   2
   1 oxygen atom   (atomic weight = 16)   16
                                            ——
                        Molecular weight = 18

B. Sodium bicarbonate ($NaHCO_3$)
   1 sodium atom   (atomic weight = 23)   23
   1 hydrogen atom   (atomic weight =   1)   1
   1 carbon atom   (atomic weight = 12)   12
   3 oxygen atoms   (atomic weight = 16)   48
                                            ——
                        Molecular weight = 84

Molecular weight, like atomic weight, is expressed simply as a figure until a scale of measurement of weight is applied to it, e.g. milligrams, grams.

Equivalent weight

If someone in a laboratory wished to synthesise water he would find that he would require to take twice as much hydrogen as oxygen and provide the appropriate conditions for their chemical combination to form water. Thus two volumes of hydrogen are *equivalent to* one volume of oxygen.

To make 36·5 grams of hydrochloric acid (HCl) it would be necessary to take 1 gram of hydrogen and 35·5 grams of chlorine. One gram of hydrogen *contains the same number of atoms as* 35·5 grams of chlorine. Provided these two gases were given the appropriate conditions for chemical combination, hydrochloric acid would be formed with no hydrogen or chlorine left. It can be said, therefore, that 1 atomic weight of hydrogen is *equivalent to* 35·5 atomic weights of chlorine.

The definition of equivalent weight is the weight of an element which will combine with *one atomic weight of chlorine* or displace *one atomic weight of hydrogen*.

Using hydrochloric acid as an example, it can be seen that the equivalent weight of both elements is their atomic weight, but if hydrochloric acid is combined with calcium it is found that one atomic weight of calcium combines with two atomic weights of chlorine releasing two atomic weights of hydrogen. The equation for this reaction is:

$$Ca + 2HCl \rightarrow CaCl_2 + H_2$$

The weight of calcium, therefore, which is equivalent to one atomic weight of chlorine is half an atomic weight of calcium (calcium: atomic weight = 40·8, therefore its equivalent weight = 20·4).

Any scale of measurement of the weight of these elements can be used provided the same scale is used for all the elements involved in the reaction. The measures commonly used are grams or milligrams or, for even smaller quantities, micrograms.

The electrolytes which are found within the human body may be expressed in terms of *milliequivalents per litre of body fluids (mEq/l)*. This is *the equivalent weight in milligrams in each litre* of body fluid.

Molar concentration

This is the method of expressing the concentration of substances present in the body fluids recommended in the *Système Internationale* which has now been adopted in the United Kingdom.

The mole (mol) is the molecular weight in grams of a substance.

A molar solution is a solution in which 1 mole in grams is dissolved in 1 litre of solvent. In the human body the solvent is water.

Molar concentration may be used to measure quantities of electrolytes, non-electrolytes, ions and atoms. For example molar solutions of the following substances mean:

1 mole of sodium chloride molecules (NaCl)	= 58·5 g per litre
1 mole of sodium ions (Na^+)	= 23 g per litre
1 mole of carbon atoms (C)	= 12 g per litre
1 mole of atmospheric oxygen (O_2)	= 32 g per litre

In physiology this system has the advantage of being a measure of the number of particles of substances present because 1 mol/litre of different substances contains the same number of particles. It has the advantage over the measure milliequivalents per litre because it can be used for non-electrolytes, in fact for any substance of known molecular weight.

Many of the chemical substances present in the body are in very low concentrations so it is more convenient to use *millimoles per litre* (mmol/l) as a biological measure. This is the molecular weight of a substance measured in milligrams (one thousandth of a gram) in 1 litre of water. Even smaller quantities such as micromoles and nanomoles are sometimes used.

Examples of normal plasma levels

Substance	Amount in S.I. units	Amount in other units
Chloride	95–105 mmol/l	95–105 mEq/l
Sodium	138–148 mmol/l	138–148 mEq/l
Calcium	2·1–2·6 mmol/l	4·3–5·3 mEq/l
Glucose	3·5–5·5 mmol/l	60–100 mg/100 ml

pH or Hydrogen Ion Concentration

The number of hydrogen ions present in a solution is a measure of the acidity of the solution. The maintenance of the normal hydrogen ion concentration within the body is an important factor in the environment of the cells.

A standard scale for the measurement of the hydrogen ion concentration in solution has been developed. All acids do not ionise completely when dissolved in water, that is, all the molecules of acid in solution do not ionise and exist in the solution as electrically charged particles. The hydrogen ion concentration is a measure, therefore, of the amount of *dissociated acid* rather than of the total amount of acid present. Strong acids dissociate more freely than weak acids, for example, hydrochloric acid dissociates freely into H^+ and Cl^-, while carbonic acid dissociates much less freely into H^+ and HCO_3^-. The number of *free hydrogen ions* in a solution is a *measure of its acidity* rather than an indication of the type of molecule from which the hydrogen ions originated.

The alkalinity of a solution is dependent upon the number of hydroxyl ions (OH^-) or other negatively charged radicals present. Water is a neutral solution because every molecule contains one hydrogen ion and one hydroxyl radical. For every molecule of water (H.OH) which dissociates, one hydrogen ion (H^+) and one hydroxyl ion (OH^-) are formed, each one neutralising the other.

The scale for measurement of pH was developed taking water as the standard. It was found by experiment that 1 molecule in 550 000 000 molecules of water ionises into a hydrogen ion and a hydroxyl ion. This is the same proportion as 1 gram hydrogen ion in 10 000 000 litres of water. Therefore 1 litre of water contains $\dfrac{1}{10\,000\,000}$ of a gram of hydrogen ion. In order to make these figures more manageable this can be written:

$$1 \text{ litre of water contains } \frac{1}{10^7} \text{ g hydrogen ion}$$
$$(10 = 10^1; 100 = 10^2; 10\,000 = 10^4)$$

The fraction $\dfrac{1}{10^7}$ may be written 10^{-7}, the negative power indicates that this is a fraction. $\dfrac{1}{10\,000\,000}$ may be written $\dfrac{1}{10^7}$ or 10^{-7}.

Later, for every-day use, only the *'power'* figure was used and the symbol pH placed before it.

Thus in a neutral solution such as water, where the number of hydrogen ions is balanced by the same number of hydroxyl ions, the pH = 7. The range of this scale is from 0 to 14. If the pH is 0 it would mean that 1 litre contained $\dfrac{1}{1} = 1$ gram of hydrogen ion; or at the other end of the scale if there were no hydrogen ions present it would be written $\dfrac{1}{10^{14}}$ or pH 14. It will be noted that a change of pH of *one* at any level in the scale means an increase or decrease by a factor of 10 in hydrogen ion concentration.

A pH reading *below* 7 indicates an *acid solution*, while readings *above* 7 indicate an *alkaline solution* (Fig. 4:5).

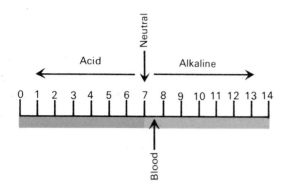

Figure 4:5   Diagram of the pH scale.

Ordinary litmus paper indicates whether a solution is acid or alkaline by colouring blue for alkaline and colouring red for acid. It is possible to obtain specially treated absorbent paper which gives an approximate measure of pH by a colour change. Laboratories which have to make accurate measurements use a *pH meter*.

pH values of the body fluids

All body fluids have pH values which must be maintained within relatively narrow limits within which the cells function normally. These pH values are not the same in every part of the body. For example, the normal range of pH values of the following secretions are:

Blood	7·4 to 7·5
Saliva	6·4 to 7·4
Gastric juice	1·5 to 1·8
Duodenal fluid	5·5 to 7·5
Gall bladder bile	5·5 to 7·7
Pancreatic juice	7·5 to 8·2
Urine	4·5 to 8·0

It can be seen that there is a wide variation of pH values within the body. The individual pH in an organ is produced by its secretion of acids or alkalis which will establish the optimum level. The highly acid pH of the gastric juice is maintained by the secretion into the stomach of hydrochloric acid produced by chemical changes which take place in special cells in the walls of the stomach. The low pH value in the stomach provides the environment best suited to the functioning of the enzyme pepsin which is present in gastric juice. Saliva has a pH of between 6·4 to 7·4—the optimum value for the action of ptyalin, the enzyme present in saliva which initiates the digestion of carbohydrates. Ptyalin action in inhibited when food containing it reaches the stomach

and is mixed with highly acid gastric juice with a pH of 1·5.

The blood has a pH value of between 7·4 and 7·5. This, therefore, is the general pH level in the body which may be altered in an individual organ, such as the stomach, to meet the needs of the special functions of that organ. The range of pH of the blood which is compatible with life is 7·0 to 7·8. The metabolic activity of the body cells produces certain acids and alkalis which tend to alter the pH of the tissue fluid and the blood. To maintain the pH within the normal range, there are substances present in the blood which act as *buffers*.

Buffers

Buffers are chemical substances such as phosphates, bicarbonates and some proteins which are able to 'bind' free hydrogen ion or hydroxyl ion and so prevent a change in pH. For example, if there is sodium hydroxide (NaOH) and carbonic acid (H_2CO_3) present, both will ionise to some extent, but they will also react together to form sodium bicarbonate ($NaHCO_3$) and water (H.OH). One of the hydrogen ions from the acid has been 'bound' in the formation of the bicarbonate radical and the other by combining with the hydroxyl radical to form water.

$$NaOH \quad + \quad H_2CO_3 \quad \rightarrow \quad NaHCO_3 \quad + \quad H.OH$$

sodium hydroxide   carbonic acid   sodium bicarbonate   water

The ability of the complex buffer systems of the blood to 'bind' hydrogen ions, or neutralise acids, is called the *alkali reserve* of the blood. If the pH of the blood falls below 7·4, the reserve of alkali has been reduced by an increase in production of hydrogen ions, and the condition of *acidosis* exists. When the reverse situation pertains and the pH is raised above 7·5, the increased alkali produced has used up the *acid reserve*, and a state of *alkalosis* exists.

The buffer system serves to prevent dramatic changes in the pH values in the blood, but it can only function effectively if there is some means by which excess acid or alkali can be excreted from the body. The two organs most active in this way are the lungs and the kidneys. When respiration is decreased there is an accumulation of carbon dioxide in the body which uses up the alkali reserve of the blood resulting in the development of acidosis. On the other hand, if there is 'over-breathing', which results in excessive excretion of carbon dioxide, the condition of alkalosis may develop.

The kidneys have the ability to form ammonia which combines with the acid products of protein metabolism and are then excreted in the urine.

The buffer and excretory systems of the body together maintain the *acid-base balance* of the body, so that the pH range of the blood remains within normal, but narrow, limits.

Body Fluids

The human body is made up of approximately 60 per cent water. For example, a man weighing 70 kg consists of about 42 litres of water. Approximately 28 litres (66 per cent) of this water is intracellular, that is, within the cells of the body, and about 14 litres (34 per cent) is extracellular. The water in the body provides the medium for the transportation of many chemical substances dissolved in it.

The extracellular fluid consists of fluid in the blood and lymph vessels, the cerebrospinal fluid and the fluid in the interstitial spaces of the body. This latter is sometimes called tissue fluid. It bathes all the cells of the body and occupies the interstitial spaces. The constituents of blood which do not pass into the tissue fluid are those which are too large in size to pass through the semipermeable walls of the capillaries. Substances which remain in the blood vessels are plasma proteins, erythrocytes, thrombocytes and leucocytes, except those which are amoeboid. Electrolytes and nutritional materials, such as glucose, amino acids, fatty acids, glycerol, carbon dioxide and oxygen are all in units which are small enough in size to pass freely across the walls of the capillaries. This means that the concentration of electrolytes in plasma is virtually the same as the concentration in the tissue fluid.

As the tissue fluid provides the external environment of the cells of the body, measurement of the electrolyte concentration in the plasma provides some indication of the intracellular concentrations of these substances. However, cell membranes are more complex structures than capillary walls, so concentrations of substances in tissue fluid are not always the same as concentrations inside cells.

Before going on to discuss the physiological factors which maintain fluid and electrolyte balance in the body, there are two physical processes which require explanation. These are *diffusion* and *osmosis*, both of which involve the movement of substances across semipermeable membranes.

A semipermeable membrane is one with 'pores' that permit the passage of small-sized substances. Although molecular weight does not always indicate the size and shape of a molecule, it can be used as a guide to the relative sizes of molecules of chemical compounds.

Diffusion

Diffusion is the physical process in which *dissolved substances* cross a membrane in order to establish equality of concentration on the two sides of the membrane. It is a fairly slow process as it takes time for equilibrium to be established. This type of movement of dissolved substances is not the deliberate movement of a specific number of particles (as shown schematically in Fig. 4:6)

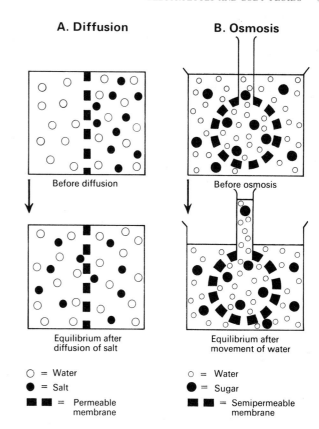

A. Diffusion **B. Osmosis**

Before diffusion

Equilibrium after diffusion of salt

Before osmosis

Equilibrium after movement of water

○ = Water
● = Salt
■ ■ = Permeable membrane

○ = Water
● = Sugar
■ ■ = Semipermeable membrane

Figure 4:6 A. Before and after diffusion. B. Before and after osmosis.

but results from the constant movement characteristic of all substances in solution. The particles bombard the dividing membrane and some pass through. The net result is equality of concentration of substances on the two sides of the membrane. In diffusion the net movement of dissolved substances is from the higher concentration to the lower, that is, down the concentration gradient.

Osmosis

Osmosis is the process of the transfer of the *water* of a solution across a semipermeable membrane. The force with which this occurs is called the *osmotic pressure*.

It would be easier to understand this phenomenon if osmotic pressure was considered as *osmotic pull*, because water crosses the semipermeable membrane from the low concentration side to the high concentration side. By taking water away from the side of lower concentration its concentration increases, and by adding water to the high concentration solution its concentration is reduced. The process will continue until the concentrations on each side of the membrane are equal. When this happens they are said to be *isotonic*.

An example may help to clarify this process. If a dilute solution of gelatine and water is prepared and placed in a beaker and a concentrated solution is placed inside a semipermeable membrane such as a cellophane bag, as shown in Figure 4:6, water passes from the *solution of lower concentration to the solution of higher concentration.* The osmotic pull which is exerted is directly related to the difference in the concentration of the dissolved substances which cannot pass through the pores in the membrane on the two sides of the membrane.

The processes of diffusion and osmosis have been described separately, but it must be understood that within the body the two processes occur concurrently.

Osmosis and diffusion cannot explain the transfer of all substances which are known to cross living cell membranes. Two further explanations offered for the transfer of such substances are:

1. Some substances can *dissolve in the fat* of the cell membrane and diffuse through it until they enter the cytoplasm of the cell.
2. Some substances can be *actively transferred* through the walls of the cell. For example, for each chemical transferred (substance X) there is a carrier substance (Y) in the cell membrane. Y combines with X to form XY. The combined substances, XY, then cross the cell wall and at its inner surface, X is released into the cell cytoplasm and Y remains in the cell membrane. From this it can be seen that the amount of X which can be transferred will depend on the amount of Y present and the rate at which the transfer occurs.

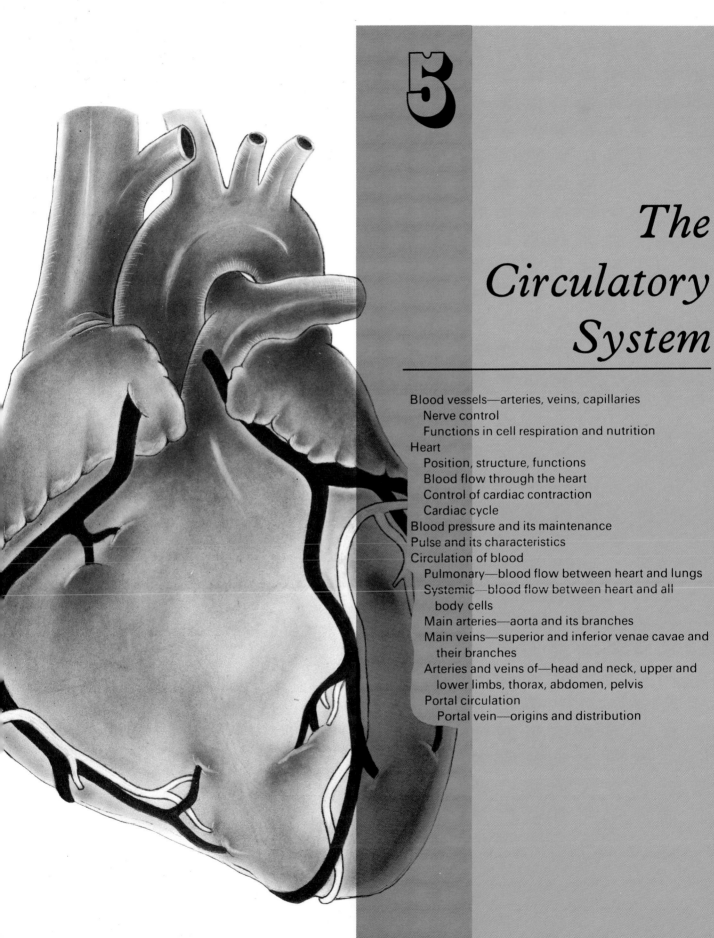

5

The Circulatory System

5. The Circulatory System

The circulatory or vascular system is divided for descriptive purposes into two main parts:

1. *The blood circulatory system*, consisting of the heart, which acts as a pump and the blood vessels through which the blood circulates. The heart and the blood vessels form a continuous system through which the blood flows.
2. *The lymphatic system*, which consists of lymph nodes and lymph vessels through which a colourless fluid known as *lymph* flows.

The two systems communicate with one another and are intimately associated.

The Blood Vessels

There are several types:
Arteries
Arterioles
Veins
Venules
Capillaries

The arteries

These are the blood vessels which transport blood away from the heart. They vary considerably in size, but have much the same structure (Fig. 5:1). They consist of three layers of tissue:

The *tunica adventitia* or outer layer consists of fibrous tissue.
The *tunica media* or middle layer consists of smooth muscle and elastic tissue.
The *tunica intima* or inner lining consists of squamous epithelium called endothelium.

The amount of muscular and elastic tissue varies in the arteries depending upon their size. In the large arteries the tunica media consists of more elastic tissue and less muscle. These proportions gradually change as the arteries become smaller until in the *arterioles* (the smallest arteries) the tunica media consists almost entirely of smooth muscle.

The veins

The veins are the blood vessels which transport blood to the heart. The walls of the veins, like those of the arteries, have three layers of tissue (Fig. 5:1):

The tunica adventitia—fibrous tissue
The tunica media—smooth muscle and elastic tissue
The tunica intima—squamous epithelium (endothelium)

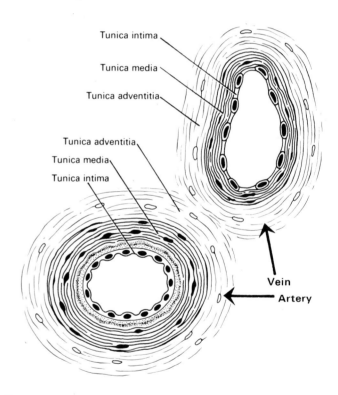

Figure 5:1 Structure of an artery and a vein.

Although composed of the same layers of tissue, the walls of the veins are much *thinner* than those of the arteries because there is less muscle and elastic tissue in the tunica media. When cut, the thin-walled veins collapse while the thicker walled arteries retain their cylindrical shape.

Some veins possess *valves*, which prevent the back flow of blood (Fig. 5:2). Valves are abundant in the veins of the limbs and are usually absent from those in the thorax

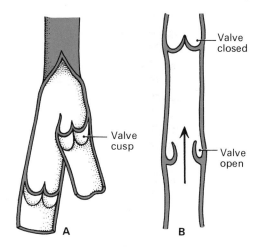

Figure 5:2 Interior of a vein. A. Showing the valves and cusps. B. Arrow showing the direction of blood flow through a valve.

and the abdomen. They are formed by a fold of tunica intima strengthened by connective tissue. In shape they are *semi-lunar* with the concavity towards the heart. The smallest veins are called *venules*.

The capillaries

The small arteries known as the arterioles break up into a number of minute vessels called *capillaries*. The wall of a capillary is composed of a single layer of endothelial cells which is very thin and permits the passage of water and other small-molecule substances. Blood cells and large-molecule substances such as plasma proteins do not normally pass through capillary walls. The capillaries form a vast network of tiny vessels which link the smallest arterioles to the smallest venules. Their diameter is approximately that of an erythrocyte (7 μm).

THE NERVOUS CONTROL OF THE BLOOD VESSELS

Both the veins and arteries are supplied by nerves of the *autonomic nervous system*. These nerves arise from the *vasomotor centre* in the *medulla oblongata* and they change the calibre of the vessels, thus controlling the amount of blood circulating to any part of the body. The changes in calibre are the result of contraction or relaxation of the muscle in the wall of the blood vessel. Medium-sized and small arteries have more muscle than elastic tissue in their walls. In large arteries, such as the aorta, the middle layer is almost entirely elastic tissue. This means that small arteries and arterioles respond to nerve stimulation whereas large arteries dilate and constrict according to the amount of blood they contain.

The nerves which reduce the lumen of the blood vessels are known as *vasoconstrictors*, and those which increase the lumen as *vasodilators* (see Ch. 12).

CELL RESPIRATION

Internal or *cell respiration* is the name given to the interchange of gases between the blood and the cells of the body.

Oxygen is carried from the lungs to the tissues in chemical combination with haemoglobin as *oxyhaemoglobin*. The exchange in the tissues takes place between the arterial end of the capillaries and the tissue fluid. The process involved is that of *diffusion from a higher concentration of oxygen in the blood to a lower concentration in the tissue fluid*.

Oxyhaemoglobin is an unstable compound and breaks up easily to liberate oxygen. One of the factors which assists the liberation of oxygen is the amount of carbon dioxide present. In active tissues there is an increased production of carbon dioxide which leads to an increased availability of oxygen. In this way oxygen is available to the tissues which most urgently require it. Oxygen *diffuses* through the capillary wall into the tissue fluid then into the cell protoplasm through its semipermeable wall (Fig. 5:3).

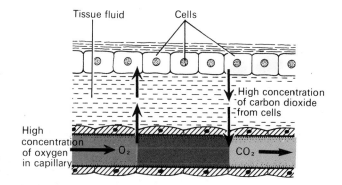

Figure 5:3 Diagram of the exchange of gases in cell respiration.

Carbon dioxide is one of the waste products of cell metabolism and, towards the venous end of the capillary, it diffuses into the blood. Blood transports carbon dioxide to the lungs for excretion by three different mechanisms:

1. Some of the carbon dioxide is dissolved in the water of the blood plasma.
2. Some is transported in chemical combination with sodium in the form of sodium bicarbonate.
3. The remainder is transported in combination with haemoglobin.

CELL NUTRITION

The nutritive materials required by the cells of the body are transported round the body in the blood plasma. In the process of passing from the blood to the cells the nutritive materials pass through the semipermeable capillary walls into the tissue fluid which bathes the cells, then through the cell wall into the cell. The mechanism of the transfer of water and other substances from the blood capillaries depends mainly upon two physical principles, diffusion and osmosis (see also pp. 53–54).

Diffusion

The walls of the blood capillaries consist of a single layer of endothelial cells constituting a *semipermeable membrane.* This membrane allows for the passage of *low molecular weight substances* through the capillary wall and the retention within the capillary of *high molecular weight substances.*

Nutrient materials which are in solution and are of low molecular weight pass through the semipermeable membrane by *diffusion*, that is, they pass from a *high concentration in the blood* to a *lower concentration in the tissue fluid*, and thus to the cells. Glucose, amino acids, fatty acids and glycerol, mineral salts, vitamins and water, which are necessary for the formation and functioning of cells, are all of relatively low molecular weight and pass out into the tissue fluid by *diffusion*.

Osmotic pressure

This is the pressure developed across a semipermeable membrane which forces water to pass from a more dilute solution to a more concentrated solution, in an attempt to establish a state of equilibrium. The extent of the osmotic pressure depends upon the *number of non-diffusable particles* in solution on each side of the membrane.

Substances of low molecular weight can diffuse freely across a semipermeable membrane and achieve the same concentration on each side of the membrane, therefore they are not involved in osmosis. The substances which influence osmotic pressure are those of molecular size too great to pass through the membrane. The *plasma proteins* in the capillaries are the substances mainly responsible for the osmotic pressure between the blood and tissue fluids.

At the arterial end of the capillaries the blood pressure is about *40 mm of mercury (mmHg)*. This is the pressure which tends to force substances into the tissue spaces. The osmotic pressure in the capillaries, exerted mainly by the plasma proteins, is about *25 mmHg*. This tends to retain water within the blood vessels. The net outward pressure is the difference between these two, that is, *15 mmHg* (Fig. 5:4).

At the venous end of the capillaries the blood pressure drops to about *10 mmHg* while the osmotic pressure remains the same at *25 mmHg*; thus there is a net pressure drawing fluid into the blood vessel of about *15 mmHg*.

This transfer of substances, including water, to the tissue spaces is a dynamic process. Blood flows slowly through the large network of capillaries from the arterial to the venous end and there is a constant change. All the water and cell waste products do not return to the blood capillaries. The excess is drained away from the tissue spaces in the minute *lymph capillaries* (Fig. 5:4).

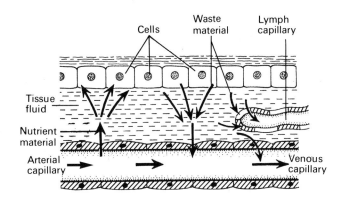

Figure 5:4   Diagram of the exchange of nutrient materials and waste products between the capillaries and the cells.

The tiny lymph capillaries originate as blind end tubes with walls similar to, but more permeable than, those of the blood capillaries. Extra tissue fluid and some of the cell waste enter the lymph capillaries and are eventually returned to the blood stream (see Ch. 6).

The Heart

The heart is a roughly cone-shaped hollow muscular organ. It is about 10 cm (4 inches) long and is about the size of the owner's fist. It weighs about 255 g (9 oz) in women and is heavier in men.

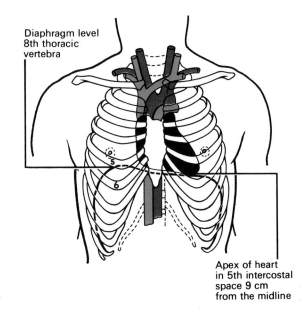

Figure 5:5   Position of the heart in the thorax.

POSITION (Fig. 5:5)

The heart is in the thoracic cavity in the middle mediastinum between the lungs. It lies obliquely, a little more to the left than the right, and presents a *base* above and an *apex* below. The apex is about 9 cm (3½ inches) to the left of the median plane at the level of the *5th intercostal space* on the *mid-clavicular line*, i.e., a little below the nipple and slightly nearer the midline. The base extends to the level of the 2nd rib.

Organs in association with the heart (Fig. 5:6)

Inferiorly— the apex rests on the central tendon of the diaphragm

Superiorly— the great blood vessels, i.e., the aorta, superior vena cava, pulmonary artery and pulmonary veins

Posteriorly— the oesophagus, trachea, left and right bronchus, descending aorta, inferior vena cava and thoracic vertebrae

Laterally— the lungs—the left lung overlaps the left side of the heart

Anteriorly— the sternum, the ribs and the intercostal muscles

STRUCTURE

The heart is composed of three layers of tissue.

1. The pericardium

The pericardium is made up of two sacs. The outer sac consists of fibrous tissue and the inner of a double layer of serous membrane.

The outer fibrous sac is continuous with the tunica adventitia of the great blood vessels above and is adherent to the diaphragm below. Its fibrous nature prevents over-distension of the heart.

The outer layer of the serous membrane, the *parietal pericardium*, lines the fibrous sac, and the inner layer, the *visceral pericardium*, is reflected on to the heart and is adherent to the heart muscle.

The serous membrane is made up of *flattened epithelial cells* which secrete serous fluid into the space between the visceral and parietal layers of pericardium. This fluid allows smooth movement between the layers when the heart beats. The space between the parietal and visceral pericardium is known as a *potential space*. In life the two

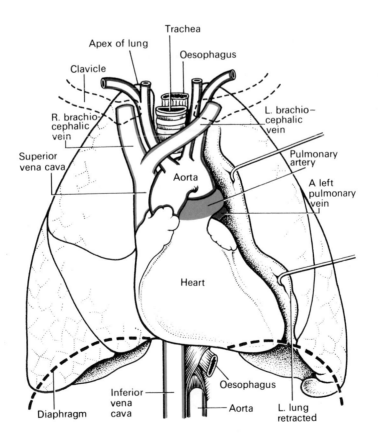

Figure 5:6 Organs in association with the heart.

layers are in close association, with only serous fluid between them.

2. The myocardium

The myocardium is composed of specialised muscle tissue known as *cardiac muscle* (Fig. 5:7). This muscle is found exclusively in the heart. It is not under the control of the will but, like voluntary muscle, cross stripes can be seen on microscopic examination. Each fibre (cell) has a nucleus and one or more branches. The ends of the cells and their branches are in very close contact with the ends and branches of adjacent cells. Microscopically these 'joints' or *intercalated discs* can be seen as thicker, darker lines than the ordinary cross stripes. This arrangement gives cardiac muscle the appearance of being a sheet of muscle rather than a very large number of individual cells. Because of the end-to-end contiguity of the fibres, each one does not need to have a separate nerve supply. When an impulse of contraction is initiated it spreads from cell to cell and thus over the whole 'sheet' of muscle.

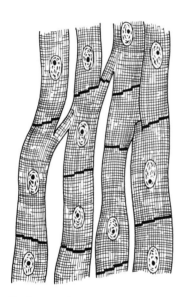

Figure 5:7 Cardiac muscle.

The myocardium is thickest at the apex and thins out towards the base. The atria and the ventricles are separated by a ring of fibrous tissue. Consequently, when a wave of contraction passes over the atrial muscle, it can only spread to the ventricles through the conducting system (see p. 62).

3. The endocardium

This forms a lining to the myocardium and is a thin, smooth, glistening membrane consisting of flattened epithelial cells continuous with the lining of the blood vessels.

INTERIOR OF THE HEART (Figs. 5:9 and 5:10)

The heart is divided into a right and left side by a partition of muscular tissue and endocardium known as the *septum*. After birth, blood cannot pass directly from the left to the right side of the heart or vice versa. Each side is divided into an upper and a lower chamber by a valve. The valves ensure that the blood flows in one direction only: from the upper chamber or *atrium* to the lower chamber or *ventricle*. The heart, therefore, has four chambers.

Right and left atria
Right and left ventricles

The valves dividing the atria from the ventricles are formed by double folds of endocardium strengthened with fibrous tissue. The valve separating the right atrium from the right ventricle is known as the *right atrioventricular valve* (tricuspid valve) and is made up of three flaps or cusps. The valve separating the left atrium from the left ventricle is called the *left atrioventricular valve* (mitral valve) and is composed of two flaps or cusps.

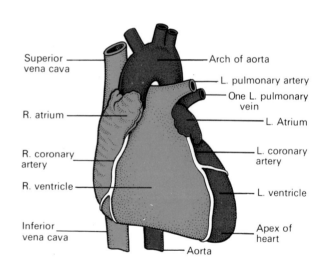

Figure 5:8 The heart and the great vessels viewed from the front.

The valves between the atria and the ventricles open and close as a result of changes in the pressure of blood within the chambers. The pressure in the ventricles rises higher than that in the atria and backward flow of blood is prevented by the closure of the valves. They are prevented from opening upwards by the *chordae tendineae* (tendinous cords), which extend from the inferior surface of the valve cusps to the walls of the ventricles. The walls of the ventricles have little muscle projections, called *papillary muscles*, to which the chordae tendineae are attached (Fig. 5:10). The papillary muscles are covered by endothelium.

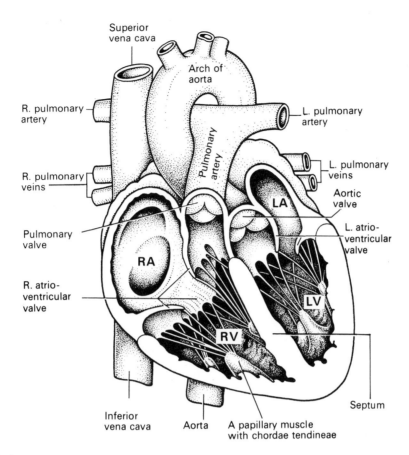

Figure 5:9 Interior of the heart.

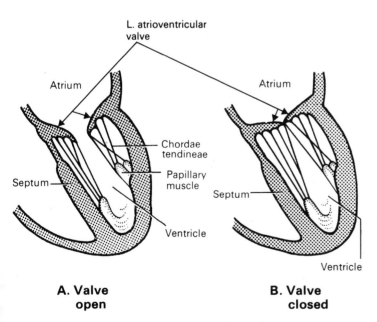

Figure 5:10 Diagram of the left atrioventricular valve.

FLOW OF BLOOD THROUGH THE HEART
(Fig. 5:11)

The two largest veins of the body, the *superior* and *inferior venae cavae*, empty their contents into the right atrium. This blood passes via the right atrioventricular valve into the right ventricle, and from the right ventricle it is pumped into the *pulmonary artery* or *trunk* (the only artery in the body which carries venous or deoxygenated blood). The opening of the pulmonary artery is guarded by a valve known as the *pulmonary valve* and is formed by three *semilunar cusps*. This valve prevents the back flow of blood into the right ventricle when the ventricular muscle relaxes. The pulmonary artery passes through the wall of the heart and divides into a *left* and a *right pulmonary artery*. These arteries carry the venous blood to the lungs where the interchange of gases occurs: carbon dioxide is excreted and oxygen is absorbed.

The arterial or oxygenated blood is carried from each lung by *two* pulmonary veins and the *four pulmonary veins* empty their contents into the left atrium of the heart. This blood passes through the left atrioventricular valve into the left ventricle, and from there it is pumped into the aorta, the first artery of the general circulation. The

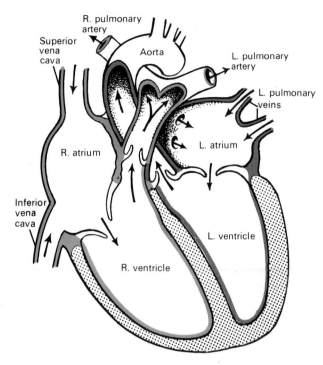

Figure 5:11   Diagram of the flow of blood through the heart.

opening of the aorta is guarded by the *aortic valve* which is formed by three *semilunar cusps* (Fig. 5:12).

From this sequence of events it can be seen that the blood passes from the right to the left side of the heart via the lungs. However, it should be noted that both atria contract at the same time followed by the simultaneous contraction of both ventricles.

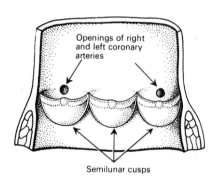

Figure 5:12   The aorta cut open to show the cusps of the semilunar valve.

The muscle layer of the walls of the atria is very thin in comparison with that of the ventricles. This is consistent with the amount of work it does. The atria propel the blood through the atrioventricular valve into the ventricles while the ventricles pump the blood to the lungs and round the whole body. The muscle layer is thickest in the wall of the left ventricle.

The pulmonary trunk leaves the heart from the upper part of the right ventricle, and the aorta leaves from the upper part of the left ventricle.

Note

The *right side* of the heart deals with *deoxygenated blood.*

The *left side* of the heart deals with *oxygenated blood.*

The vessels *carrying blood* to the heart are *veins.*

The vessels *carrying blood away* from the heart are *arteries.*

BLOOD SUPPLY TO THE HEART

The heart is supplied with arterial blood by the *right* and *left coronary arteries*. These are the first branches from the aorta immediately distal to the aortic valve.

The venous return is by the *coronary sinus* which empties into the right atrium.

THE CONDUCTING SYSTEM OF THE HEART
(Fig. 5:13)

The heart has an intrinsic system whereby the muscle is stimulated to contract without the need for nerve supply from the brain. However, the intrinsic system can be stimulated or depressed by nerve impulses initiated in the brain.

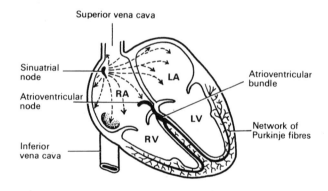

Figure 5:13   Diagram of the conducting system in the heart.

There are small groups of specialised neuromuscular cells in the myocardium which initiate and conduct impulses of contraction over the heart muscle.

The sinuatrial node (SA node)

This small mass of specialised cells is in the wall of the right atrium near the opening of the superior vena cava. The sinuatrial node is often described as the '*pace-maker*' of the heart because it is capable of initiating impulses which stimulate the myocardium to contract without any outside influence from the nervous system.

The atrioventricular node (AV node)

This mass of neuromuscular tissue is situated in the wall

of the atrial septum near the atrioventricular valves. Normally the atrioventricular node is stimulated by the contraction which sweeps over the atrial myocardium. However, it too is capable of initiating impulses of contraction but at a slower rate than the SA node, if there is no stimulation by the nervous system.

The atrioventricular bundle (AV bundle)

This consists of a mass of specialised fibres which originates from the atrioventricular node and passes downwards in the septum that separates the right and left ventricles. This bundle of fibres (called the bundle of His) then divides into two branches, one going to each ventricle. Within the myocardium of the ventricles the branches break up into a network of fine filaments or fibres known as the *fibres of Purkinje*. The atrioventricular bundle and the Purkinje fibres convey the impulse of contraction from the atrioventricular node to the myocardium of the ventricles.

The impulses of contraction initiated by the sinuatrial node stimulate the myocardium of the atria to contract. This wave of contraction stimulates the atrioventricular node to produce impulses which pass to the apex of the heart in the Purkinje fibres then over the muscle of the ventricles. In this way the ventricular wave of contraction begins at the apex of the heart and blood is forced into the pulmonary artery and into the aorta which leave the heart near its base.

NERVE SUPPLY TO THE HEART

In addition to the intrinsic stimulation of the myocardium described above the heart is influenced by nerves originating in the *cardiac centre* in the *medulla oblongata* which reach it through the autonomic nervous system. These are the *parasympathetic* and *sympathetic nerves* and they are antagonistic to one another.

The *vagus nerves* (parasympathetic) tend to slow the rate at which impulses are produced by the sinuatrial node therefore *decreasing* the rate and force of the heart beat.

The *sympathetic nerves* tend to speed up the rate of impulse production by the sinuatrial node thus *increasing* the rate and force of the heart beat.

The rate at which the heart beats is the result of a fine balance of sympathetic and parasympathetic effects. It is usually decreased during rest and increased during excitement, exercise and when the blood volume is decreased.

FUNCTION

The function of the heart is to maintain a constant circulation of blood throughout the body. The heart acts as a pump and its action consists of a series of events known as the *cardiac cycle*.

The cardiac cycle (Fig. 5:14)

In a human being, when the heart is beating normally the cardiac cycle occurs about 74 times per minute. Thus each cycle lasts about *0·8 of a second*. The cardiac cycle consists of:

    Atrial systole—contraction of the atria
    Ventricular systole—contraction of the ventricles
    Complete cardiac diastole—relaxation of the atria and
      ventricles

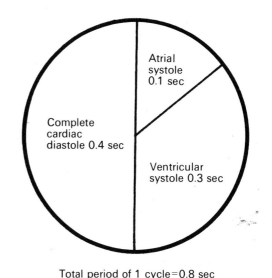

Total period of 1 cycle=0.8 sec

Figure 5:14 Diagram of the stages of one cardiac cycle.

It does not matter at which stage of the cardiac cycle a description starts. For convenience the period when the atria are filling has been chosen.

The superior vena cava and the inferior vena cava pour deoxygenated blood into the right atrium *at the same time* as the four pulmonary veins pour oxygenated blood into the left atrium. The sinuatrial node emits an impulse of contraction. This stimulates the myocardium to contract and the contraction spreads like a wave over both atria, pushing the blood through the atrioventricular valves into the ventricles (atrial systole 0·1 sec). When this wave of contraction reaches the atrioventricular node it is stimulated to emit an impulse of contraction, which spreads to the ventricular muscle via the atrioventricular bundle and the Purkinje fibres. This results in a wave of contraction which sweeps upwards from the apex of the heart and pushes the blood into the pulmonary artery and the aorta (ventricular systole 0·3 sec).

After contraction of the ventricles the heart rests for *0·4 of a second*, and this period is known as *complete cardiac diastole*. After complete cardiac diastole the cycle begins again with atrial systole (Fig. 5:15).

The valves of the heart and of the great vessels open and close according to the pressure within the chambers of the heart. The atrioventricular valves are open while the ventricular muscle is relaxed during atrial systole.

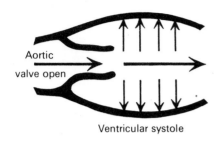

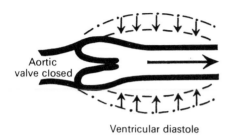

Figure 5:15   Diagram showing the elasticity of the walls of the aorta.

When the ventricles contract there is a gradual increase in the pressure in these chambers, and when it rises above atrial pressure the atrioventricular valves close. When the ventricular pressure rises above that in the pulmonary artery and in the aorta, the pulmonary and aortic valves open and blood flows into these vessels. When the ventricles relax and the pressure within the ventricles falls, the reverse process occurs. First the pulmonary and aortic valves close, then the atrioventricular valves open and the cycle begins again. This sequence of opening and closing valves ensures that the blood flows in only one direction.

HEART SOUNDS

The individual is not usually conscious of his heart beat, but if the ear or the receiver of a stethoscope is placed upon the chest wall at about the midclavicular line in the fifth intercostal space the beat of the heart can be heard.

Two sounds, which are separated by a short pause, can be clearly distinguished. They are described in words as '*lubb dup*'. The first sound, '*lubb*', is fairly loud and is due to the contraction of the ventricular muscle and the closure of the atrioventricular valves.

The second sound, '*dup*', is softer and is due to the closure of the aortic and pulmonary valves.

ELECTRICAL CHANGES IN THE HEART

When muscles contract there is a change in the electrical potential across the membrane of muscle fibres. As the body fluids and tissues are good conductors of electricity, the electrical changes which occur in the contracting myocardium can be detected by attaching electrodes to the surface of the body. The pattern of electrical activity

may be displayed on an oscilloscope screen or printed out on paper. This tracing is called an *electrocardiogram* (ECG).

The normal ECG tracing shows five waves which, by convention, have been named P, Q, R, S and T (Fig. 5:16).

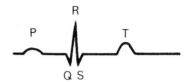

Figure 5:16   Electrocardiograph tracing of one cardiac cycle.

The P wave is caused by the impulse of contraction which sweeps over the atria.

The Q, R, S wave is indicative of the spread of the impulse of contraction from the atrioventricular node through the atrioventricular bundle and the Purkinje fibres and the contraction of the ventricular muscle.

The T wave represents the relaxation of the ventricular muscle.

By examining the pattern of waves and the time interval between them the physician obtains valuable information about the state of the myocardium and the conducting system within the heart.

Blood Pressure

Blood pressure may be defined as the force or pressure which the blood exerts on the walls of the blood vessels in which it is contained. As there is some delay in the movement of blood through the arteriolar and capillary systems, the blood pressure in the arteries is higher than that in the veins. This means that the arteries are always full and their walls are continuously subjected to stretch.

The arterial blood pressure is the result of the discharge of blood from the left ventricle into the *already full aorta*.

When the left ventricle contracts and pushes blood into the aorta the pressure produced is known as the *systolic blood pressure*, which is found in an adult to be about 120 mmHg (millimetres of mercury) or 16 kPa (kilopascals).

When *complete cardiac diastole* occurs and the heart is resting following the ejection of blood, the pressure within the arteries is termed the *diastolic blood pressure*. In an adult this is about 80 mmHg or 11 kPa. These figures vary according to the time of day, the posture, sex and age of the individual. During bedrest at night the blood pressure tends to be lower. It increases with age and in women it is usually higher than in men.

The blood pressure is measured by the use of a

sphygmomanometer and is usually expressed in the following manner:

$$BP = \frac{120}{80} \, mmHg \quad or \quad BP = \frac{16}{11} \, kPa$$

MAINTENANCE OF NORMAL BLOOD PRESSURE

A number of factors are involved in the maintenance of the blood pressure:

  The cardiac output
  The blood volume
  The peripheral resistance
  The elasticity of the artery walls
  The venous return

Cardiac output

The cardiac output may be considered as the amount of blood ejected from the heart by each contraction of the ventricles (stroke volume), or the amount ejected each minute (minute volume). The minute volume takes into consideration the rate *and* force of cardiac contraction. An increase in minute volume raises both the systolic and diastolic pressure but an increase in the stroke volume increases the systolic pressure more than it does the diastolic.

Blood volume

A sufficient amount of blood must be circulating in the vessels to maintain the normal blood pressure. If, for example, a haemorrhage has occurred and a large amount of blood is lost, there is an accompanying fall in the blood pressure.

The peripheral or arteriolar resistance

The arterioles are very small blood vessels. They have a tunica media composed almost entirely of smooth muscle which responds to nerve stimulation. Nerve impulses reach the muscle layer in blood vessel walls via vaso-constrictor nerves (sympathetic nerves), which originate in the vasomotor centre in the medulla oblongata (see Ch. 12). Continuous nerve stimulation maintains a state of slight contraction or *tone*, but an increase in the activity of the vasomotor centre results in constriction of the arterioles and an increase in the blood pressure. A decrease in vasomotor activity results in vasodilatation and a fall in the blood pressure. Dilatation and constriction of arterioles occurs selectively around the body, resulting in changes in the blood flow through organs according to their needs. The highest priorities are the blood supply to the brain and the heart muscle, and in an emergency, supplies to other parts of the body are reduced in order to ensure an adequate supply to these organs. Generally, changes in the amount of blood flowing to any organ depend on how active it is. A very active organ needs more oxygen and nutritional materials than a resting organ and it produces more waste materials for excretion.

The elasticity of the artery walls

There is a considerable amount of elastic tissue in the arterial walls, especially in the walls of large blood vessels. Therefore, when the left ventricle ejects blood into the already full aorta, it distends then recoils pushing the blood onwards. This distension and recoil occurs all through the arterial system. During cardiac diastole the elastic recoil of the arteries maintains the diastolic pressure (Fig. 5:15).

Venous return

The amount of blood returned to the heart through the superior and inferior venae cavae plays an important part in cardiac output. The force of contraction of the left ventricle ejecting blood into the aorta and subsequently into the arteries, arterioles and capillaries is not sufficient to return the blood through the veins back to the heart. Other factors are involved in assisting the venous return.

The position of the body. Gravity assists the venous return from the head and neck when the individual is standing or sitting.

Muscular contraction. The contraction of muscles, particularly skeletal muscles, puts pressure on the veins. This squeezing or milking action has the effect of pushing the blood towards the heart. Backward flow of blood is prevented by the valves in the veins (Fig. 5:17). This

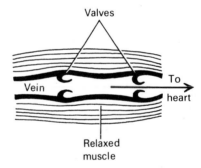

A. Muscle relaxed, valves open

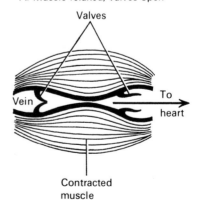

B. Muscle contracted squeezing vein—proximal valve open, distal valve closed

Figure 5:17 Diagram of the flow of blood through a vein aided by the contraction of skeletal muscle.

milking action affects the deep veins and as the pressure in them is lowered blood flows into them from the superficial veins.

Effects of respiratory movements. During inspiration the expansion of the chest creates a negative pressure within the thorax. This has the effect of assisting the flow of blood towards the heart. In addition, when the diaphragm descends during inspiration, the intra-abdominal pressure rises and squeezes blood towards the heart.

The Pulse

The pulse is described as a wave of distension and elongation felt in an artery wall due to the contraction of the left ventricle forcing about 70 to 90 millilitres of blood into the already full aorta. When the aorta is distended a wave passes along the walls of the arteries and can be felt at any point where an artery can be pressed gently against a bone. The waves occur on average about 74 times per minute in health and represent the number of heart beats.

It must be appreciated that this wave is felt in the artery wall before the blood ejected into the aorta could possibly reach the area. The wave travels about 10 to 15 times more rapidly than does the blood and is quite independent of it.

Some information may be obtained by taking the pulse of an individual.

1. *The rate* at which the heart is beating.
2. *The rhythm*, that is, the regularity with which the heart beats occur: the length of time between each beat should be the same.
3. *The volume or strength* of the beat. It should be possible to compress the artery with moderate pressure thus stopping the flow of blood. The compressibility of the blood vessel gives some indication of the blood pressure and the state of the blood vessel wall.
4. *The tension*. The artery should feel soft and pliant under the fingers, not hard and tortuous.

Factors affecting the pulse rate

Position. When the individual is standing up the pulse rate will be more rapid than when he is lying down.

Age. The pulse rate in children is more rapid than in adults.

Sex. The pulse rate tends to be more rapid in the female than in the male. The difference is usually about five beats per minute.

Exercise. Any exercise—walking, running or playing games—will increase the rate of the pulse.

Emotion. When any strong emotion is experienced the pulse rate is increased, for example, excitement, fear, anger, grief.

The Circulation of the Blood

Three systems are discussed when considering the circulation of blood throughout the body:

The pulmonary circulation
The systemic or general circulation
The portal circulation

THE PULMONARY CIRCULATION (Fig. 5:18)

The pulmonary circulation consists of the circulation of blood from the right ventricle of the heart to the lungs and back to the left atrium. In the lungs carbon dioxide is excreted and oxygen is absorbed.

The pulmonary artery or trunk, carrying deoxygenated blood, leaves the upper part of the right ventricle of the heart. It passes upwards through the walls of the heart and at the level of the fifth thoracic vertebra it divides into a left and right pulmonary artery.

The left pulmonary artery runs to the root of the left lung where it divides into two branches, one passing into each lobe.

The right pulmonary artery passes to the root of the right lung and divides into two branches. The larger branch carries blood to the middle and lower lobes, and the smaller branch to the upper lobe.

Within the lung these arteries divide and subdivide into smaller arteries, subsequently becoming arterioles and capillaries. It is between the capillaries and the lung tissue that the interchange of gases occurs. In each lung the capillaries containing oxygenated blood join up and eventually form two veins.

Two pulmonary veins leave each lung, therefore *four pulmonary veins* return oxygenated blood to the left atrium of the heart. During atrial systole this blood passes into the left ventricle, and during ventricular systole it is forced into the aorta and then into the general circulation.

THE SYSTEMIC OR GENERAL CIRCULATION

The blood pumped out from the left ventricle is carried by the branches of the aorta around the body and is returned to the heart by the superior and inferior venae cavae. A general impression of the positions of the aorta and the main arteries to the limbs is given in Figure 5:19. Figure 5:20 provides an overview of the venae cavae and the veins of the limbs.

The circulation of blood to the different parts of the body will be described in the order in which their arteries branch off the aorta.

THE AORTA (Fig. 5:21)

The aorta begins at the upper part of the left ventricle and after passing upwards for a short distance it arches backwards and to the left. It then descends behind the

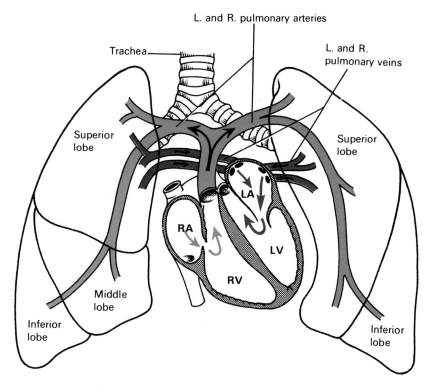

Figure 5:18 Diagram showing the flow of blood between the heart and the lungs.

heart through the thoracic cavity a little to the left of the thoracic vertebrae. At the level of the 12th thoracic vertebra it passes behind the diaphragm then downwards to the level of the 4th lumbar vertebra, where it divides into the two *common iliac arteries.*

Throughout its length the aorta gives off numerous branches. Some of the branches are *paired,* i.e., there is a right and left branch of the same name; others are single or *unpaired* branches.

For descriptive purposes the aorta is divided into two parts:

The thoracic aorta
The abdominal aorta

THE THORACIC AORTA

This part of the aorta is described in three portions:

The ascending aorta
The arch of the aorta
The descending aorta

The ascending aorta

The ascending aorta is about 5 cm (2 inches) in length and lies behind the sternum.

The right and left coronary arteries (Fig. 5:22) arise from the aorta just above the level of the aortic valve. They supply the tissues of the heart with oxygenated

blood. As they traverse the heart they break up, eventually becoming capillaries. The venous blood is collected into several small veins which join up to form the *coronary sinus* which opens into the right atrium.

The arch of the aorta

The arch of the aorta is a continuation of the ascending aorta. It begins behind the manubrium of the sternum and runs upwards, backwards and to the left in front of the trachea. It then passes downwards to the left of the trachea and is continuous with the descending aorta.

Three branches are given off from the upper aspect of the arch of the aorta (Fig. 5:23):

The brachiocephalic artery or trunk
The left common carotid artery
The left subclavian artery

The brachiocephalic artery is about 4 to 5 cm ($1\frac{1}{2}$ to 2 inches) long and passes obliquely upwards, backwards and to the right. At the level of the sternoclavicular joint it divides into the *right common carotid artery* and the *right subclavian artery.*

CIRCULATION OF BLOOD TO THE HEAD AND NECK

The *right common carotid artery* is a branch of the brachio-

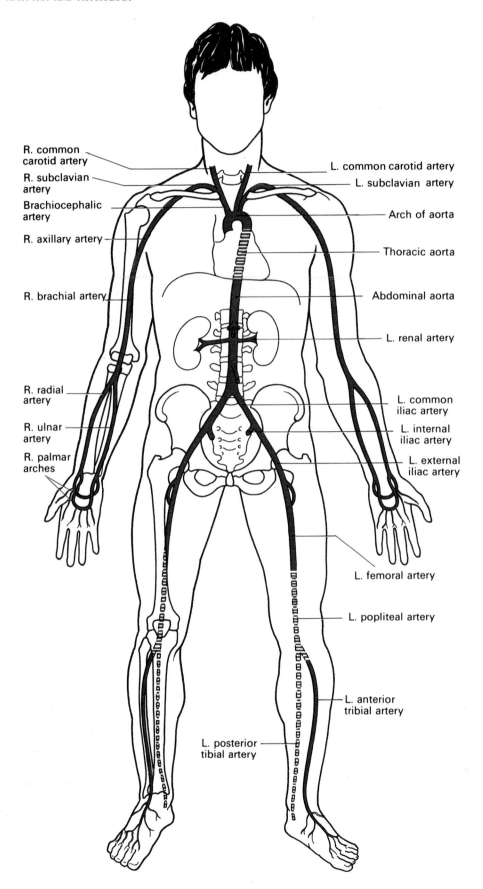

R. common
carotid artery

R. subclavian
artery

Brachiocephalic
artery

R. axillary artery

R. brachial artery

R. radial
artery

R. ulnar
artery

R. palmar
arches

L. common carotid artery

L. subclavian artery

Arch of aorta

Thoracic aorta

Abdominal aorta

L. renal artery

L. common
iliac artery

L. internal
iliac artery

L. external
iliac artery

L. femoral artery

L. popliteal artery

L. anterior
tribial artery

L. posterior
tibial artery

Figure 5:19   Aorta and the main arteries of the limbs.

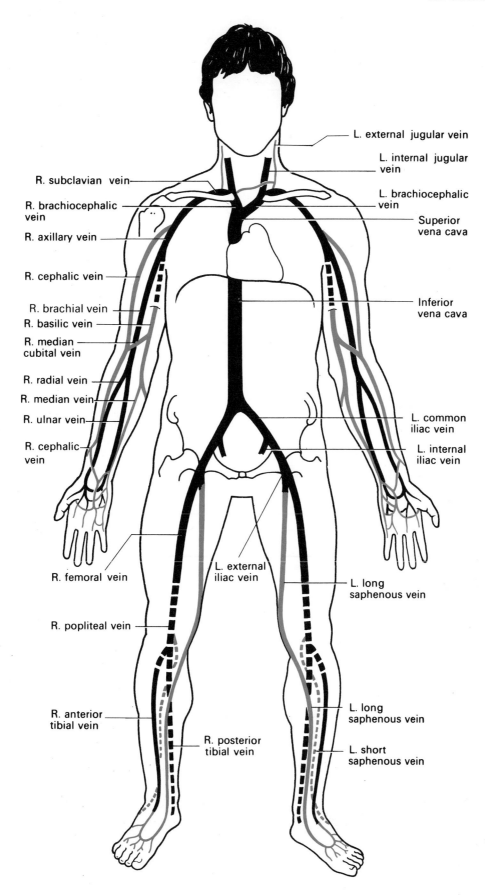

Figure 5:20 Venae cavae and the main veins of the limbs. Deep veins in black, superficial veins in blue.

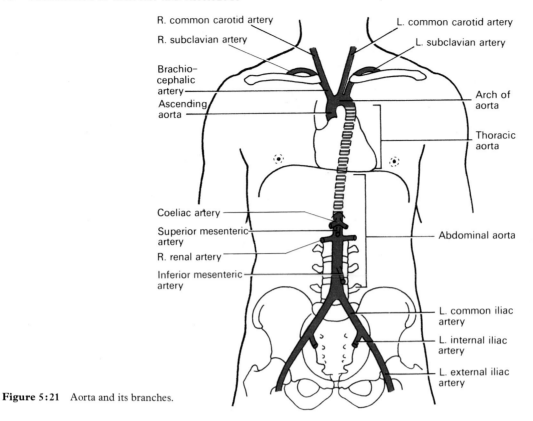

Figure 5:21 Aorta and its branches.

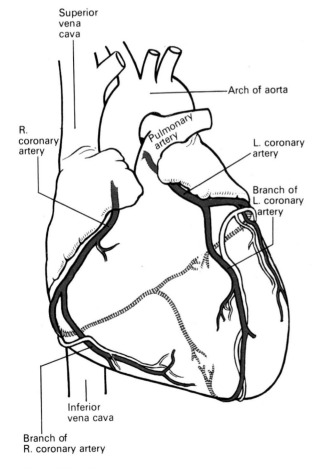

Figure 5:22 Coronary arteries.

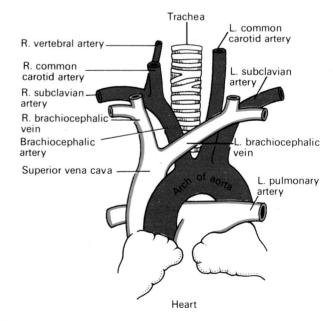

Figure 5:23 The arch of the aorta and its branches.

cephalic artery. The *left common carotid artery* arises directly from the arch of the aorta. They pass upwards on either side of the neck and they have the same distribution on each side. The common carotid arteries are embedded in fascia, known as the *carotid sheath*. At the level of the upper border of the thyroid cartilage they divide into:

The external carotid artery
The internal carotid artery

Branches of the external carotid artery (Fig. 5:24)

This artery supplies the superficial tissues of the head and neck, giving off a number of branches:

The superior thyroid artery supplies the thyroid gland and adjacent muscles.

The lingual artery supplies the tongue, the lining membrane of the mouth, the structures in the floor of the mouth, the tonsil and the epiglottis.

The facial artery passes outwards over the mandible just in front of the angle of the jaw and supplies the muscles of facial expression and structures in the mouth.

The occipital artery supplies the posterior part of the scalp.

The temporal artery passes upwards over the zygomatic process in front of the ear and supplies the frontal, temporal and parietal parts of the scalp.

The maxillary artery supplies the muscles of mastication and a branch of this artery, the *middle meningeal artery*, runs deeply to supply structures in the interior of the skull.

Branches of the internal carotid artery

The internal carotid artery supplies the greater part of the brain, the eye, the forehead and the nose. It ascends to the base of the skull and passes through the carotid foramen in the temporal bone. Many branches arise from the internal carotid artery; four of these branches are:

The ophthalmic artery supplies the eye
The anterior cerebral artery
The middle cerebral artery
The posterior communicating artery } supply the brain

The circulus arteriosus (the circle of Willis) (Fig. 5:25)

The greater part of the brain is supplied with arterial blood by a striking arrangement of arteries known as the *circulus arteriosus* or the *circle of Willis*.

The *two anterior cerebral arteries* arise from the internal carotid arteries and are joined together by an artery known as the *anterior communicating artery*.

Posteriorly two *vertebral arteries* which rise from the subclavian arteries pass through the foramina in the transverse processes of the cervical vertebrae and enter

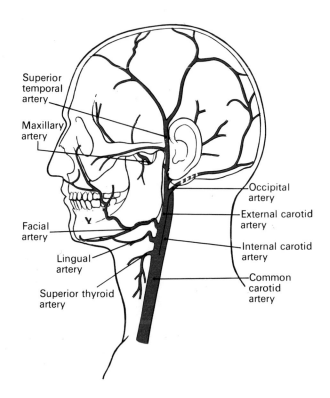

Figure 5:24 Main arteries of the left side of the head and neck.

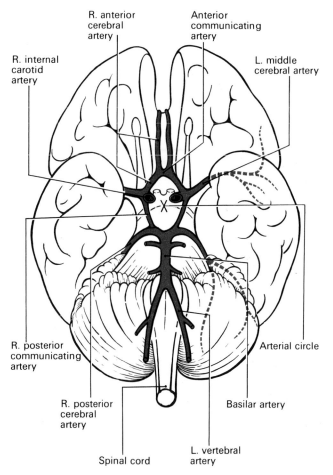

Figure 5:25 Arteries forming the circulus arteriosus (circle of Willis) and its main branches to the brain.

the skull through the foramen magnum (Fig. 5:26). Just inside the skull they join together to form the *basilar artery*. After travelling for a short distance the basilar artery divides to form the two *posterior cerebral arteries*. Each of these arteries is joined to the corresponding internal carotid artery by a *posterior communicating artery*. The *circulus arteriosus* is therefore formed by:

  2 anterior cerebral arteries
  1 anterior communicating artery
  2 posterior communicating arteries
  2 posterior cerebral arteries

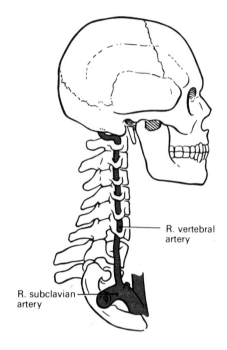

Figure 5:26   Right vertebral artery.

It is from this circle that the anterior cerebral arteries pass forward to supply the anterior part of the brain. The middle cerebral arteries pass laterally to supply the sides of the brain, and the posterior cerebral arteries supply the posterior part of the brain.

Branches of the basilar artery supply parts of the brain stem.

THE VENOUS RETURN FROM THE HEAD AND NECK (Fig. 5:27)

The venous blood from the head and neck is returned by *deep and superficial veins*.

It will be remembered that from the external carotid artery several branches supply arterial blood to the superficial parts of the face and scalp. *Superficial veins* with the same names return the venous blood from these areas and unite to form the external jugular vein.

The external jugular vein begins in the neck at the level of the angle of the jaw. It passes downwards in front of

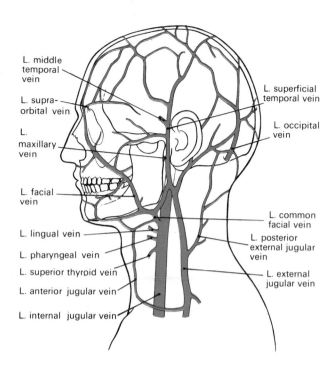

Figure 5:27   Veins of the left side of the head and neck.

the sternocleidomastoid muscle, then behind the clavicle to enter the *subclavian vein*.

The venous blood from the deep areas of the brain is collected into channels which are known as *sinuses*.

The venous sinuses of the brain
(Figs. 5:28 and 5:29)

The walls of the venous sinuses are formed by layers of *dura mater* lined with endothelium. The dura mater is

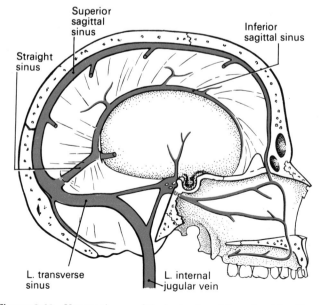

Figure 5:28   Venous sinuses of the brain viewed from the right side.

the name given to the outer protective covering of the brain (see p. 171). The main venous sinuses are:

1 superior sagittal sinus
1 inferior sagittal sinus
1 straight sinus
2 transverse or lateral sinuses

The superior vena cava, which drains all the venous blood from the head, neck and upper limbs, is about 7 cm (2¾ inches) long. It passes downwards along the right border of the sternum and ends in the right atrium of the heart.

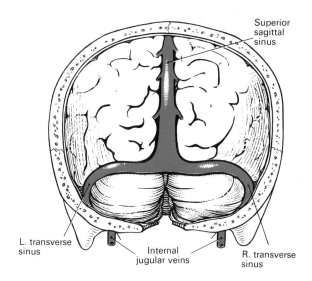

Figure 5:29 Venous sinuses of the brain viewed from above.

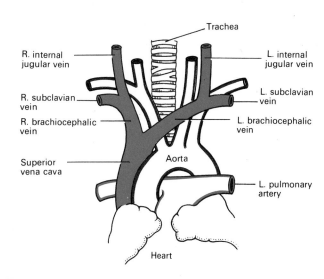

Figure 5:30 Superior vena cava and the veins which form it.

The superior sagittal sinus carries the venous blood from the superior part of the brain. It commences in the frontal region in folds of dura mater and passes directly backwards in the midline of the skull to the occipital region. In the occipital region it turns to the right side and continues as the *right transverse sinus*.

The inferior sagittal sinus lies deep within the brain and passes backwards to form the *straight sinus*.

The straight sinus runs backwards and downwards to become the *left transverse sinus*.

The transverse sinuses commence in the occipital region, one as a continuation of the superior sagittal sinus and the other of the straight sinus. They run, in a curved groove of the skull, forwards and medially to become continuous with the *internal jugular veins* in the middle cranial fossa.

The *internal jugular veins* begin at the jugular foramina in the middle cranial fossa and each is the continuation of a transverse sinus. They run downwards in the neck behind the sternocleidomastoid muscles. Behind the clavicle they unite with the *subclavian veins* to form the *brachiocephalic veins*.

The brachiocephalic veins are two in number and are situated one on each side in the root of the neck. Each is formed by the internal jugular and the subclavian vein. The left brachiocephalic vein is longer than the right and passes obliquely behind the manubrium of the sternum, where it joins the right brachiocephalic vein to form the *superior vena cava* (Fig. 5:30).

CIRCULATION OF BLOOD TO THE UPPER LIMB

Arterial supply (Figs. 5:19 and 5:31)

The subclavian arteries. The right subclavian artery arises from the brachiocephalic artery, the left directly from the arch of the aorta. They are slightly arched and pass behind the clavicles and over the first ribs before entering the axillae, where they continue as the *axillary arteries*.

Before entering the axilla each subclavian artery gives off two branches: the *vertebral artery*, which passes upwards to supply the brain; the *internal mammary artery*, which supplies the breast and a number of structures in the thoracic cavity.

The axillary artery is a continuation of the subclavian artery and lies in the axilla. The first part lies deeply, then it runs more superficially to become the *brachial artery*.

The brachial artery is a continuation of the axillary artery. It runs down the medial aspect of the upper arm then passes to the front and extends to about 1 cm below the elbow joint, where it divides into a *radial* and an *ulnar artery*.

The radial artery passes down the radial or lateral side of the forearm to the wrist. In its lower part, just above the wrist, it lies superficially and can be felt in front of the radius. It is here that the radial pulse is palpable. The artery then passes between the first and second metacarpal bones and enters the palm of the hand.

The ulnar artery runs downwards on the ulnar or

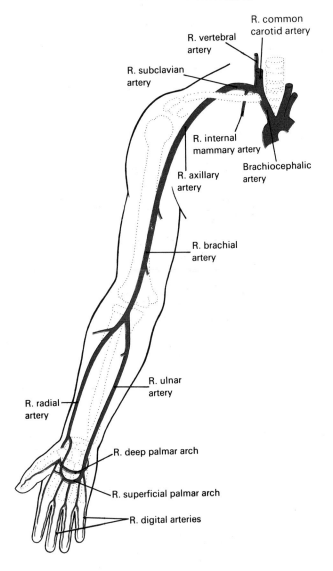

Figure 5:31   Main arteries of the right arm.

medial aspect of the forearm to cross the wrist and pass into the hand.

Together the radial and ulnar arteries form two arterial arches in the hand, *the deep and superficial palmar arches*. From these arches *palmar metacarpal* and *palmar digital arteries* arise to supply the structures in the hand and fingers.

Branches from the axillary, brachial, radial and ulnar arteries supply all the structures in the upper limb.

Venous return from the upper limb (Figs. 5:20 and 5:32)

The veins of the upper limb are divided into two groups: deep and superficial veins.

The deep veins follow the course of the arteries and have the same names:

Palmar metacarpal veins

Deep palmar venous arch

Ulnar and radial veins

Brachial vein

Axillary vein

Subclavian vein

The superficial veins begin in the hand and consist of the following:

Cephalic vein

Basilic vein

Median vein

Median cubital vein

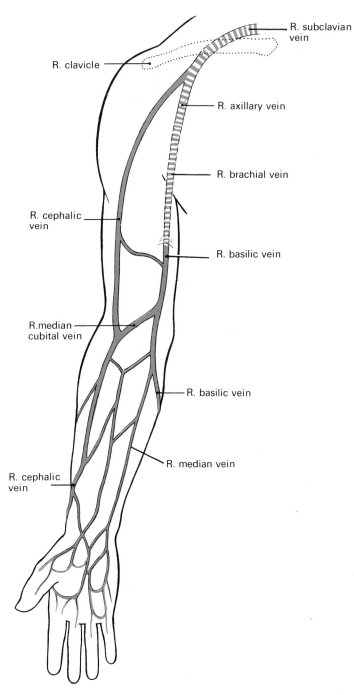

Figure 5:32   Main veins of the right arm. Dotted lines indicate deep veins.

The cephalic vein begins at the back of the hand where it collects blood from a complex of superficial veins, many of which can be easily seen. It then winds round the radial side to the anterior aspect of the foramen. In front of the elbow it gives off a large branch, the *median cubital vein*, which slants upwards and medially to join the *basilic vein*. After crossing the elbow joint the cephalic vein passes up the lateral aspect of the arm to end in the axillary vein. Throughout its length it receives blood from the superficial tissues on the lateral aspects of the hand, forearm and arm.

The basilic vein begins at the back of the hand on the ulnar aspect. It ascends on the ulnar side of the forearm and the medial aspect of the upper arm. It then passes into the deeper structures to join the axillary vein. It receives blood from the medial aspect of the hand, forearm and arm. There are many small veins which link the cephalic and basilic veins.

The median vein is a small vein which is not always present. It begins at the palmar surface of the hand, ascends on the front of the forearm and ends in the basilic vein or the median cubital vein.

The brachiocephalic vein is formed when the subclavian and internal jugular veins unite. There is one on each side.

The superior vena cava is formed when the two brachiocephalic veins unite. It drains all the venous blood from the head, neck and upper limbs and terminates in the right atrium. It is about 7 cm ($2\frac{3}{4}$ inches) long and passes downwards along the right border of the sternum.

The descending aorta (Figs. 5:21 and 5:33)

The descending aorta is continuous with the arch of the aorta and begins at the level of the 4th thoracic vertebra. It extends downwards on the anterior surface of the bodies of the thoracic vertebrae to the level of the 12th thoracic vertebra, where it passes behind the diaphragm to become the abdominal aorta.

The descending aorta in the thorax gives off many *paired branches* which supply the walls of the thoracic cavity and the organs within the cavity.

The bronchial arteries supply the bronchi and their branches, connective tissue in the lungs and the lymph nodes at the root of the lungs.

The oesophageal arteries supply the oesophagus.

The intercostal arteries. One intercostal artery runs along the inferior border of each rib and supplies the intercostal muscles, some muscles of the thorax, the ribs, the skin and its underlying connective tissues.

Venous return from the thoracic cavity (Fig. 5:34)

Most of the venous blood from the organs in the thoracic cavity is drained into the *azygos vein* and the *hemiazygos vein*. Some of the main veins which join them are the *bronchial, oesophageal* and *intercostal veins*. The azygos

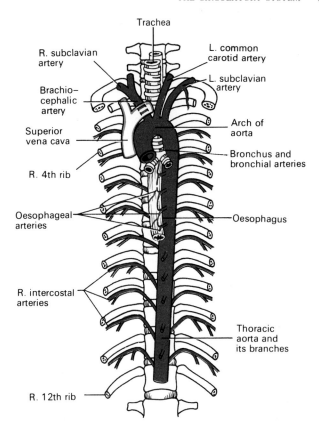

Figure 5:33 Aorta and its main branches in the thorax.

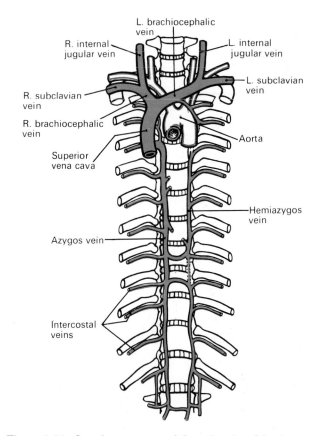

Figure 5:34 Superior vena cava and the main veins of the thorax.

vein joins the superior vena cava and the hemiazygos vein joins the left brachiocephalic vein.

THE ABDOMINAL AORTA (Fig. 5:35)

The abdominal aorta is a continuation of the thoracic aorta. The name changes when the aorta enters the abdominal cavity by passing behind the diaphragm at the level of the 12th thoracic vertebra. It descends in front of the bodies of the vertebrae to the level of the 4th lumbar vertebra, where it divides into the *right* and *left common iliac arteries*.

When a branch of the abdominal aorta supplies an organ it is only mentioned here and is described in more detail in association with the anatomy and physiology of the organ. However, illustrations showing the distribution of blood from the coeliac, superior and inferior mesenteric arteries are presented here (Figs. 5:36 and 5:37).

Many branches arise from the abdominal aorta some of which are paired and some unpaired.

Paired branches

The inferior phrenic arteries, which supply the diaphragm.

The renal arteries, which supply the kidneys and give off branches to supply the adrenal glands.

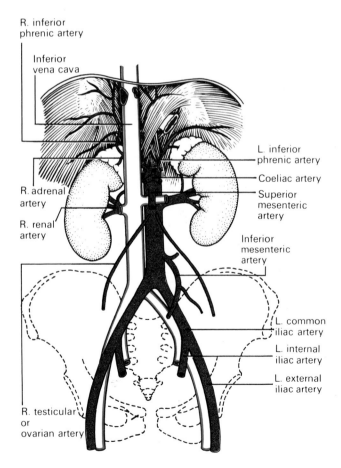

R. inferior
phrenic artery

Inferior
vena cava

L. inferior
phrenic artery

Coeliac artery

Superior
mesenteric
artery

R. adrenal
artery

R. renal
artery

Inferior
mesenteric
artery

L. common
iliac artery

L. internal
iliac artery

L. external
iliac artery

R. testicular
or
ovarian artery

Figure 5:35   Abdominal aorta and its branches.

The testicular arteries, which supply the testes in the male.

The ovarian arteries, which supply the ovaries in the female.

The testicular and ovarian arteries are much longer than the other paired branches. This is because the testes and the ovaries begin their development in the region of the kidneys. As they grow they descend into the scrotum and the pelvis respectively and are accompanied by their blood vessels.

Unpaired branches (Figs. 5:36 and 5:37)

Coeliac artery. This is a short thick artery about 1·25 cm (½ inch) in length. It arises immediately below the diaphragm and divides into three branches:

The left gastric artery, which supplies the stomach

The splenic artery, which supplies the pancreas and the spleen

The common hepatic artery, which supplies the liver, gall bladder and parts of the stomach, duodenum and pancreas

Superior mesenteric artery branches from the aorta between the coeliac artery and the renal arteries. It supplies the whole of the small intestine and the proximal half of the large intestine.

Inferior mesenteric artery arises from the aorta about 4 cm (1½ inches) above its division into the common iliac arteries. It supplies the distal half of the large intestine and part of the rectum.

Venous return from the abdominal organs

The inferior vena cava is formed when *right* and *left common iliac veins* join at the level of the body of the 5th lumbar vertebra. This, the largest vein in the body, conveys blood from all parts of the body below the diaphragm to the right atrium of the heart. It passes through the central tendon of the diaphragm at the level of the 8th thoracic vertebra.

Paired veins from the testes, ovaries, kidneys and adrenal glands join the inferior vena cava.

Blood from the remaining organs in the abdominal cavity passes through the liver before entering the inferior vena cava. This is called the *portal circulation*.

THE PORTAL CIRCULATION

In all the parts of the circulation which have been described previously, venous blood passes from the tissues to the heart by the most direct route. In the portal circulation blood passes from *the abdominal part of the digestive system and the spleen via the liver* and the inferior vena cava to the heart. In this way blood with a high concentration of nutrient materials goes to the liver first, where certain modifications take place including the regulation of their supply to other parts of the body.

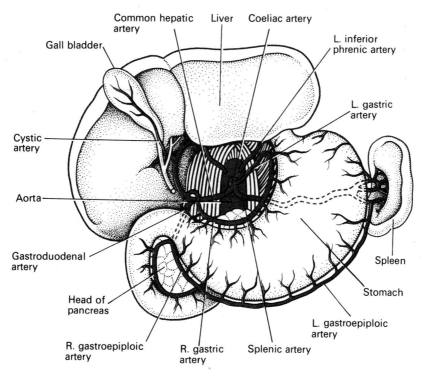

Figure 5:36 Coeliac artery and its branches and the inferior phrenic arteries.

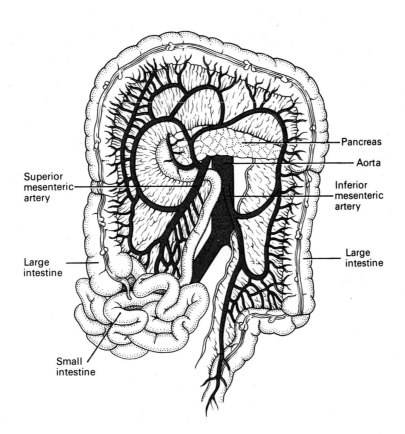

Figure 5:37 Superior and inferior mesenteric arteries and their branches.

The portal vein (Figs. 5:38 and 5:39)

This is formed by the following veins joining together:

    Splenic vein
    Inferior mesenteric vein
    Superior mesenteric vein
    Gastric veins
    Cystic vein

The splenic vein drains blood from the spleen, the pancreas and part of the stomach.

The inferior mesenteric vein returns the venous blood from the rectum, pelvic and descending colon of the large intestine. It joins the splenic vein.

The superior mesenteric vein returns venous blood from the small intestine. This vein also drains the caecum, ascending and transverse colon of the large intestine. The *superior mesenteric vein* unites with the *splenic vein* to form the *portal vein.*

The gastric veins join the portal vein.

The cystic vein which drains venous blood from the gall bladder joins the portal vein.

CIRCULATION OF BLOOD TO THE PELVIS AND LOWER LIMB
(Figs. 5:19, 5:20, 5:40, 5:41 and 5:42)

Arterial supply

The common iliac arteries. The right and left common iliac arteries are formed when the abdominal aorta divides at the level of the 4th lumbar vertebra. In front of the sacroiliac joint each divides into:

    An internal iliac artery
    An external iliac artery

The internal iliac artery runs medially to supply the

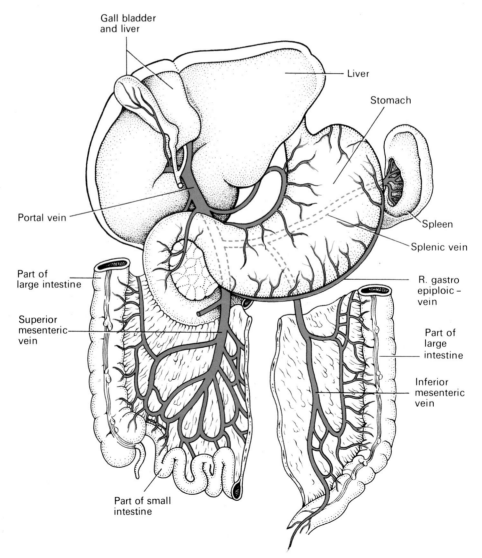

Figure 5:38 Venous drainage from the abdominal organs and the formation of the portal vein.

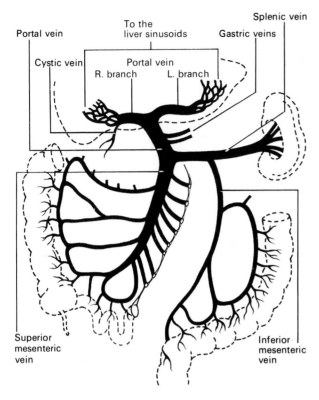

Figure 5:39   Portal vein—the veins which form it and its termination in the liver.

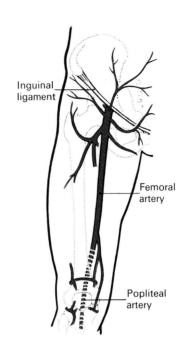

Figure 5:40   Femoral artery and its main branches.

organs within the pelvic cavity. In the female one of the main branches is the *uterine artery* which supplies the uterus.

The external iliac artery runs obliquely downwards to the level of the inguinal ligament where it becomes the femoral artery.

The femoral artery (Fig. 5:40) begins at the mid-point of the inguinal ligament and extends downwards, at first in front of the thigh, then medially, and eventually passes round the medial aspect of the femur to enter the popliteal space where it becomes the *popliteal artery*. It supplies blood to the structures of the thigh and some superficial structures in the pelvis and in the inguinal region.

The popliteal artery (Fig. 5:41) passes through the popliteal fossa behind the knee. It supplies the structures in this area and the knee joint. At the lower border of the popliteal fossa it divides into the anterior and posterior tibial arteries.

The anterior tibial artery (Fig. 5:41) passes forwards between the tibia and fibula and supplies the structures in the front of the leg. It lies on the tibia, runs in front of the ankle joint and continues over the dorsum (top) of the foot as the *dorsalis pedis artery*.

The dorsalis pedis artery is a continuation of the anterior tibial artery and passes over the dorsum of the foot supplying arterial blood to the structures in this area. It ends by passing between the first and second metatarsal

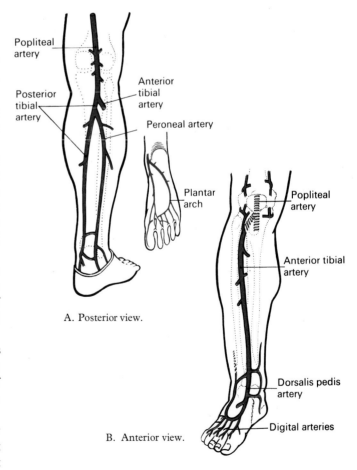

A. Posterior view.

B. Anterior view.

Figure 5:41   Popliteal artery and its main branches.

bones into the sole of the foot where it contributes to the formation of the plantar arch.

The posterior tibial artery (Fig. 5:41) runs downwards and medially on the back of the leg. Near its origin it gives off a large branch called the *peroneal artery* which supplies the lateral aspect of the leg. In the lower part it becomes superficial and passes medial to the ankle joint to reach the sole of the foot where it continues as the *plantar artery.*

The plantar artery supplies the structures in the sole of the foot. This artery and its branches form an arch similar to that of the hand, from which the *digital branches* arise to supply the toes.

Venous return from the lower limb and pelvis

The veins returning venous blood from the lower limb are, like those of the upper limb, divided into deep and superficial veins (Fig. 5:20).

The deep veins

The deep veins accompany the arteries and their branches and have the same names. They are:

    Digital veins
    Plantar venous arch
    Posterior tibial vein
    Anterior tibial vein
    Popliteal vein
    Femoral vein
    External iliac vein
    Internal iliac vein
    Common iliac vein

The femoral vein ascends in the thigh to the level of the inguinal ligament where it becomes the external iliac vein.

The external iliac vein is the continuation of the femoral vein where it enters the pelvis lying close to the femoral artery. It passes along the brim of the pelvis and at the level of the sacroiliac joint it is joined by the *internal iliac vein* to form the *common iliac vein.*

The internal iliac vein receives tributaries from several veins which drain the organs of the pelvic cavity.

The two common iliac veins begin at the level of the sacroiliac joints. They ascend obliquely and end a little to the right of the body of the 5th lumbar vertebra by uniting to form the inferior vena cava.

The superficial veins (Fig. 5:42)

The two main superficial veins draining blood from the superficial structures of the lower limbs are:

    The short saphenous vein
    The long saphenous vein

The short saphenous vein begins behind the ankle joint where many small veins which drain the dorsum of the foot join together. It ascends superficially along the back

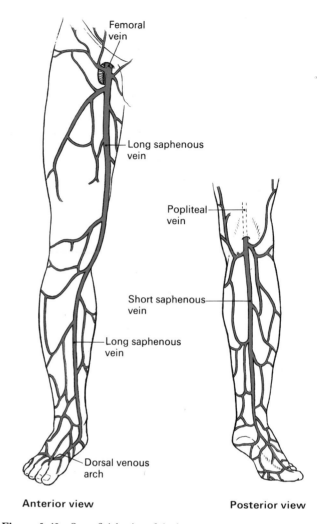

Figure 5:42 Superficial veins of the leg.

of the leg and in the popliteal space it becomes the *popliteal vein*—a deep vein.

The long saphenous vein is the longest vein in the body. It begins at the medial half of the dorsum of the foot and runs upwards, crossing the medial aspect of the tibia and up the inner side of the thigh. Just below the inguinal ligament it joins the *femoral vein.*

Summary of the Systemic Circulation

THE THORACIC AORTA

The ascending aorta
    Coronary arteries—paired

The arch

Brachiocephalic { right common carotid artery
{ right subclavian artery
Left common carotid artery
Left subclavian artery

The descending aorta
    Bronchial arteries
    Oesophageal arteries   }   paired arteries
    Intercostal arteries

THE ABDOMINAL AORTA
(branches in the order in which they leave the aorta)
    Phrenic arteries—paired

Coeliac artery { gastric arteries / splenic artery / hepatic artery } unpaired

    Superior mesenteric artery—unpaired
    Renal arteries—paired
    Ovarian or testicular arteries—paired
    Inferior mesenteric artery—unpaired

Common iliac arteries { external iliac artery / internal iliac artery } paired

ARTERIES SUPPLYING THE HEAD AND NECK

Common carotid arteries—paired
  External carotid artery:
    thyroid artery
    lingual artery
    facial artery
    occipital artery
    temporal artery
    maxillary artery
  Internal carotid artery:
    ophthalmic artery
    anterior cerebral artery
    middle cerebral artery
    posterior communicating artery
  Circulus arteriosus (circle of Willis):
    anterior cerebral arteries
    middle cerebral arteries
    posterior communicating arteries
    posterior cerebral arteries, which are branches
    from the basilar arteries
  Vertebral arteries—paired
  Basilar artery—unpaired

VENOUS RETURN FROM THE HEAD AND NECK

Superficial veins
  Thyroid vein
  Facial vein } empty into the external jugular vein
  Occipital vein

Deep sinuses
  Superior sagittal sinus
  Inferior sagittal sinus } unpaired
  Straight sinus
  Transverse sinuses—paired

ARTERIES SUPPLYING THE UPPER EXTREMITY
  Subclavian artery
  Axillary artery
  Brachial artery
  Radial artery
  Ulnar artery
  Deep and superficial palmar arches
  Palmar metacarpal arteries
  Palmar digital arteries

VENOUS RETURN FROM THE UPPER EXTREMITY

Deep veins
  Palmar digital veins
  Palmar metacarpal veins
  Palmar venous arch
  Radial vein
  Ulnar vein
  Brachial vein
  Axillary vein
  Subclavian vein
  Brachiocephalic vein
  Superior vena cava

Superficial veins
  Digital veins
  Palmar venous arch
  Cephalic vein
  Basilic vein
  Median cubital vein
  Median vein

ARTERIES SUPPLYING THE PELVIS AND LOWER EXTREMITY
  Common iliac artery
  Internal iliac artery
  External iliac artery
  Femoral artery
  Popliteal artery
  Anterior tibial artery
  Posterior tibial artery
  Plantar arch
  Digital arteries

VENOUS RETURN FROM THE PELVIS AND LOWER EXTREMITY

Deep veins
  Digital veins
  Plantar veins
  Anterior tibial veins
  Posterior tibial veins
  Popliteal vein
  Femoral vein
  External iliac vein
  Internal iliac vein
  Common iliac vein
  Inferior vena cava

Superficial veins
  Long saphenous vein
  Short saphenous vein

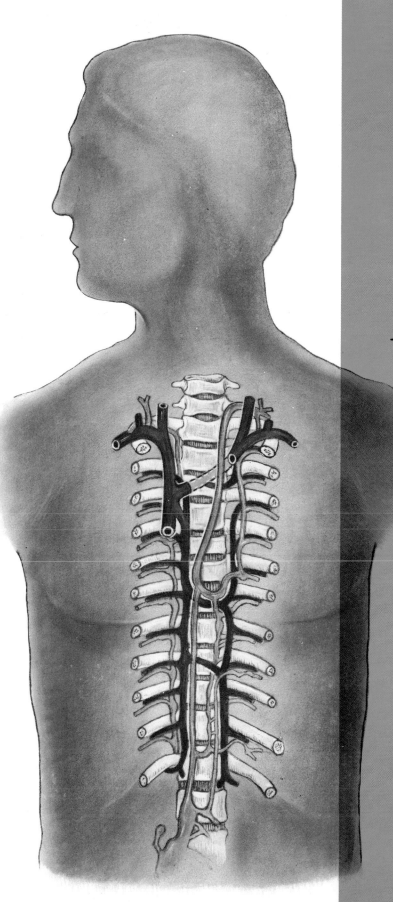

The Lymphatic System

6. *The Lymphatic System*

The lymphatic system communicates with the blood circulatory system and is closely associated with it.

As described previously the tissue fluid is derived from the blood plasma (p. 53). A certain amount of this fluid and waste products from the cells are returned to the blood capillaries, but within the tissue spaces fine capillary vessels known as *lymph capillaries* begin, which help to drain water and particles in the water from the interstitial spaces. These vessels have blind ends. Their walls are only one cell thick and are more permeable than blood capillaries. This means that larger particles can pass through the walls of the lymph capillaries than can enter the blood capillaries.

Figure 6:1 illustrates the movement of nutrient materials, water and oxygen from the blood capillaries into the tissue fluid then to the cells. Waste products and water excreted by the cells have two routes by which they return to the circulation: one is by the blood capillaries and the other is by the lymph capillaries. It is easier for water and dissolved substances to cross the wall of the lymph capillary because it is more permeable and the pressure inside is lower than in the blood capillary.

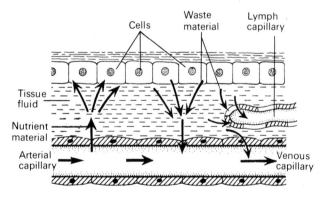

Figure 6:1 Diagram of the beginning of a lymph capillary.

The fluid within the lymph capillaries and vessels is known as *lymph*. The composition of lymph is very like that of the blood plasma, but the dissolved substances are in different concentrations. Lymph also contains materials which may be damaging to the body. Because of the greater permeability of the lymph capillaries sub-

stances of larger size can enter these vessels and be removed from the interstitial spaces. For example, if infection is present and phagocytosis has occurred the neutrophils and monocytes with their ingested micro-organisms pass through the walls of the lymph capillaries and are drained away in the lymph.

The Structures of the Lymphatic System

Lymph capillaries
Lymph vessels
The thoracic duct
The right lymphatic duct
Lymph nodes
Other lymphatic tissue

LYMPH CAPILLARIES

Structure
The lymph capillaries are composed of a single layer of *endothelial cells* and originate in the tissue spaces as minute *blind end tubes*. These minute capillaries join with one another to form the *lymph vessels* until eventually *two main lymphatic ducts* are formed.

LYMPH VESSELS (Fig. 6:2)

Structure
1. An outer coat consisting of *fibrous tissue* which acts as a protective covering.
2. A middle coat of *muscular and elastic tissue*.
3. An inner lining composed of a single layer of *endothelial cells*.
4. Many *valves* consisting of a double layer of lining membrane give the vessels a knotted or beaded appearance. The valves prevent the backward flow of the lymph.

THE LYMPHATIC DUCTS (Fig. 6:3)
There are two lymphatic ducts which collect lymph from the whole body and return it to the blood. Their structure is the same as that of the smaller lymph vessels.

THE THORACIC DUCT (Fig. 6:4)
This duct begins at the *cysterna chyli*, which is a sac-like

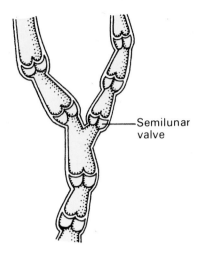

Figure 6:2   A lymph vessel cut open to show semilunar valves.

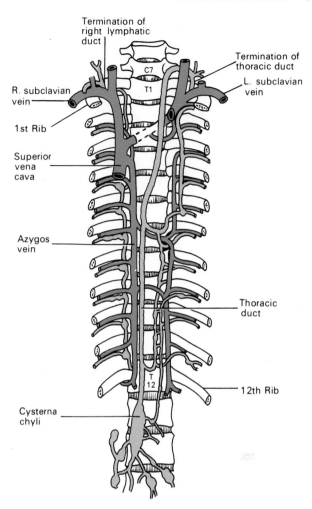

Figure 6:4   Origin and position of the thoracic duct.

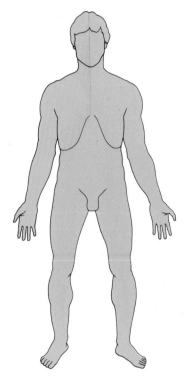

Figure 6:3   Diagram of lymph drainage.
Green area drained by the thoracic duct.
Orange area drained by the right lymphatic duct.

dilatation situated in front of the bodies of the first and second lumbar vertebrae to the right of the abdominal aorta.

The thoracic duct is the largest lymph vessel in the body and contains several valves. It is approximately 40 cm (16 inches) in length and extends from the 2nd lumbar vertebra to the root of the neck where it opens into the *left subclavian vein.*

This duct drains lymph from the lower limbs, pelvic cavity, abdominal cavity, *left side* of the chest, head and neck and the *left* arm (Figs. 6:3 and 6:4).

THE RIGHT LYMPHATIC DUCT
The right lymphatic duct is about 1 cm (less than ½ an inch) in length. It lies in the root of the neck and terminates by emptying its contents into the *right subclavian vein.* The right lymphatic duct receives all the lymph which has drained from the *right side* of the chest, head and neck and the *right* arm.

LYMPH NODES (Fig. 6:5)
All the small and medium-sized lymph vessels open into *lymph nodes* which are situated in strategic positions throughout the body. The lymph drains through at least one node before returning to the blood. These nodes vary considerably in size: some are as small as a pin head and the largest are about the size of an almond.

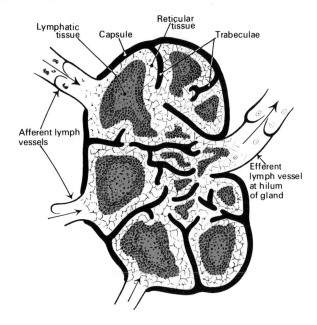

Figure 6:5   Diagram of the inside of a lymph node. Arrows show the direction of flow of lymph.

Structure

Lymph nodes have a *surrounding capsule of fibrous tissue* which dips down into the node substance forming partitions known as *trabeculae*. The main substance of the node consists of *reticular and lymphatic tissue* containing many *lymphocytes*.

As many as four or five lymph vessels, known as afferent vessels, may enter a lymph node while only one, the efferent vessel, carries lymph away from the node. Each node has a concave surface called the hilum. At the hilum the blood vessels supplying the node enter and leave and the efferent lymph vessel leaves.

There is a vast number of lymph nodes situated in strategic positions throughout the body. They are arranged in *deep* and *superficial groups*. Those through which lymph from the head and neck pass are the deep and superficial *cervical nodes* (Fig. 6:6).

The lymph from the *upper limbs* passes through nodes situated in the elbow region then through the deep and superficial *axillary nodes* (Fig. 6:7).

Lymph from the organs and tissues in the *thoracic cavity* is drained through many nodes; the more important groups are the *tracheobronchial* and the *intercostal nodes*. Most of the lymph from the breast passes through the

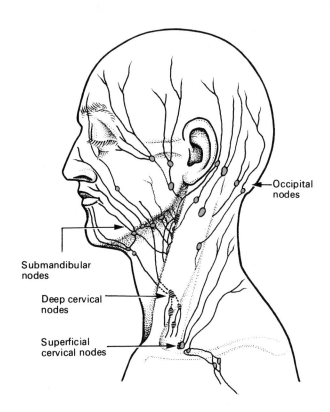

Figure 6:6   Some lymph nodes of the face and neck.

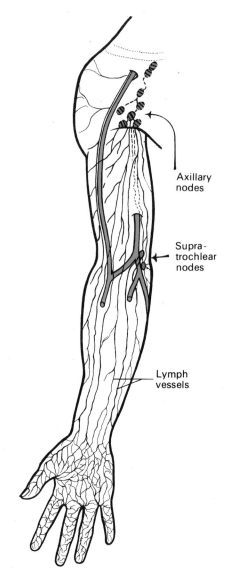

Figure 6:7   Some lymph nodes of the upper limb and their relationship to the veins.

nodes in the axilla before being returned to the venous circulation.

Lymph from the *pelvic and abdominal cavities* passes through many lymph nodes before entering the cysterna chyli. The abdominal and pelvic nodes are situated mainly in association with the blood vessels supplying the organs and close to the main arteries, i.e. the aorta and the external and internal iliac arteries.

The lymph from the *lower limbs* is drained through deep and superficial nodes behind the knee (popliteal nodes) and in the groin (inguinal nodes) (Fig. 6:8).

The lacteals are the lymph capillaries which drain lymph from the small intestine. A proportion of the fat absorbed from the small intestine passes into the lymph capillaries and this high concentration of fat gives the lymph a milky appearance. Because of this, lymph entering the thoracic duct is known as *chyle*.

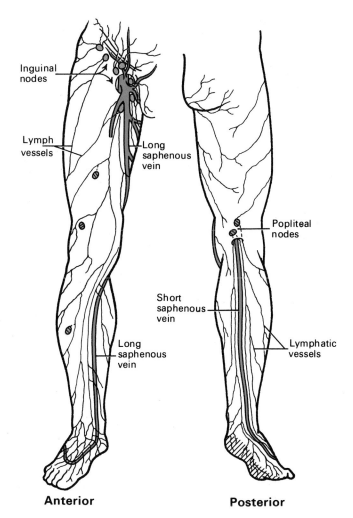

Anterior   **Posterior**

Figure 6:8 Some lymph nodes of the lower limb and their relationship to the veins.

Functions

1. Lymph is filtered as it passes through the lymph nodes leaving behind any particles which would not normally be found in serum.

2. The lymphoid tissue in the nodes breaks down material which has been filtered off, for example, micro-organisms, phagocytes, tumour cells and cells which have been damaged by inflammation. The nodes are not always successful in destroying this type of material but, on the whole, they provide an effective barrier against the spread of noxious particulate material.

3. Lymphocytes which develop in the bone marrow are of two types (see Ch. 5). After they have been *activated* they pass to the lymph nodes and other collections of lymphoid tissue, where they multiply. From there they enter the blood where they perform their specific protective activities.

LYMPHATIC TISSUE

Lymphatic tissue is found in a number of situations in the body in addition to the lymph nodes:

The palatine tonsil—	between the mouth and the oral part of the pharynx
The pharyngeal tonsil—	on the wall of the nasal part of the pharynx
The solitary lymphatic follicles The aggregated lymphatic follicles (Peyer's patches)	in the wall of the small intestine
The vermiform appendix—	in the first part of the large intestine, the caecum

These structures are discussed in the sections on the respiratory and digestive systems (Chs. 7 and 9).

THE SPLEEN (Fig. 6:9)

The spleen is formed partly by lymphatic tissue and will be described here as its functions are associated with the cirulatory system.

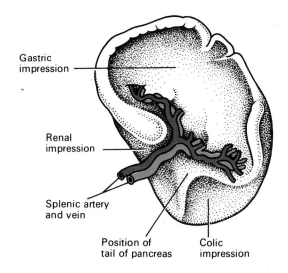

Figure 6:9 Spleen.

The spleen lies in the *left hypochondriac region* of the abdominal cavity between the fundus of the stomach and the diaphragm. It is purplish in colour and varies in size in different individuals, but is usually about 12 cm (4¾ inches) long, 7 cm (2¾ inches) wide and 2·5 cm (1 inch) thick. It weighs about 200 grams (7 ounces).

Organs in association with the spleen

Superiorly and posteriorly— diaphragm

Inferiorly— left colic flexure of the large intestine

Anteriorly— fundus of the stomach

Medially— pancreas and the left kidney

Laterally— separated from the 9th, 10th and 11th ribs and the intercostal muscles by the diaphragm

Structure

The spleen is slightly ovoid in shape and is covered anteriorly by *peritoneum*. Under the peritoneum there is a *fibroelastic* capsule which surrounds the gland. Fibrous tissue spreads into the organ substance to form *trabeculae*. The substance of the organ is known as the *splenic pulp*. It is composed of lymph vessels and a large number of blood capillaries (Fig. 6:10).

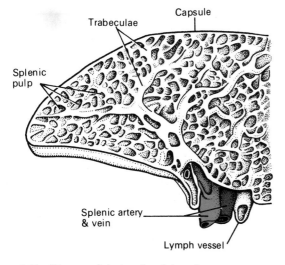

Figure 6:10 Diagram of the interior of the spleen.

The lower medial border of the organ is concave and is known as the *hilum*. The structures entering and leaving at the hilum are:

The splenic artery
The splenic vein
Lymph vessels
Nerves

Functions

Phagocytosis. As described previously (p. 39) erythrocytes are destroyed in the spleen and the breakdown pro-

ducts, bilirubin and iron, are passed to the liver via the splenic and portal veins. Other cellular material such as micro-organisms, leucocytes and platelets are phagocytosed in the spleen.

A reservoir for blood. It is known that the spleen stores blood in some animals, but the extent to which it does so in man is uncertain. However, it is believed to store erythrocytes and in states of emergency such as hypoxia they pass into the circulation and increase the oxygen-carrying capacity of the blood.

The formation of lymphocytes. The spleen forms lymphocytes some of which pass into the circulating blood while others remain in the spleen.

The formation of antibodies and antitoxins. The spleen produces considerable amounts of specific antibodies and antitoxins which are passed into the blood.

THE THYMUS

The thymus gland lies in the upper part of the mediastinum behind the sternum and extends upwards into the root of the neck. It weighs about 12 g at birth and grows until the individual reaches puberty, when it begins to atrophy. Its maximum weight, at puberty, is between 30 and 40 g and by middle age it has returned to approximately its weight at birth (Fig. 6:11).

Organs in association with the thymus

Anteriorly— sternum and upper four costal cartilages

Posteriorly— aortic arch and its branches, brachiocephalic vein, trachea

Laterally— lungs

Superiorly— structures in the root of the neck

Inferiorly— heart

Structure

The gland consists of two lobes of unequal size and shape which are joined near their upper borders. The lobes are enclosed by a fibrous capsule, which dips into the substance of the lobes dividing it into lobules. The cells making up the lobules are divided into an inner medulla and an outer cortex.

The thymus receives its blood supply from adjacent blood vessels.

Functions

The thymus gland is the site of activation of T-lymphocytes which originate in red bone marrow (see also p. 40). In the child it is very active and may influence the development of other lymphoid tissue through secretions as yet unidentified. Its activity, like that of all lymphoid tissue, declines with age.

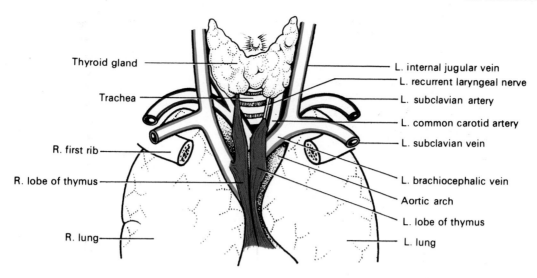

Thyroid gland

Trachea

R. first rib

R. lobe of thymus

R. lung

L. internal jugular vein

L. recurrent laryngeal nerve

L. subclavian artery

L. common carotid artery

L. subclavian vein

L. brachiocephalic vein

Aortic arch

L. lobe of thymus

L. lung

Figure 6:11   Thymus gland in the adult, and related structures.

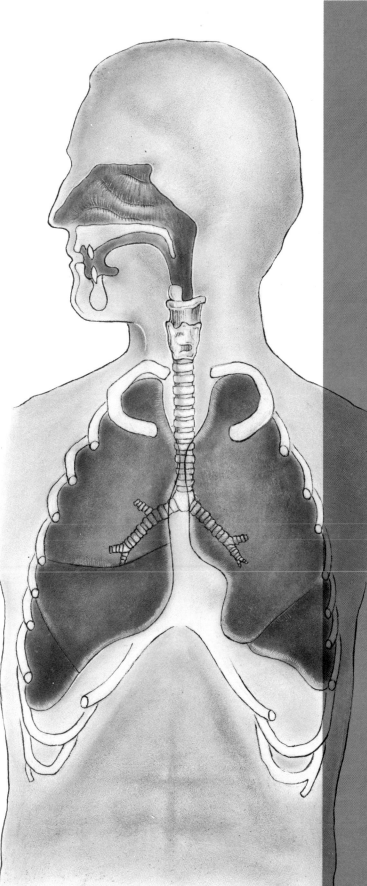

7

The Respiratory System

7. *The Respiratory System*

Most of the energy required by the cells of the body is derived from chemical reactions which can only take place in the presence of oxygen (O_2). The main waste product of these reactions is carbon dioxide (CO_2). The respiratory system provides the route by which the supply of oxygen present in the atmospheric air gains entry to the body and it provides the route of excretion of carbon dioxide.

The condition of the atmospheric air entering the body varies considerably according to the external environment, e.g., it may be dry, cold and contain dust particles or it may be moist and hot. As the air breathed in moves through the air passages to reach the lungs, it is warmed or cooled to body temperature, moistened to become saturated with water vapour and 'cleaned' as particles of dust stick to the mucus which coats the lining membrane. The blood provides the transport system for these gases between the lungs and the cells of the body. The exchange of gases between the blood and the lungs is called *external respiration* and that between the blood and the cells *internal respiration*.

The organs of the respiratory systems are:

The nose
The pharynx
The larynx
The trachea
Two bronchi (one bronchus to each lung)
The bronchioles and smaller air passages.
The two lungs and their coverings—the pleura
The muscles of respiration—the intercostal muscles and the diaphragm

A general view of the organs of the respiratory system is given in Figures 7:1 and 7:2.

The Nose and Nasal Cavity

POSITION AND STRUCTURE

The nasal cavity is the first of the respiratory organs and consists of a large irregular cavity divided into two equal parts by a *septum* situated in the midline. The posterior bony part of the septum is formed by the perpendicular plate of the ethmoid bone and the vomer. Anteriorly it consists of hyaline cartilage (Fig. 7:3).

The roof is formed by the cribriform plate of the ethmoid bone, the sphenoid bone, the frontal bone and the nasal bones.

The floor is formed by the roof of the mouth, which consists of the hard palate in front and the soft palate behind. The hard palate is composed of the maxilla and palatine bones and the soft palate consists of unstriped muscle.

The medial wall is formed by the *septum*.

The lateral walls are formed by the maxilla, the ethmoid bone and the inferior conchae (Fig. 7:4).

The posterior wall is formed by the posterior wall of the pharynx.

OPENINGS INTO THE NASAL CAVITY

The anterior nares are the openings from the exterior to the nasal cavity.

The posterior nares are the openings from the nasal cavity into the pharynx.

The sinuses are cavities in the bones of the face and the cranium which contain air. There are tiny openings between the air sinuses and the nasal cavities. The main sinuses are:

The maxillary sinuses in the lateral walls
The frontal and sphenoidal sinuses in the roof
The ethmoidal sinuses in the upper part of the lateral walls

The nasolacrimal ducts extend from the lateral walls of the nose to the conjunctival sacs of the eye (see p. 217).

THE LINING OF THE NOSE

The nose is lined with *ciliated columnar epithelium* (ciliated mucous membrane) which contains mucus-secreting goblet cells (Figs. 7:5 and 7:6).

At the anterior nares this ciliated mucous membrane blends with the skin and posteriorly it extends into the nasal part of the pharynx.

RESPIRATORY FUNCTION OF THE NOSE

The nose is the first of the respiratory passages through which the incoming air passes. The function of the nose

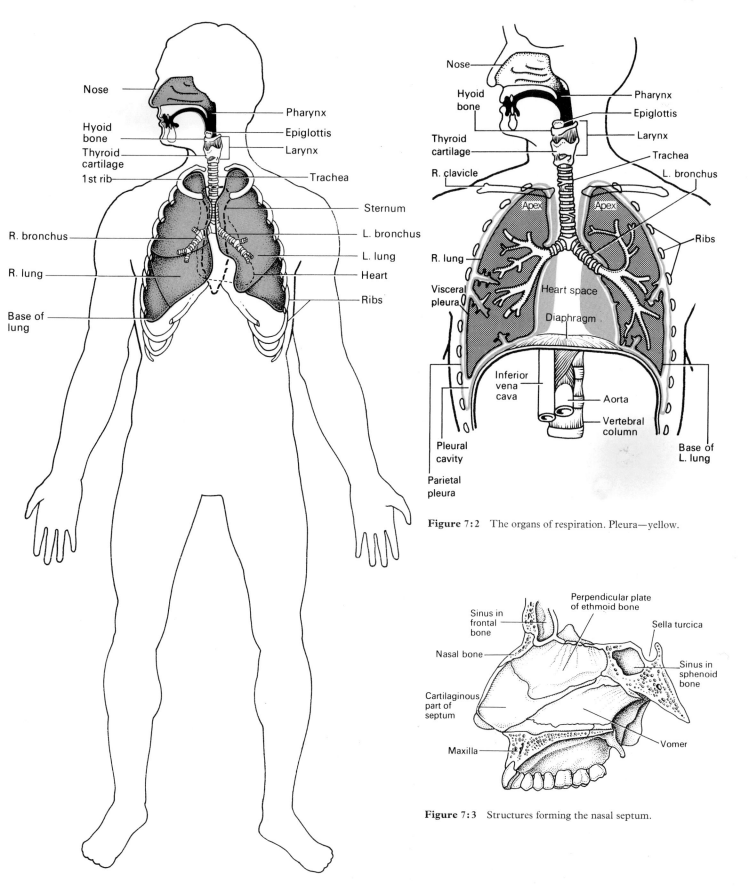

Figure 7:1 The parts of the respiratory system and related structures.

Figure 7:2 The organs of respiration. Pleura—yellow.

Figure 7:3 Structures forming the nasal septum.

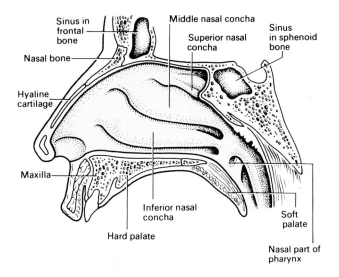

Figure 7:4   Lateral wall of right nasal cavity. Epithelium—yellow.

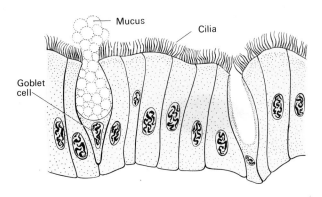

Figure 7:5   Ciliated columnar epithelium with goblet cells.

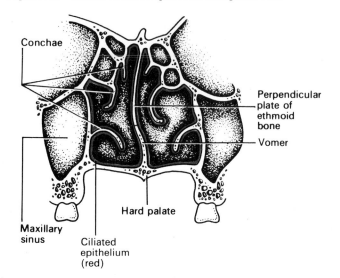

Figure 7:6   Interior of the nose viewed from the front.

is to begin the process by which the air is *warmed, moistened and filtered.*

The air is warmed as it passes over the surface of the nose; it is moistened by contact with the moist mucus; and it is filtered in the sense that particles of dust and other impurities such as bacteria in the air stick to the mucus.

The cilia of the mucous membrane waft the mucus towards the throat and it is swallowed. Figure 7:7 illustrates the pathway of air through the nose.

OLFACTORY FUNCTION OF THE NOSE

The nose is the organ of the sense of smell. In the roof of the nose, in relation to the cribriform plate of the ethmoid bone and the superior conchae, there are nerve endings and fibres of the sense of smell. These are stimulated by chemical substances which are given off by odorous materials. The resultant nerve impulses are conveyed to the brain by the *olfactory nerves* where the sensation of smell is perceived (for further explanation see p. 190).

The Pharynx

POSITION

The pharynx is a tube 12 to 14 cm (about 5 inches) in length which extends from the base of the skull to the sixth cervical vertebra. It lies behind the nose, the mouth and the larynx and is wider at its upper end.

Structures in association with the pharynx

Superiorly—   the inferior surface of the base of the skull

Inferiorly—   it is continuous with the oesophagus

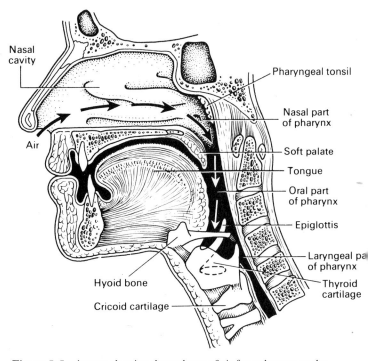

Figure 7:7   Arrows showing the pathway of air from the nose to the larynx.

Anteriorly— the wall is incomplete because of the openings into the nose, the mouth and the larynx

Posteriorly— areolar tissue, involuntary muscle and the bodies of the first six cervical vertebrae

For descriptive purposes the pharynx is divided into three parts: *nasal, oral* and *laryngeal.*

The nasal part of the pharynx lies behind the nose above the level of the soft palate. On its lateral walls are the two openings of the *auditory tubes* which lead to the middle ear. On the posterior wall there is the *pharyngeal tonsil* (adenoid) which consists of lymphoid tissue. It is most prominent in children up to approximately the age of 7 years and thereafter it gradually atrophies.

The oral part of the pharynx lies behind the mouth extending from below the level of the soft palate to the level of the upper part of the body of the third cervical vertebra. The walls of the pharynx blend with the soft palate to form two folds on each side. Between each pair of folds there is a collection of lymphoid tissue called the *palatine tonsil.*

The laryngeal part of the pharynx extends from the oral part above and continues as the oesophagus below, i.e., from the level of the third to the sixth cervical vertebrae.

STRUCTURE

The pharynx is composed of three layers of tissue:

1. *Mucous membrane* which lines the pharynx. The type varies slightly in the different parts. In the nasal part it is continuous with the lining of the nose and consists of *ciliated columnar epithelium*; in the oral and laryngeal parts it is *stratified squamous epithelium* which is continuous with the lining of the mouth.

2. *Fibrous tissue* which forms the intermediate layer. It is thicker in the nasal part, where there is little muscle, and becomes thinner towards the lower end, where the muscle layer is thicker.

3. *Muscle tissue.* This layer consists of several muscles known as the *constrictor muscles* of the pharynx. The muscle fibres are unstriped and they play an important part in the mechanism of swallowing (deglutition) which, in the pharynx, is not under voluntary control.

The blood supply to the pharynx is through several arteries which are branches of the facial artery. The venous return is into the facial and internal jugular veins.

The nerve supply is from the pharyngeal plexus, which is formed by parasympathetic and sympathetic nerves. Parasympathetic supply is mainly through the vagus and glossopharyngeal nerves. Sympathetic supply is by nerves from the superior cervical ganglion (see p. 192).

FUNCTIONS

The pharynx is an organ involved in both the respiratory and the digestive systems: air passes through the nasal and oral parts and food through the oral and laryngeal parts. By the same methods as in the nose the air is further warmed and moistened as it passes through the pharynx.

The auditory tubes pass between the nasal part of the pharynx and the middle ear, and through these tubes air passes to the middle ear. The presence of air in the middle ear at atmospheric pressure is essential for satisfactory hearing (see p. 199).

The lymphatic tissue of the palatine and laryngeal tonsils helps to prevent micro-organisms from entering the body.

The Larynx

POSITION

The larynx or 'voice box' extends from the root of the tongue and the hyoid bone to the trachea. It lies in front of the laryngeal part of the pharynx at the level of the third, fourth, fifth and sixth cervical vertebrae but is slightly higher in the female than in the male. Until puberty there is little difference in the size of the larynx in the different sexes. Thereafter it grows larger in the male, which explains the prominence of the 'Adam's apple' and the generally deeper voice.

Structures in association with the larynx

Superiorly— the hyoid bone and the root of the tongue

Inferiorly— it is continuous with the trachea

Anteriorly— the muscles attached to the hyoid bone and the muscles of the neck

Posteriorly— laryngeal part of the pharynx and cervical vertebrae

Laterally— the lobes of the thyroid gland

STRUCTURE

The larynx is composed of several irregularly shaped cartilages attached to each other by ligaments and membranes. The main cartilages are:

1 thyroid cartilage ⎫
1 cricoid cartilage   ⎬ hyaline cartilage
2 arytenoid cartilages ⎭
1 epiglottis— elastic fibrocartilage

The thyroid cartilage is the most prominent and consists of two flat pieces of cartilage known as the *laminae*, which are fused together anteriorly to form the *laryngeal prominence* (Adam's apple). Immediately above the laryngeal prominence the laminae are separated by a V-shaped notch known as the *thyroid notch.* The thyroid cartilage is incomplete posteriorly and the posterior border of each lamina is extended to form two processes called the *superior* and *inferior cornu* (Fig. 7:8).

The upper part of the thyroid cartilage is lined with stratified squamous epithelium like the larynx, and the lower part with ciliated columnar epithelium like the trachea. There are many muscles attached to its outer surface.

larynx. They give attachment to the vocal cords and to muscles and are lined with ciliated columnar epithelium.

The epiglottis is a leaf-shaped cartilage attached to the inner surface of the anterior wall of the thyroid cartilage immediately below the thyroid notch. It rises obliquely upwards behind the tongue and the body of the hyoid bone. It is covered with stratified squamous epithelium.

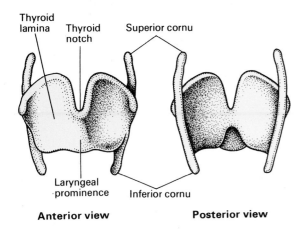

Figure 7:8 Thyroid cartilage.

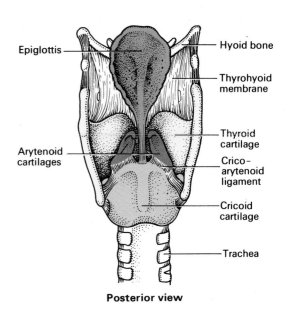

Figure 7:10 Larynx.

The cricoid cartilage lies below the thyroid cartilage. It is shaped like a signet ring, completely encircling the larynx with the narrow part in front and the broad part behind. The broad posterior part articulates with the arytenoid cartilages above and with the inferior cornu of the thyroid cartilage below. It is lined with ciliated columnar epithelium and there are muscles and ligaments attached to its outer surface (Fig. 7:9)

LIGAMENTS AND MEMBRANES

The hyoid bone and the cartilages which form the larynx are attached to one another by ligaments and membranes.

The thyrohyoid membrane is a broad flat membrane composed of fibroelastic tissue attached to the lower border of the hyoid bone above and to the thyroid cartilage below (Fig. 7:11).

The cricoarytenoid ligaments attach the arytenoid cartilages to the cricoid cartilage (Fig. 7:10).

The cricothyroid ligament and cricovocal membrane are composed mainly of yellow elastic tissue and extend from the thyroid cartilage above to the cricoid cartilage below (Fig. 7:12).

The thyroepiglottic ligament and hyoepiglottic ligaments attach the epiglottis to the thyroid cartilage and hyoid bone respectively.

BLOOD AND NERVE SUPPLY

Blood is supplied to the larynx through the superior and inferior laryngeal arteries and drained by the thyroid vein, which joins the internal jugular vein.

The parasympathetic nerve supply is from the superior laryngeal and recurrent laryngeal nerves, which are branches of the vagus nerves, and by sympathetic nerves from the superior cervical ganglion (see p. 194).

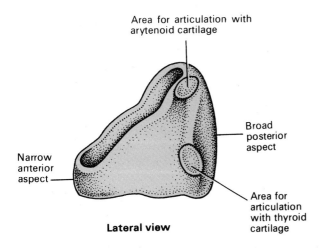

Figure 7:9 Cricoid cartilage.

The arytenoid cartilages are two roughly pyramid-shaped cartilages situated on top of the broad part of the cricoid cartilage forming part of the posterior wall of the

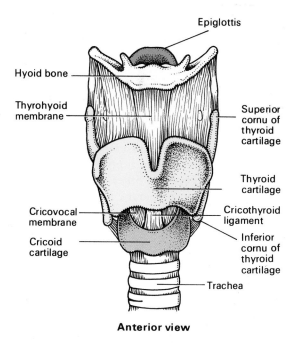

Anterior view

Figure 7:11   Larynx.

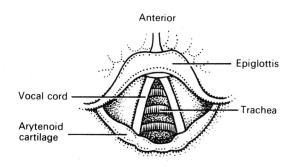

Figure 7:12   Interior of the larynx viewed from above.

These provide motor nerve supply to the muscles of the larynx and sensory fibres to the lining membrane.

INTERIOR

Within the larynx there are the *vocal cords* (Fig. 7:12), thus the term 'voice box'. They consist of two pale folds of mucous membrane with cord-like free edges which extend from the inner wall of the thyroid prominence anteriorly to the arytenoid cartilages posteriorly.

When the muscles of the arytenoid cartilages contract, the cartilages adduct and rotate medially, pulling the vocal cords together and narrowing the gap between them, thus forming the *chink of the glottis*. If air is forced through this chink it causes vibration of the cords and sound is produced. When the muscles relax the cartilages rotate laterally and abduct, separating the cords, and no sound is produced (Fig. 7:13).

Sound has the properties of *pitch, loudness* and *quality*.

The pitch of the voice depends upon the *length* and *tightness* of the cords. In adults the vocal cords are longer in the male than in the female, thus the male voice has a deeper pitch than that of the female.

The loudness of the voice depends upon the *force* with which the cords vibrate. The greater the force of expired air the more vibration of the cords and the louder the sound.

The quality and resonance of the voice depend upon the shape of the mouth; the position of the tongue and the lips; the facial muscles; the air sinuses in the bones of the face and the skull.

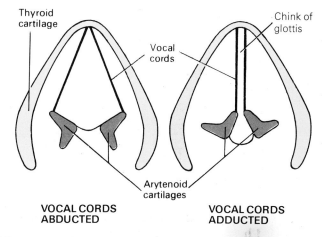

Figure 7:13   Diagram of the extreme positions of the vocal cords.

FUNCTIONS

1. The larynx provides a passageway for air between the pharynx and the trachea. As the air from the outside passes through, it is further moistened, filtered and warmed thus continuing the process started in the nose.

2. The vocal cords produce sound of varying loudness and pitch.

3. During swallowing (deglutition) the larynx moves upwards occluding the opening into it from the pharynx. This ensures that food passes into the oesophagus and not into the lower respiratory passages.

The Trachea

POSITION

The trachea or windpipe is a continuation of the larynx and extends to about the level of the fifth thoracic vertebra where it divides (bifurcates) into the right and left bronchi, one bronchus going to each lung. It is approximately 10 to 11 cm (about 4 inches) long and lies mainly in the median plane in front of the oesophagus.

Structures in association with the trachea (Fig. 7:14)

Superiorly—	the larynx
Inferiorly—	the right and left bronchi
Anteriorly—	upper part—the isthmus of the thyroid gland; lower part—the arch of the aorta and the sternum
Posteriorly—	the oesophagus separates the trachea from the vertebral column
Laterally—	the lobes of the thyroid gland and the lungs

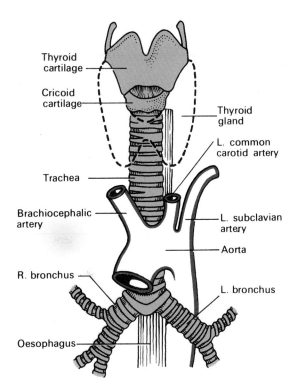

Figure 7:14 The trachea and some of its associated structures.

STRUCTURE

The trachea is composed of from 16 to 20 incomplete (C-shaped) rings of hyaline cartilages situated one above the other. The cartilages are incomplete posteriorly. Connective tissue and involuntary muscle join the cartilages and form the posterior wall where they are incomplete. The soft tissue posterior wall is in contact with the oesophagus.

There are three layers of tissue which 'clothe' the cartilages of the trachea.

The outer layer consists of fibrous and elastic tissue and encloses the cartilages.

The middle layer is composed of connective tissue, involuntary muscle, hyaline cartilage and areolar tissue and contains blood and lymph vessels and nerves.

The inner lining consists of ciliated columnar epithelium containing goblet cells which secrete mucus.

BLOOD AND NERVE SUPPLY

The arterial blood supply is mainly by the inferior thyroid and bronchial arteries and the *venous return* is through the inferior thyroid veins into the brachiocephalic vein.

The *nerve supply* is by parasympathetic and sympathetic fibres. Parasympathetic supply is through the recurrent laryngeal nerves and other branches of the vagi. Sympathetic supply is by nerves from the sympathetic ganglia (see p. 194).

The Bronchi

The two bronchi commence when the trachea divides, that is, at about the level of the 5th thoracic vertebra (Fig. 7:15).

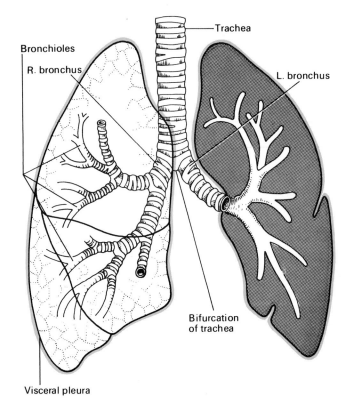

Figure 7:15 The bronchial tree and the lungs. Anterior part of the left lung removed. Visceral pleura—yellow.

The right bronchus is a wider, shorter tube than the left bronchus and it lies in a more vertical position. It is approximately 2·5 cm (1 inch) in length. After entering the right lung at the hilum it divides into three branches one of which passes to each lobe. Each branch then subdivides into numerous smaller branches.

The left bronchus is about 5 cm (2 inches) long and is

narrower than the right. After entering the lung it divides into two branches, one of which goes to each lobe. Each branch then subdivides into progressively smaller tubes within the lung substance.

STRUCTURE

The bronchi are composed of the same tissues as the trachea. As they become smaller by subdividing, the cartilages become less well defined and more irregular in shape. The bronchi are lined with ciliated columnar epithelium.

The Bronchioles, Smaller Air Passages and Alveoli

STRUCTURE

There is no clear anatomical division between bronchi and bronchioles. As the air passages get smaller they lose their cartilages. The largest air passages with no cartilages in their walls are the *bronchioles*.

The larger bronchioles are composed of muscle tissue, fibrous tissue and elastic tissue with an inner lining of ciliated columnar epithelium. As the tubes subdivide and become still smaller the fibrous and muscle tissue disappear, and the columnar epithelium changes to a single layer of flattened epithelial cells. The minute bronchioles, known as the *terminal bronchioles*, branch to form *respiratory bronchioles* which branch still more to form *alveolar ducts*. The alveolar ducts then lead into minute sac-like structures known as the *alveoli* (Fig. 7:16). It

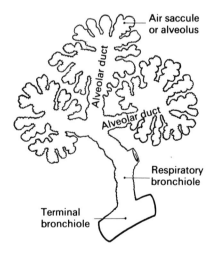

Figure 7:16   Diagram of small air passages and alveoli.

is here that the interchange of gases takes place between the air in the alveoli and the blood in the capillaries, which are also composed of a single layer of flattened epithelial cells.

BLOOD AND NERVE SUPPLY, LYMPH DRAINAGE

The arterial blood supply to the walls of the bronchi and smaller air passages is by branches of the *right and left bronchial arteries* and the *venous return* is mainly by the bronchial veins. On the right side they empty into the azygos vein and on the left into the superior intercostal vein.

The nerve supply is by the vagus and sympathetic nerves. The vagus nerves (parasympathetic) stimulate constriction of the bronchial tree and sympathetic stimulation causes dilatation.

The lymphatic vessels and lymph nodes. Lymph is drained from the walls of the air passages in a network of lymph vessels. It passes through lymph nodes situated around the trachea and bronchial tubes then into the thoracic duct on the left side and the right lymphatic duct on the other.

FUNCTIONS: TRACHEA, BRONCHI, BRONCHIOLES

1. Because of the presence of the cartilages in the larger air passages they remain permanently open thus allowing for the unobstructed passage of air between the outside atmosphere and the alveoli of the lungs.

2. The mucus which coats the lining membrane is of a sticky consistency to which particles present in the air adhere, for example, dust and bacteria. This ensures that much of the particulate matter suspended in inspired air is removed before it reaches the alveoli.

3. The wave motion of the cilia of the lining membrane wafts mucus and any adherent particles towards the throat. When the mucus reaches the pharynx it is usually swallowed but it may be expectorated. The process may be aided by coughing.

4. The diameter of the respiratory passages may be altered by contraction or relaxation of the involuntary muscle in their walls. The amount of change in the calibre of air passages which contain cartilages is limited, but it is sufficient to affect the volume of air entering the lungs. These changes are regulated by the autonomic nerve supply: parasympathetic stimulation constricts the air passage and sympathetic stimulation dilates them (see p. 196).

The Lungs

POSITION AND ASSOCIATED STRUCTURES
(Fig. 7:17)
The lungs are two in number, one lying on each side of the midline in the thoracic cavity. They are cone-shaped and are described as having an *apex*, a *base*, a *costal surface* and a *medial surface*.

The apex is rounded like the narrow end of a cone and rises into the root of the neck, about 25 mm (1 inch)

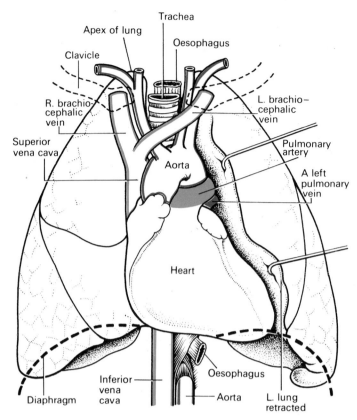

Figure 7:17   Organs in association with the lungs.

above the level of the middle third of the clavicle. The structures associated with it are the first rib and the blood vessels and nerves in the root of the neck.

The base is concave and semilunar in shape and is closely associated with the thoracic surface of the diaphragm.

The costal surface is convex and is closely associated with the costal cartilages, the ribs and the intercostal muscles.

The medial surfaces are concave and are separated from each other by a space called the *mediastinum* which is occupied by the heart, the great vessels, the trachea, the right and left bronchus, the oesophagus, lymph nodes, lymph vessels and nerves. The medial surface of each lung has a roughly triangular-shaped area known as the *hilus* and the structures which form the *root of the lung* enter and leave at the hilus. The roots of the lungs are situated at the level of the fifth, sixth and seventh thoracic vertebrae.

Structures entering and leaving each lung at the hilus are (Fig. 7:18):

A bronchus
A pulmonary artery
Two pulmonary veins
A bronchial artery
The bronchial veins

Lymph vessels
Parasympathetic and sympathetic nerves

STRUCTURE

THE LOBES

The *right lung* is divided into three distinct lobes: superior, middle and inferior.

The *left lung* is divided into only two lobes: superior and inferior.

THE PLEURA

The pleura consists of a closed sac of serous membrane (one for each lung) which contains a small amount of serous fluid. The lung is invaginated into this sac so that it forms into two layers: one is closely associated with the lung and the other lines the thoracic cavity (see Fig. 7:2).

The visceral pleura is the layer which is adherent to the lung, investing each lobe and passing into the fissures which separate the lobes.

The parietal pleura is adherent to the inside of the chest wall and the thoracic surface of the diaphragm. It is reflected off the adjacent structures in the mediastinum and is continuous with the visceral pleura round the edges of the hilus.

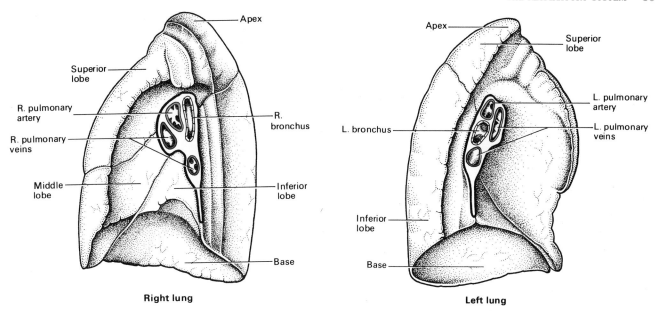

Figure 7:18 The parts of the lungs and the structures entering at the hilum.

The pleural cavity is only a potential space because in health the two layers of pleura are separated by only a thin film of serous fluid, sufficient to prevent friction between them during breathing. The serous fluid is secreted by the epithelial cells of the membrane.

THE INTERIOR OF THE LUNGS

The lungs are composed of the bronchi and smaller air passages, alveoli, connective tissue, blood vessels, lymph vessels and nerves. As stated earlier the left lung is divided into two lobes and the right lung into three lobes. Each lobe is made up of a large number of *lobules*.

The lobules are composed of tiny bronchioles which subdivide into terminal bronchioles, respiratory bronchioles, alveolar ducts and many thousands of alveoli.

The pulmonary blood supply (Fig. 7:19)

The *pulmonary artery* divides into a right and left branch,

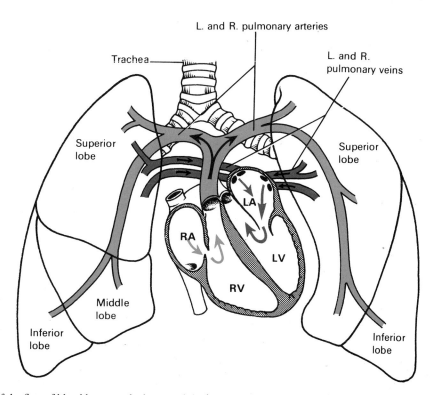

Figure 7:19 Diagram of the flow of blood between the heart and the lungs.

one of which conveys *deoxygenated blood* to each lung. On entering the lung it divides into many branches which accompany the bronchi and bronchioles. They subsequently end in a dense capillary network in the walls of the alveoli (Fig. 7:20). The walls of the alveoli and those of the capillaries consist of only one layer of flattened

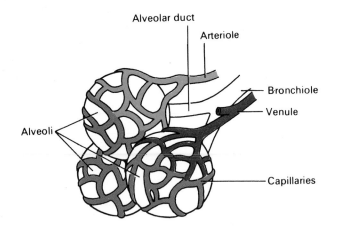

Figure 7:20 Diagram of the capillary network surrounding the alveoli.

epithelial cells. The exchange of gases between the air in the alveoli and the blood in the capillaries takes place across these two very fine membranes. The pulmonary capillaries join up, eventually becoming *two main pulmonary veins* in each lung. They leave the lungs at the hilus and convey *oxygenated blood* to the left atrium of the heart. The innumerable blood capillaries and blood vessels in the lungs are supported by connective tissue.

The blood supply to the respiratory passages, lymphatic drainage and nerve supply have already been described.

External Respiration

Expansion and contraction of the lungs ensure that a regular exchange of gases takes place between the alveoli and the external air. This is dependent upon the arrangement of the pleura and the contraction and relaxation of the muscles of respiration.

THE MECHANISM OF RESPIRATION
This is the process by which the lungs expand to take in air then contract to expel it. The cycle of respiration, which occurs about 15 times per minute, consists of three phases:
1. Inspiration
2. Expiration
3. Pause

The expansion of the chest during inspiration occurs as a result of muscular activity which is partly voluntary and partly involuntary. The main muscles of respiration in normal quiet breathing are the *intercostal muscles* and the *diaphragm*. During difficult or deep breathing they are assisted by the muscles of the abdomen, the neck and the shoulders.

The intercostal muscles
There are 11 pairs of intercostal muscles which occupy the spaces between the 12 pairs of ribs. They are arranged in two layers and are called the external intercostal muscles and the internal intercostal muscles (Fig. 7:21).

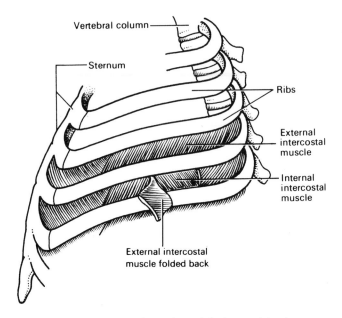

Figure 7:21 The intercostal muscles and the bones of the thorax.

The external intercostal muscle fibres extend in a *downwards and forwards* direction from the lower border of the rib above to the upper border of the rib below.

The internal intercostal muscle fibres extend in a *downwards and backwards* direction from the lower border of the rib above to the upper border of the rib below, crossing the external intercostal muscle fibres at right angles.

The first rib is fixed. Therefore, when the intercostal muscles contract they *pull* all the other ribs towards the first rib. Because of the shape of the ribs they *move outwards* when they are *pulled upwards*. In this way the thoracic cavity is enlarged *anteroposteriorly and laterally*. The intercostal muscles are stimulated to contract by the *intercostal nerves* (see Fig. 7:23).

The diaphragm
When *relaxed* the diaphragm is a dome-shaped structure which separates the thoracic from the abdominal cavity. It forms the floor of the thoracic cavity and the roof of the abdominal cavity and consists of a central tendon from which muscle fibres radiate to be attached to the vertebral

column, the lower ribs and the sternum. When the muscle of the diaphragm is relaxed the central tendon is at the level of the 8th thoracic vertebra (Fig. 7:22).

When the diaphragm *contracts*, its muscle fibres shorten and the central tendon is *pulled downwards*, enlarging the thoracic cavity in *length*. This increases the pressure in the abdominal and pelvic cavities. The diaphragm is supplied by the *phrenic nerves*.

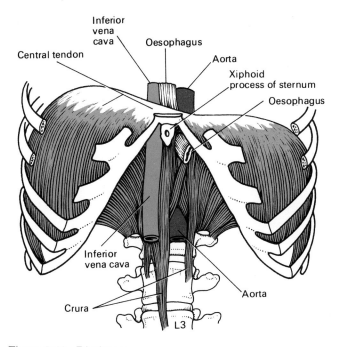

Figure 7:22   Diaphragm.

It is important to appreciate that the intercostal muscles and the diaphragm contract *simultaneously*, thus ensuring the enlargement of the thoracic cavity in all directions, that is from back to front, side to side and from top to bottom (Fig. 7:23).

THE CYCLE OF RESPIRATION

As described previously the visceral pleura is adherent to the lungs and the parietal pleura to the inner wall of the thorax and to the diaphragm. There is a potential space between these two layers of serous membrane called the pleural cavity which contains a very small amount of serous fluid.

When the capacity of the thoracic cavity is increased by simultaneous contraction of the intercostal muscles and the diaphragm, the parietal pleura moves with the walls of the thorax and the diaphragm. This reduces the pressure in the pleural cavity to a level considerably lower than atmospheric pressure and the visceral pleura tends to follow the parietal pleura. During this process the lungs are stretched and the pressure within the alveoli and in the air passage is reduced. This results in air being drawn into the lungs in an attempt to equalise the atmospheric and alveolar air pressures.

This is the process of *inspiration* which is described as *active* because it is the result of muscle contraction. When the diaphragm and intercostal muscles *relax* the ribs glide back to their original position, the diaphragm ascends, the lungs recoil and *expiration occurs*. This is a *passive* process. After expiration there is a short pause, then the cycle begins again.

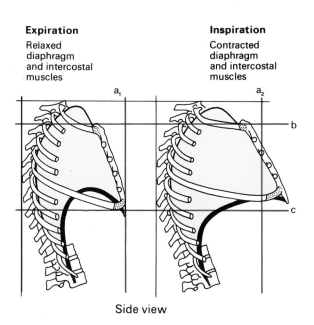

Side view

1. Outward movement of ribs shown by lines a_1 & a_2.
2. Upward movement of ribs & sternum shown by lines b & c.
3. Lowering of diaphragm shown by line c.

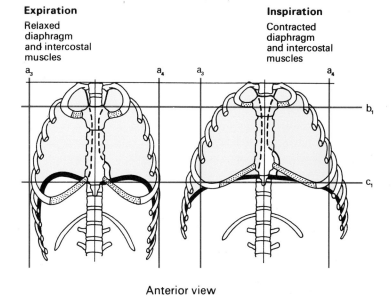

Anterior view

1. Outward movement of ribs shown by lines a_3 & a_4.
2. Upward movement of ribs shown by lines b_1 & c_1.
3. Lowering of diaphragm shown by line c_1.

Figure 7:23   Diagram of the changes in the capacity of the thoracic cavity (and the lungs) during breathing.

In normal quiet breathing there are about 15 complete respiratory cycles per minute. The lungs and the air passages are never empty and, as the exchange of gases takes place across only the alveolar walls, the remaining capacity of the respiratory passages is termed the *dead space* (about 150 ml).

The *tidal volume* (about 400 ml) is the name given to the amount of air which passes into and out of the lungs during each cycle of quiet breathing.

Alveolar ventilation is the amount of air which moves into and out of the alveoli each minute.

> Alveolar ventilation = respiratory rate (tidal volume − dead space volume)
> = 15 (400 − 150) ml
> = 3·75 litres per minute

THE INTERCHANGE OF GASES

The interchange of gases in the lungs occurs between the blood in the capillary network which surrounds the alveoli and the air in the alveoli of the lungs.

Some *properties of gases* are:

1. The molecules of gases are always in motion.
2. Gases always tend to diffuse from an area of *higher concentration* to one of *lower concentration*, i.e., down the concentration gradient.
3. Gases always exert pressure upon all the walls of their container. Unlike liquids, gases always fill their container. If it is not enclosed on all sides a gas will escape.

The atmospheric pressure at sea level is 101·3 kilopascals (kPa) or 760 millimetres of mercury (mmHg).* This pressure is exerted by the mixture of gases which make up the inspired air in the following proportions:

Oxygen	21 per cent
Carbon dioxide	0·04 per cent
Nitrogen and other inert gases	79 per cent
Water vapour	variable

During respiration the lungs and the respiratory passages are never empty of air. Instead there is a *tidal volume* of air (about 400 ml) which passes into and out of the lungs and air passage during each cycle or respiration in quiet breathing. The ebb and flow of air results in inspired air being mixed with the air already in the lungs. When it reaches the alveoli the air is saturated with water vapour and because of the tidal movement the *concentration of gases* in the alveoli remains fairly constant.

The total pressure exerted on the walls of the alveoli by the mixture of gases is the same as atmospheric pressure: 101·3 kPa (760 mmHg). Each gas in the mixture exerts a part of the total pressure proportional to its concentration which is known as the *partial pressure* (see Table 7:1).

*1 mmHg = 133·3 Pa = 0·1333 kPa
1 kPa = 7·5 mmHg

Table 7:1   Partial pressures of gases

Gas	Alveolar air kPa	Alveolar air mmHg	Deoxygenated blood kPa	Deoxygenated blood mmHg	Oxygenated blood kPa	Oxygenated blood mmHg
Oxygen	13·3	100	5·3	40	13·3	100
Carbon dioxide	5·3	40	5·8	44	5·3	40
Nitrogen and other inert gases	76·4	573	76·4	573	76·4	573
Water vapour	6·3	47				
	101·3	760				

The partial pressure of nitrogen (PN_2) is the same in the alveoli as it is in the blood. This stable state is maintained because nitrogen as a gas is not used by the body but it can diffuse across the walls of the alveoli and the capillaries.

The partial pressure of oxygen (Po_2) in the alveoli is higher than that in the deoxygenated blood in the capillaries of the pulmonary arteries (see Table 7:1). As gases diffuse from a higher to a lower concentration, the movement of oxygen is from the alveoli to the blood.

The reverse is the case in relation to carbon dioxide. The Pco_2 is higher in deoxygenated blood than in alveolar air, so carbon dioxide passes across the walls of the capillaries and the alveoli into the alveolar air (see Table 7:1 and Fig. 7:24).

The partial pressure of each gas in the blood when leaving the lungs via the pulmonary veins is the same as in the alveolar air.

The slow movement of blood through the capillaries surrounding the alveoli allows time for the interchange of gases to take place and for the uptake of oxygen by the erythrocytes in the blood. Oxygen is transported round the body in solution in the blood water and in combination with haemoglobin in the erythrocytes.

THE CONTROL OF RESPIRATION

This activity is partly *chemical* and partly *nervous*, but the aspects are too closely linked to be described separately.

Respiration is controlled by nerve cells in the brain stem: the *respiratory centre* in the *medulla oblongata* and the *pneumotaxic centre* in the *pons varolii* (see Ch. 12). The cells in the respiratory centre are concerned with inspiration and those in the pneumotaxic centre with the inhibition of inspiration, which results in expiration. Nerve impulses which originate in the respiratory centre pass to the diaphragm in the phrenic nerves and to the intercostal muscles in the intercostal nerves. This results in contraction of these muscles, and inspiration occurs (Fig. 7:25).

There are nerve endings in the lungs which are sensitive to stretch and which are stimulated when the lungs are in-

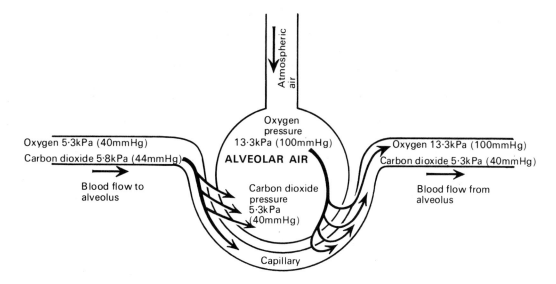

Oxygen
pressure
13·3kPa (100mmHg)
ALVEOLAR AIR

Atmospheric air

Oxygen 5·3kPa (40mmHg)
Carbon dioxide 5·8kPa (44mmHg)

Blood flow to
alveolus

Carbon dioxide
pressure
5·3kPa
(40mmHg)

Oxygen 13·3kPa (100mmHg)
Carbon dioxide 5·3kPa (40mmHg)

Blood flow from
alveolus

Capillary

Figure 7:24   Diagram of the interchange of gases between air in the alveoli and the blood capillaries.

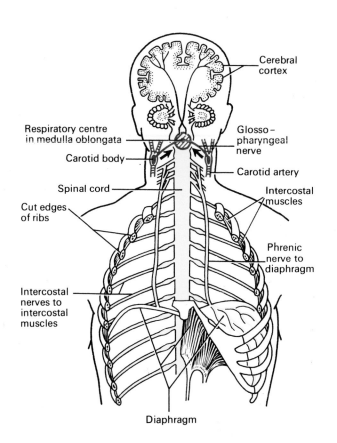

Cerebral
cortex

Respiratory centre
in medulla oblongata

Glosso-
pharyngeal
nerve

Carotid body

Spinal cord

Carotid artery

Intercostal
muscles

Cut edges
of ribs

Phrenic
nerve to
diaphragm

Intercostal
nerves to
intercostal
muscles

Diaphragm

Figure 7:25   Diagram of some of the nerves involved in the control of
respiration.

flated. The nerve impulses produced are passed to the pneumotaxic centre in the afferent fibres of the vagus nerves, and expiration occurs.

In the walls of the arch of the aorta and the carotid artery there are groups of cells which are sensitive to changes in the Pco_2 and Po_2 in the blood. They are called the *aortic* and *carotid bodies* and are described as *chemoreceptors*. The nerve impulses which originate in these cells are transmitted to the respiratory centre in the *glossopharyngeal* and the *vagus nerves*.

The chemoreceptors and the respiratory centre are stimulated by an increase in the Pco_2 in the blood which results in increased ventilation of the lungs. A small reduction in the Po_2 has the same effect but a substantial reduction tends to have a depressing effect.

Normally, quiet breathing is sufficient to maintain a balance between the blood Pco_2 and Po_2 while the individual is at rest or taking light exercise. During strenuous exercise breathing becomes deeper and more rapid in response to the needs of the muscles for more oxygen and to excrete the excess carbon dioxide produced.

In normal quiet breathing the intercostal muscles and the diaphragm are the only muscles involved, but in deep or forced breathing other muscles come into play. They are termed the *accessory muscles of respiration* and include the sternocleidomastoid, the pectoralis major, the platysma, and the latissimus dorsi (Figs. 17:6 and 18:1). The contraction of these muscles in addition to the diaphragm and intercostal muscles ensures the maximum increase in the capacity of the thoracic cavity.

INTERNAL OR CELL RESPIRATION

This is the name given to the interchange of gases

which takes place between the blood and the cells of the body.

Oxygen is carried from the lungs to the tissues dissolved in plasma and in chemical combination with haemoglobin as *oxyhaemoglobin*. The exchange in the tissues takes place between the arterial end of the capillaries and the tissue fluid. The process involved is the same as that which occurs in the lungs, that is, *diffusion from a higher concentration to a lower concentration* (Fig. 7:26). In this case the higher concentration of oxygen is in the blood and the lower concentration is in the tissue fluid. The cells obtain their oxygen from the tissue fluid.

Oxyhaemoglobin is an unstable compound which breaks up easily to liberate oxygen. As the cells of the body require a constant supply of oxygen, the process of diffusion of oxygen from the blood across the capillary wall to the tissue fluid and then into the cells is continuous. The rate at which this process is carried on is increased in the presence of a higher than usual concentration of carbon dioxide, which occurs when cells in a particular area are more than usually active. The higher P_{CO_2} assists the release of oxygen from oxyhaemoglobin. In this way cells receive a supply of oxygen which is consistent with their activity and the supply changes as the amount of activity changes.

Carbon dioxide is one of the waste products of carbohydrate and fat metabolism in the cells. The method of transfer of carbon dioxide from the cells into the blood at the *venous end of the capillary* is also by *diffusion*. Blood transports carbon dioxide by three different mechanisms:

1. Some of the carbon dioxide is dissolved in the water of the blood plasma.
2. Some is transported in chemical combination with sodium in the form of sodium bicarbonate.
3. The remainder is transported in combination with haemoglobin.

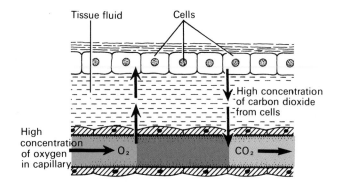

Figure 7:26   Diagram of the interchange of gases during internal respiration.

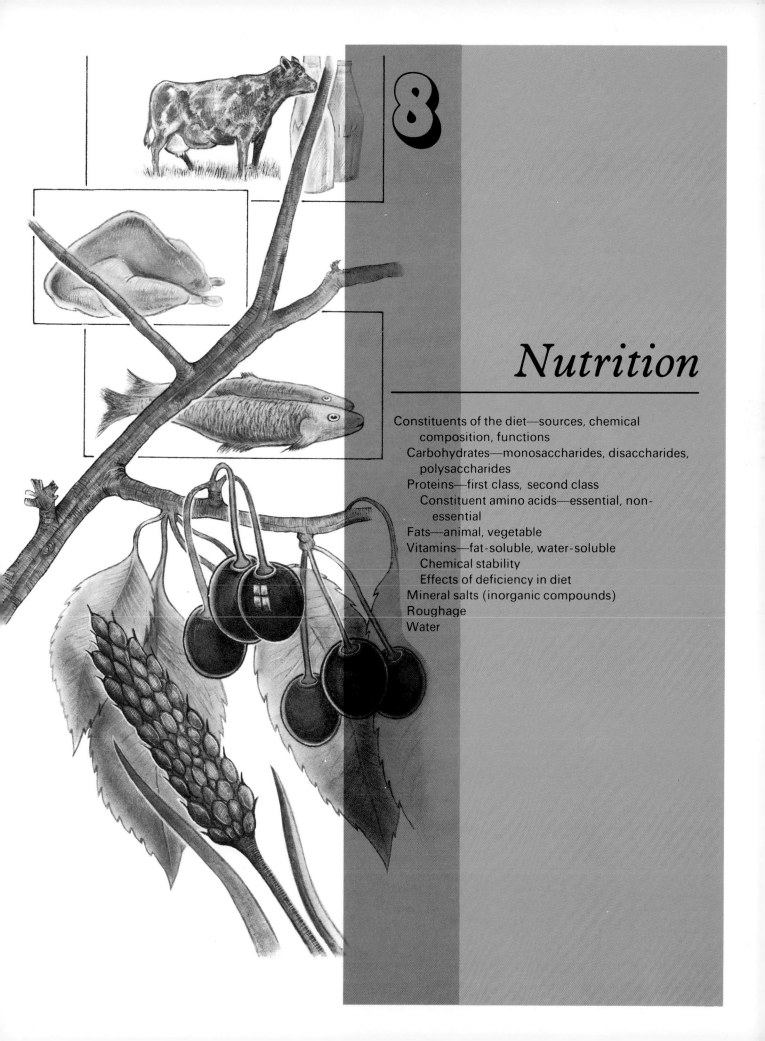

Nutrition

8. Nutrition

Before discussing the digestive system it is necessary to have an understanding of the needs of the body regarding diet.

The essentials of diet include:

Carbohydrates
Proteins
Fats
Vitamins
Mineral salts
Water
Roughage

If the cells of the body are to be able to function efficiently these nutritional substances must be available in the *correct proportions*.

Many of the foods which are eaten contain a proportion of a number of the essential dietary components. For example, potatoes and bread are considered carbohydrates but both contain some protein and some vitamins.

Foods are described as carbohydrate or protein because they are composed mainly of one or other of these nutrients.

Carbohydrates

These are found in sugar, jam, cereals, bread, biscuits, potatoes, fruit and vegetables. They consist of carbon, hydrogen and oxygen, the hydrogen and oxygen being in the same proportion as in water.

Carbohydrates are classified according to the complexity of the chemical substances of which they are formed.

MONOSACCHARIDES
These are, chemically, the simplest form in which a carbohydrate can exist. They are made up of single units or molecules which, if they were broken down further, would cease to be monosaccharides. Carbohydrates are absorbed from the alimentary canal as monosaccharides and more complex carbohydrates are broken down to this form by digestion (see Ch. 9).

Examples of monosaccharides are: glucose, fructose, galactose.

DISACCHARIDES
These consist of two monosaccharide molecules chemically combined.

Examples of disaccharides or sugars are: sucrose, maltose, lactose.

POLYSACCHARIDES
These consist of complex molecules made up of large numbers of monosaccharide molecules in chemical combination.

Examples of polysaccharides are: starches, glycogen, cellulose, dextrins.

Not all polysaccharides can be digested by human beings, for example cellulose present in vegetables passes through the alimentary canal almost unchanged.

Functions
1. To provide energy and heat.
2. To act as a protein sparer, that is, when there is an adequate supply of carbohydrate in the diet, protein does not need to be used to provide energy and heat.
3. If carbohydrate is eaten which is in excess of the body's needs it is converted to fat and deposited in the fat depots, for example, under the skin.

Proteins or Nitrogenous Foods

Proteins are made up of a number of units called *amino acids* which are linked together chemically. Each protein consists of a specific number of different amino acids arranged in a way which is characteristic of that protein.

Proteins cannot be absorbed as such. They are broken down into their constituent amino acids and it is in this form that they are absorbed into the blood.

AMINO ACIDS
These are composed of the elements carbon, hydrogen, oxygen, nitrogen, sulphur and phosphorus. They are divided into two categories, *essential* and *non-essential*.

Essential amino acids are so-called because they cannot

be synthesised in the body and therefore must be included in the diet. These are:

isoleucine	methionine	tryptophan
leucine	phenylalanine	valine
lysine	threonine	

Non-essential amino acids are those which can be synthesised in the body. They are:

alanine	cystine	hydroxyproline
arginine	glutamic acid	proline
asparagine	glutamine	serine
aspartic acid	glycine	tyrosine
cysteine	histidine	

FIRST CLASS PROTEIN

This is the name given to protein foods which contain all the essential amino acids in the correct proportions. They are derived almost entirely from animal sources and include:

meat	fish	soya beans
milk	eggs	

SECOND CLASS PROTEIN

These do not contain all the essential amino acids in the correct proportions; they are mainly of vegetable origin. Examples are peas, beans and lentils, which are known as pulses. A small proportion of protein is to be found in other vegetables and in some of the mainly carbohydrate foods, such as bread and potatoes.

Functions

1. To provide the amino acids required for the formation, growth and repair of body cells and for the formation of some of the secretions the cells produce, e.g., hormones and enzymes.

2. To provide the amino acids required for the formation of blood proteins, i.e. albumen, globulin, fibrinogen and prothrombin.

3. To provide energy and heat. Normally, this is a secondary function and becomes important only when there is not enough carbohydrate in the diet and fat stores are depleted.

4. When protein is eaten in excess of the body's needs, the nitrogenous part is detached and excreted by the kidneys and the remainder is converted to fat for storage in the fat depots, for example, in the fat cells of adipose tissue.

Fats

Fats consist of carbon, hydrogen and oxygen, but they differ from carbohydrates in that the hydrogen and oxygen are not in the same proportions as in water. Fats are divided into two groups, *animal* and *vegetable*.

Animal fat is found in milk, cheese, butter, eggs, meat and oily fish such as herring, cod and halibut. All the animal sources of protein contain some animal fat.

Vegetable fat is found in margarine and in vegetable oils. Nuts of various kinds are the best natural source of vegetable fat.

Functions

1. To produce energy and heat.

2. To support certain organs of the body, for example, the kidneys and the eyes.

3. To transport the fat-soluble vitamins A, D, E and K.

4. It is present in the nerve sheaths and in the secretions of the sebaceous glands in the skin.

5. It is used in the formation of cholesterol.

6. When eaten in excess of that required by the body it is stored in the fat depots.

Vitamins

Vitamins are chemical compounds which are essential for health. They are found widely distributed in food and are divided into two main groups:

Fat-soluble vitamins—	A, D, E, and K
Water-soluble—	B complex, C

FAT-SOLUBLE VITAMINS

VITAMIN A (RETINOL)

This vitamin is found in such foods as cream, egg yolk, fish oil, milk, cheese and butter. It is absent from vegetable fats and oils but is added to margarine during manufacture. It can be formed in the body from certain carotenes of which the main dietary sources are green vegetables and carrots.

Vitamin A is only absorbed from the small intestine satisfactorily if fat absorption is normal.

Functions

1. It is necessary for the regeneration of the visual purple in the retina of the eye which is bleached by bright light. If there is insufficient vitamin A adaptation to seeing in dim light is delayed.

2. It influences the nutrition of epithelial cells and tends to reduce the severity of micro-organism infection. Because of this, it is sometimes known as the *anti-infective vitamin*.

3. It is necessary to maintain the cornea of the eye in a healthy state.

VITAMIN D

Vitamin D_3 is sometimes termed the *antirachitic vitamin*. It is found mainly in animal fats such as eggs, butter, cheese, cod and halibut liver oils.

Man and animals can synthesise cholicalciferol (vitamin

D_3) by the action of the ultraviolet rays of the sun on a form of cholesterol in the skin (7-dehydrocholesterol).

Calciferol (vitamin D_2) is formed in plants and is used widely in therapeutics.

Functions

This is the vitamin which regulates calcium and phosphorus metabolism and is therefore associated with the calcification of bones and teeth.

VITAMIN E (TOCOPHEROL)

The sources of this vitamin are peanuts, lettuce, egg yolk, wheat germ, whole cereal, milk and butter.

Functions

Lack of this vitamin in animals causes muscle wasting and failure in reproduction, but it is not quite certain whether it has the same functions in human beings.

VITAKIN K

The sources of vitamin K are fish, liver, leafy green vegetables and fruit. Bile salts must be present in the small intestine before it can be absorbed.

Functions

It is necessary for the formation by the liver of prothrombin and several other factors essential for the clotting of blood.

WATER-SOLUBLE VITAMINS

VITAMIN B COMPLEX

This consists of a group of water soluble vitamins which are more or less closely associated.

VITAMIN B_1 (THIAMINE)

This vitamin is present in the germ of cereals, nuts, yeast, egg yolk, liver and legumes.

Functions

It is essential for the normal carbohydrate metabolism.
It stimulates appetite.
It helps to regulate the functioning of the nervous system.
It is associated with the control of water balance in the body.

VITAMIN B_2 (RIBOFLAVIN)

This is found in yeast, leafy vegetables, milk, liver, eggs, kidney, cheese, roe.

Functions

It is concerned with the oxidation of all foods.
It is associated in some way with the physiology of vision.

It is necessary for the growth of all tissues in man and animals.

FOLIC ACID

This is found in liver, kidney, fresh leafy green vegetables and yeast. It is synthesised by bacteria in the large intestine, and significant amounts derived from this source are absorbed.

Functions

It is associated with the development of the erythrocytes in the red bone marrow.

NICOTINIC ACID (NIACIN)

This is found in liver, cheese, yeast, whole cereal, eggs, fish, peanuts and bemax.

Functions

It is necessary for:
The metabolism of carbohydrates
The normal functioning of the gastrointestinal tract
The satisfactory functioning of the nervous system

VITAMIN B_6 (PYRIDOXINE)

This is found in egg yolk, peas, beans, soya bean, yeast, meat, liver.

Functions

It is believed to be necessary for satisfactory protein and fat metabolism.

VITAMIN B_{12} (CYANOCOBALAMIN)

This is found in liver, meat, eggs, milk and fermented liquors.

Functions

It is essential for the maturation of erythrocytes in the red bone marrow.

PANTOTHENIC ACID

This is found in eggs, liver and yeast.

Functions

It is associated with healthy skin and hair.
It is required for the normal functioning of the adrenal glands.

BIOTIN

This is found in egg yolk, liver and tomatoes.

Functions

It is required for healthy skin and conjunctiva.

VITAMIN C (ASCORBIC ACID)

This is found in fresh fruit, especially blackcurrants,

oranges, grapefruit and lemons, and also in rose-hips and green vegetables.

Functions
It is necessary for:
- The maintenance of the strength of the walls of the blood capillaries
- The development and maintenance of healthy bones and teeth
- The formation of red blood cells
- The production of antibodies
- The formation of connective tissue

Mineral Salts

Mineral salts (inorganic compounds) are necessary within the body for all body processes. They are usually required in small quantities.

They consist of soluble compounds of:

calcium	phosphorus	sodium
iron	iodine	potassium

CALCIUM
This is found in milk, cheese, eggs, green vegetables and some fish. The normal daily requirement for an adult is about 1 gram. An adequate supply should be obtained in a normal, well-balanced diet.

Functions
In association with vitamin D it is essential for the hardening of bones and teeth. Therefore an adequate supply in young people is important.

It plays an important part in the coagulation of blood and is associated with the mechanism of muscle contraction.

PHOSPHORUS
Sources of phosphorus include cheese, oatmeal, liver and kidney. If there is sufficient calcium in the diet it is unlikely that there will be a deficiency of phosphorus.

Functions
It is associated with calcium in the hardening of bones and teeth and helps to maintain the constant composition of the body fluids.

SODIUM
Sodium is supplied to the body in fish, meat, eggs, milk, and as cooking and table salt. The normal intake of sodium chloride per day varies from 5 to 20 g and the daily requirement is between 2 and 5 g. Excess is excreted in the urine.

Functions
It is the most commonly occurring *extra-cellular cation* and is associated with:
The contraction of muscle
The transmission of nerve impulses in nerve fibres
The maintenance of the eletrolyte balance in the body

POTASSIUM
This substance is to be found widely distributed in all foods. The normal intake of potassium chloride varies from 5 to 7 g per day and this meets the potassium requirements.

Functions
It is the most commonly occurring *intra-cellular cation*. It is involved in:
- Many chemical activities inside cells
- The contraction of muscles
- The transmission of nerve impulses
- The maintenance of the electrolyte balance in the body

IRON
Iron, as a soluble compound, is found in liver, kidney, beef, egg yolk, whole meal bread and green vegetables. In normal adults about 1 mg of iron is lost from the body daily. The normal daily diet contains more, but the amount absorbed is limited to the amount lost.

Functions
Iron is essential for the formation of *haemoglobin* in the red blood cells.

It is necessary for tissue oxidation.

IODINE
Iodine is found in salt water fish and in vegetables which have grown in soil containing iodine. In some parts of the world where iodine is deficient in soil very small quantities are added to table salt. The daily requirement of iodine depends upon the individual's metabolic rate. Some people have a higher normal metabolic rate than others and their iodine requirements are greater. The minimum daily requirement is 20 μg.

Function
It is essential for the formation of *thyroxine* and *tri-iodothyronine*, the hormones secreted by the thyroid gland.

Roughage

Roughage is the undigestable part of the diet, for example, cellulose of fruit and vegetables and connective tissue of meat and fish.

Functions

It gives bulk to the diet and helps to satisfy the appetite.

It stimulates peristalsis, that is, the muscular activity of the alimentary tract.

It stimulates bowel movement.

Water

Water is a liquid compound of hydrogen and oxygen, made up by the chemical combination of two parts of hydrogen and one part of oxygen (H_2O). Water makes up about 60 per cent of the body weight in men and about 50 per cent in women.

Functions

1. It provides the moist environment which is required by all living cells in the body, i.e., all the cells of the body except the superficial layers of the skin, the nails, the hair and the outer hard layer of the teeth.

2. It participates in all the chemical reactions which occur inside and outside the body cells.

3. It dilutes and moistens food.

4. It assists in the regulation of body temperature as a constituent of sweat, which is secreted on to the skin. The evaporation of sweat cools the body.

5. As a major component of blood and tissue fluid it transports some substances in solution and some in suspension round the body.

6. It dilutes waste products and poisonous substances in the body.

7. It contributes to the formation of urine and faeces.

Tables 8:1 and 8:2 summarise the vitamins—their chemical names, sources, stability, functions, deficiency diseases and daily adult requirements.

Table 8:1 Summary—fat-soluble vitamins

Vitamin (letter)	Chemical name	Source	Stability	Functions	Deficiency diseases	Daily requirement (adults)
A	Retinol (carotene provitamin in plants)	Milk, butter, cheese, egg-yolk, fish, liver, oils, green and yellow vegetables	Some loss at high temperatures and long exposure to light and air	Maintains healthy epithelial tissues and cornea. Formation of visual purple	Keratinisation Xerophthalmia Stunted growth Night blindness	4000– 5000 I.U.
D	Choli-calciferol and calciferol	Fish liver oils, milk, cheese and egg-yolk, irradiated 7-dehydrocholesterol in human skin	Very stable	Facilitates the absorption and utilisation of calcium and phosphorus = healthy bones and teeth	Rickets Osteomalacia	400 I.U.
E	Tocopherol	Egg-yolk, milk, butter, green vegetables, nuts	Destroyed by rancid fat and iron salts	Maintains healthy muscular system	Very rare muscular dystrophies	10–15 I.U.
K	Menaphthone	Leafy vegetables, fish, liver, fruit	Destroyed by light, strong acids and alkalis	Formation of prothrombin in the liver	Slow blood clotting Haemorrhages in the newborn	

Bile is necessary for the absorption of these vitamins.   I.U. = International Units
Mineral oils interfere with absorption.

Table 8:2   Summary—water-soluble vitamins

Vitamin (letter)	Chemical name	Source	Stability	Functions	Deficiency diseases	Daily requirement (adults)
B_1	Thiamin	Yeast, liver, germ of cereals, nuts, pulses, rice polishings, egg yolk, liver, legumes	Stable	Metabolism of carbo-hydrates and nutri-tion of nerve cells Efficient water ex-change in the body	General fatigue and loss of muscle tone Ultimately leads to beri beri	1–1·5 mg
B_2	Riboflavin	Liver, yeast, milk, eggs, green vegetables, kidney	Destroyed by light and alkalis	Necessary for tissue oxidation and growth	Angular stomatitis Cheilosis Dermatitis Eye lesions	1–2 mg
B_6	Pyridoxine	Meat, liver, vegetables, bran of cereals, egg-yolk, beans, soya beans	Stable	Protein metabolism Formation of RBCs and WBCs	Very rare	1–2 mg
B_{12}	Cyano-cobalamin	Liver, milk, moulds, fermenting liquors, egg	Destroyed by heat	Maturation of RBCs	Pernicious anaemia Degeneration of nerve fibres of the spinal cord	2–3 μg
B	Folic acid	Dark green vegetables, liver, kidney, eggs Synthesised in colon	Destroyed by heat and moisture	Formation of RBCs	Anaemia	300–400 μg
B	Nicotinic acid (Niacin)	Yeast, offal, fish, pulses, whole meal cereals. Synthesised in the body from tryptophan	Fairly stable	Necessary for tissue oxidation	Prolonged deficiency causes pellagra, i.e. dermatitis, diarrhoea, dementia	16–20 mg
B	Pantothenic acid	Liver, yeast, egg-yolk, fresh vegetables	Destroyed by ex-cessive heat and freezing	Probably required for formation of RBCs	Dermatitis, adrenal insufficiency	Unknown
B	Biotin	Yeasts, liver, kidney, pulses, nuts	Stable	Carbohydrates and fat metabolism Growth of bacteria	Dermatitis con-junctivitis	Unknown
C	Ascorbic acid	Citrus fruits, currants, berries, green vege-tables, potatoes, liver and glandular tissue in animals	Destroyed by heat, aging, acids, alkalis, chopping, salting and drying	Formation of inter-cellular matrix Maturation of red blood cells	Multiple haemorrhages Slow wound healing Anaemia Gross deficiency causes scurvy	40–60 mg

Paraaminobenzoic acid and Inositol are two vitamins of the B group about which little is known.

9

The Digestive System

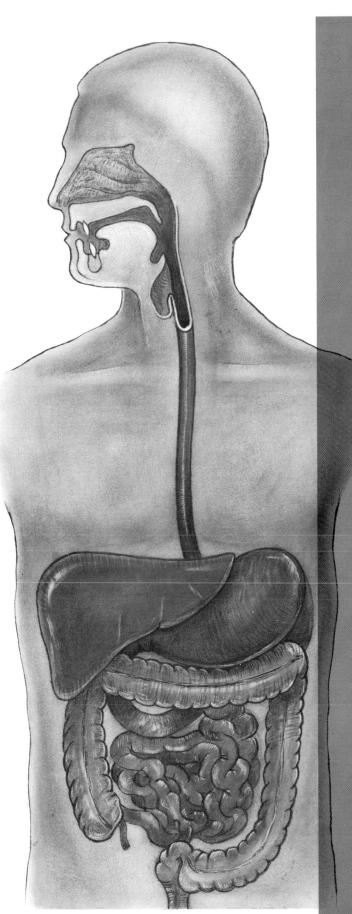

9. *The Digestive System*

The digestive system is the collective name used to describe the *alimentary canal*, some *accessory organs* and a variety of *digestive processes* which take place at different levels in the canal to prepare food eaten in the diet for absorption. The alimentary canal begins at the mouth, passes through the thorax, abdomen and pelvis and ends at the anus. It has a general structure which is modified at different levels to provide for the processes occurring at each level. The complex of digestive processes gradually simplify the foods eaten until they are in a form suitable for absorption.

For example, meat, even when cooked, is chemically too complex to be absorbed from the alimentary canal. It therefore goes through a series of changes which release its constituent nutrients: some amino acids, some mineral salts and some vitamins. Chemical substances or *enzymes*★ which effect these changes are secreted into the canal by special glands, some of which are in the walls of the canal and some outside the canal but with ducts which lead into it.

After they are absorbed the nutrient materials are used in the synthesis of the constituents of the body. They provide the raw materials for the manufacture of new cells, hormones and enzymes, and the energy needed for these processes and for the disposal of waste materials.

The activities in the alimentary canal can be grouped under four main headings.

Ingestion, which means taking food into the alimentary tract.

Digestion, which can be divided into the *mechanical* breakdown of food by, for example, *mastication* (chewing) and *chemical* breakdown by *enzymes* present in secretions produced by glands of the digestive system. These secretions include:

Saliva from the salivary glands
Gastric juice from the stomach
Intestinal juice from the small intestine
Pancreatic juice from the pancreas
Bile from the liver

Absorption is the process by which digested food substances pass through the walls of some organs of the

---

★An enzyme is a chemical substance which causes or speeds up a chemical change in other substances without itself being changed.

alimentary canal into the blood and lymph capillaries for circulation round the body.

Elimination. Food substances which have been eaten but cannot be digested and absorbed are excreted by the bowel as faeces.

The Organs of the Digestive System (Fig. 9:1)

The alimentary tract
This is a long tube through which food passes. It commences at the mouth and terminates at the anus, and the various parts are given separate names, although structurally they are remarkably similar.

The parts of the alimentary tract are:

The mouth
The pharynx
The oesophagus
The stomach
The small intestine
The large intestine
The rectum and anal canal

Accessory organs
Various secretions are poured into the alimentary tract, some by glands in the lining membrane of the organs, for example, gastric juice by the lining of the stomach, and some by glands situated outside the tract. The latter are the accessory organs of digestion and their secretions pass through ducts to enter the tract. They consist of:

Three pairs of salivary glands
The pancreas
The liver and the biliary tract

From these lists it can be seen that this is a large system involving a considerable number of organs and glands. They are linked physiologically as well as anatomically in that digestion and absorption occur in stages, each stage being dependent upon the previous stage or stages.

For descriptive purposes the system will be dealt with in six sections each of which will include the appropriate anatomy and physiology. The sections are:

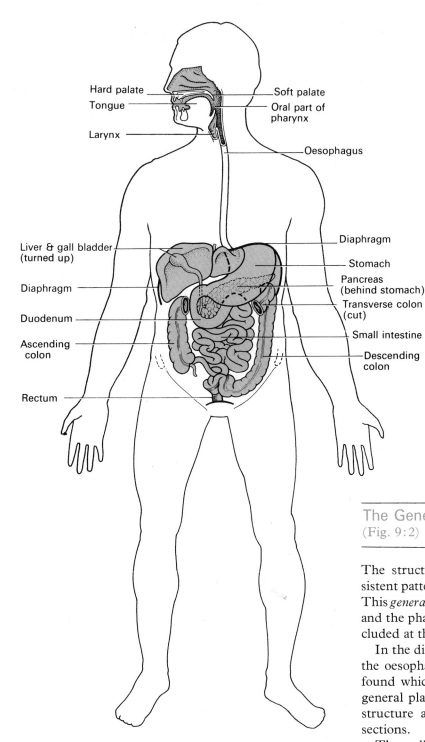

Hard palate

Tongue

Larynx

Soft palate

Oral part of pharynx

Oesophagus

Liver & gall bladder (turned up)

Diaphragm

Duodenum

Ascending colon

Rectum

Diaphragm

Stomach

Pancreas (behind stomach)

Transverse colon (cut)

Small intestine

Descending colon

Figure 9:1 The organs of the digestive system.

1. The general plan of the alimentary tract
2. Mouth, pharynx, oesophagus, salivary glands
3. Stomach
4. Small intestine, pancreas, liver, biliary tract
5. Large intestine, rectum and anal canal
6. Metabolism

The General Plan of the Alimentary Tract
(Fig. 9:2)

The structure of the alimentary canal follows a consistent pattern from the level of the oesophagus onwards. This *general plan* does not apply so obviously to the mouth and the pharynx so these parts of the tract have been excluded at this stage.

In the different organs which make up the tract, from the oesophagus onwards, modifications of structure are found which are associated with special functions. The general plan is described here and the modifications in structure and function are described in the different sections.

The walls of the alimentary tract are made up of the following four layers of tissue:

Adventitia or outer covering
Muscle layer
Submucous layer
Mucous membrane lining

ADVENTITIA OR OUTER COVERING
In the thorax this consists of *loose fibrous tissue* and in the abdomen the organs are covered by a serous membrane called *peritoneum*.

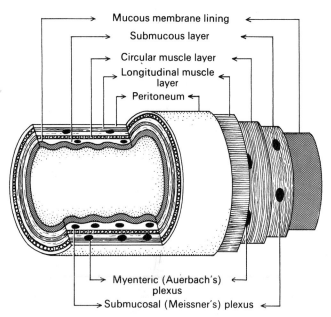

Figure 9:2   General plan of the alimentary canal.

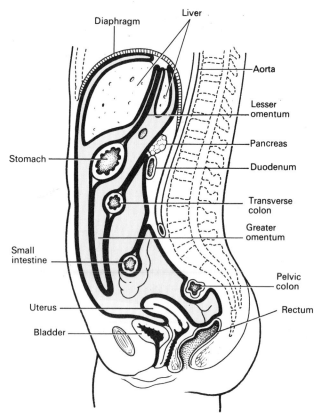

Figure 9:3   The peritoneum (coloured) and its association with the abdominal organs of the digestive system and the pelvic organs.

THE PERITONEUM (Fig. 9:3)

The peritoneum is the largest serous membrane of the body. It consists of a closed sac within the abdominal cavity and has two layers:

The parietal layer, which lines the abdominal wall
The visceral layer, which covers the organs or viscera within the abdominal and pelvic cavities

The arrangement of the peritoneum is complicated; it is as though the organs had been invaginated into it from below, behind and above. This means that the pelvic organs are covered only superiorly, the intestine is surrounded by it and is attached by a double layer to the posterior abdominal wall. The liver is almost completely covered with peritoneum which attaches it to the under surface of the diaphragm. Most of the abdominal organs are invaginated into the peritoneum from behind, thus the parietal layer lines the anterior abdominal wall.

The two layers of peritoneum are actually in contact and friction between them is prevented by the presence of serous fluid which is secreted by the peritoneal cells, thus the *peritoneal cavity* is only a *potential cavity*. In the male it is completely closed but in the female the uterine tubes open into it (see Ch. 15).

MUSCLE LAYER

With some exceptions this consists of two layers of *smooth* muscle. The muscle fibres of the outer layer are arranged longitudinally, and those of the inner layer encircle the wall of the tube. Between these two muscle layers there are blood vessels, lymph vessels and a plexus of nerves called the *myenteric* or *Auerbach's plexus* which consists of sympathetic and parasympathetic nerves. The nerves of the myenteric plexus supply the adjacent smooth muscle and blood vessels.

The contraction of these muscle layers occurs in waves which push the contents of the tract onwards. This type of contraction of smooth muscle is called *peristalsis*. The other main effect of muscle contraction is to mix the contents of the tract with the digestive juices.

Onward movement of the contents is prevented by *sphincters* situated at various points on the tract. These consist of an increased number of circular muscle fibres. They delay onward movement allowing time for digestion and absorption to take place.

SUBMUCOUS LAYER

This layer consists of loose connective tissue with some elastic fibres. Within this layer there are plexuses of blood vessels and nerves, lymph vessels and varying amounts of lymphoid tissues. The blood vessels consist of arterioles, venules and capillaries. The nerve plexus is called the *submucosal* or *Meissner's plexus* and it contains sympathetic and parasympathetic nerves which supply the mucous membrane lining.

MUCOUS MEMBRANE

This layer has three main functions: protective, secretory and absorptive. In parts of the tract which are subject to mechanical injury this layer consists of *stratified squamous epithelium* with mucus secreting glands just below the surface. In areas where the food eaten is already soft and moist and where the secretion of digestive juices and absorption occur, the mucous membrane consists of *columnar epithelial cells* interspersed with goblet cells which secrete mucus (Fig. 9:4). Below the surface in the parts lined by columnar epithelium there are collections of specialised cells called glands which pour their secretions into the lumen of the tract. These are the *digestive juices* and they contain the enzymes which chemically simplify the foods eaten. Under the epithelial lining there are varying amounts of lymphoid tissue.

NERVE SUPPLY

The alimentary tract is supplied by nerves from both parts of the autonomic nervous system, i.e. parasympathetic and sympathetic, and in the main their actions are antagonistic (Fig. 9:5). In the normal healthy state one influence may outweigh the other according to the needs of the body as a whole at a particular time.

The parasympathetic supply to most of the alimentary tract is provided by two cranial nerves, the *vagus nerves*. Stimulation causes muscle contraction and the secretion of digestive juices. The most distal part of the tract is supplied by pelvic nerves.

The sympathetic supply is provided by numerous nerves which emerge from the spinal cord in the thoracic and lumbar regions. These nerves form *plexuses* in the thorax, abdomen and pelvis, and from them nerves

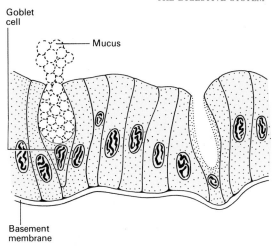

Figure 9:4 Columnar epithelium with goblet cells.

pass to the organs of the alimentary tract. Their action is to reduce muscle contraction and glandular secretion.

Within the walls of the canal there are two nerve plexuses from which both sympathetic and parasympathetic fibres are distributed. The *myenteric of Auerbach's plexus* lies between the two layers of muscle tissue and supplies the muscle, and the *submucosal or Meissner's*

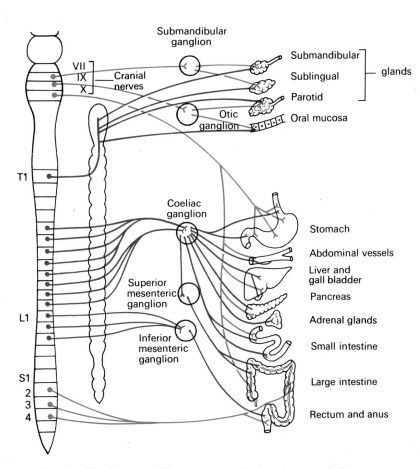

Figure 9:5 Autonomic nerve supply to the digestive system. Blue—parasympathetic; red—sympathetic.

plexus lies in the submucosa and supplies the mucous membrane and the secretory glands.

BLOOD SUPPLY

ARTERIAL BLOOD SUPPLY

In the thorax

The oesophagus is supplied by paired oesophageal arteries from the thoracic aorta.

In the abdomen and pelvis

The alimentary tract, the pancreas, the liver and the biliary tract are supplied by the unpaired *coeliac artery* and the *superior and inferior mesenteric arteries* (Figs. 9:6 and 9:7).

The coeliac artery leaves the aorta just below the diaphragm and divides into three branches which supply the stomach, duodenum, pancreas, spleen, liver, gall bladder and bile ducts. The branches are:

The right gastric artery
The splenic artery
The hepatic artery

The superior mesenteric artery leaves the abdominal aorta and supplies the whole of the small intestine, the caecum, the ascending colon and most of the transverse colon.

The inferior mesenteric artery leaves the abdominal aorta and supplies a small part of the transverse colon, the descending colon, the pelvic colon and most of the rectum.

The distal part of the recum and the anus are supplied by the *middle* and *inferior rectal arteries* which are branches of the internal iliac arteries.

VENOUS DRAINAGE

In the thorax

Venous blood from the oesophagus enters the *azygos* and *hemiazygos veins*. The azygos vein joins the superior vena cava near the heart, and the hemiazygos joins the left brachiocephalic vein.

Some blood from the lower part of the oesophagus drains into the *left gastric vein.*

In the abdomen and pelvis

The veins which drain blood from the lower part of the oesophagus, the stomach, pancreas, small intestine, large intestine and most of the rectum join to form the *portal vein* (Fig. 9:8). This blood, containing a high concentration of nutritional materials, is conveyed to the liver then to the inferior vena cava. The circulation of blood in the liver is described later (see p. 134).

Blood from the lower part of the rectum and the anal canal drains into the *internal iliac veins.*

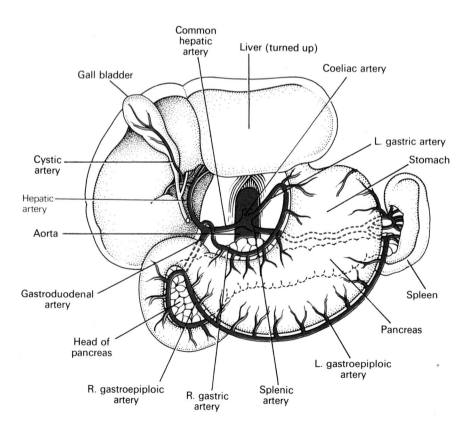

Figure 9:6   Branches of the coeliac artery and the organs they supply.

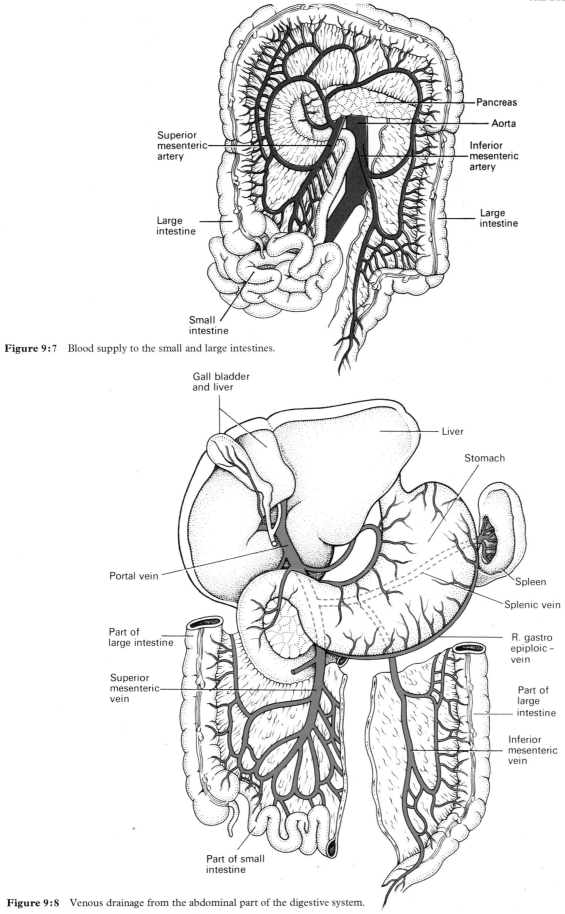

Figure 9:7 Blood supply to the small and large intestines.

Figure 9:8 Venous drainage from the abdominal part of the digestive system.

Mouth, Pharynx, Salivary Glands, Oesophagus

THE MOUTH (Fig. 9:9)

The mouth or oral cavity is bounded by muscles and bones:

Anteriorly— by the lips

Posteriorly— it is continuous with the oral part of the pharynx

Laterally— by the muscles of the cheeks

Superiorly— by the bony hard palate and muscular soft palate

Inferiorly— by the muscular tongue and the soft tissues of the floor of the mouth

The oral cavity is lined throughout with *mucous membrane* which consists of *stratified squamous epithelium* containing small mucus secreting glands.

The part of the mouth outside the gums and teeth is called the *vestibule* and the remainder of the cavity the *mouth proper*. The mucous membrane lining of the cheeks and the lips is reflected on to the gums or *alveolar ridges*.

The *palate* is divided into the anterior part which is called the *hard palate* and the posterior part called the *soft palate*. The bones forming the hard palate are the maxilla and the palatine bones. The soft palate is muscular, curves downwards from the posterior end of the hard palate and blends with the walls of the pharynx at the sides.

The *uvula* is a curved fold of muscle covered with mucous membrane which hangs down from the middle of the free border of the soft palate. Originating from the upper end of the uvula there are four folds of mucous membrane, two passing downwards at each side to form membranous arches. The posterior folds, one on each side, are called *palatopharyngeal arches* and the two anterior folds are called the *palatoglossal arches*. On each side, between the pair of arches, there is a collection of lymphoid tissue called the *palatine tonsil*.

THE TONGUE

The tongue is a voluntary muscular structure which occupies the floor of the mouth. It is attached by its base to the *hyoid bone* and by a fold of its mucous membrane covering, called the *frenulum*, to the floor of the mouth (Fig. 9:10). The superior surface consists of stratified squamous epithelium, with numerous *papillae* (little projections) which contain the nerve endings of the sense of taste; these are sometimes called the *taste buds*. There are three varieties of papillae (Fig. 9:11).

Vallate papillae are usually about 8 to 12 in number and are arranged in an inverted V shape towards the base of the tongue. These are the largest of the papillae and are the most easily seen.

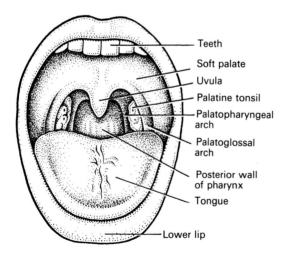

Figure 9:9   Structures seen in the widely open mouth.

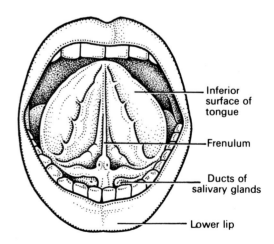

Figure 9:10   The inferior surface of the tongue.

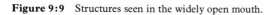

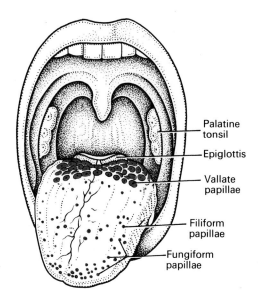

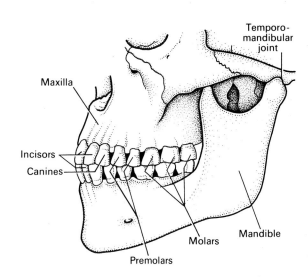

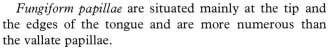

Figure 9:11   Diagram of the papillae of the tongue and related structures.

Figure 9:12   The permanent teeth and the jaw bones.

Fungiform papillae are situated mainly at the tip and the edges of the tongue and are more numerous than the vallate papillae.

Filiform papillae are the smallest of the three types and are found to be most numerous on the surface of the anterior two-thirds of the tongue.

Blood supply
The main arterial blood supply to the tongue is by the *lingual branch* of the *external carotid artery*. Venous drainage is by the *lingual vein* which joins the internal jugular vein.

Nerve supply
This is as follows:

The hypoglossal nerves supply the voluntary muscle tissue.

The lingual branch of the mandibular nerves are the nerves of ordinary sensation, that is, pain, temperature and touch.

The facial and glossopharyngeal nerves are the nerves of the special sensation of taste.

Functions of the tongue
The tongue plays an important part in mastication (chewing), deglutition (swallowing) and speech. It is the organ of taste and the nerve endings of the sense of taste are present in the papillae.

THE TEETH
The teeth are embedded in the alveoli or sockets of the alveolar ridges of the mandible and the maxilla (Fig. 9:12). Each individual has two sets of teeth, the *temporary* or *deciduous teeth* and the *permanent teeth* (Figs. 9:13 and 9:14). At birth the teeth of both dentitions

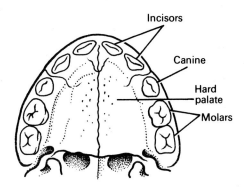

Figure 9:13   The roof of the mouth and the deciduous teeth. Viewed from below.

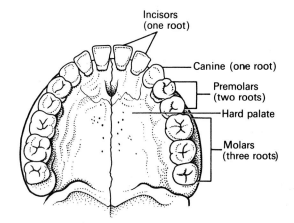

Figure 9:14   The roof of the mouth and the permanent teeth. Viewed from below.

are present in immature form in the mandible and maxilla.

The temporary teeth are 20 in number, 10 in the upper jaw and 10 in the lower jaw. The incisors and canine teeth each have one root, the molars in the upper jaw have three roots and those in the lower jaw have two roots.

Deciduous teeth						
Jaw	Molars	Canine	Incisors	Incisors	Canine	Molars
Upper	2	1	2	2	1	2
Lower	2	1	2	2	1	2

These teeth begin to erupt when the child is about *6 months* old, and should all be present by the end of *24 months*.

The permanent teeth begin to replace the deciduous teeth in the *6th year* of age and this dentition, consisting of 32 teeth, is usually complete by the *24th year*. The incisors and canines have one root; the upper premolars have two roots and the lower usually only one. Like the deciduous teeth, the upper molars have three roots and the lower two.

Permanent teeth								
Jaw	Molars	Premolars	Canine	Incisors	Incisors	Canine	Premolars	Molars
Upper	3	2	1	2	2	1	2	3
Lower	3	2	1	2	2	1	2	3

The *incisor* and *canine* teeth are the cutting teeth and are used for biting off pieces of food, whereas the *premolar* and *molar* teeth, with broad, flat surfaces, are used for grinding or chewing food (Fig. 9:15).

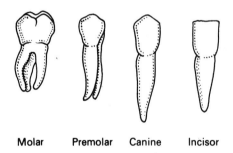

Molar Premolar Canine Incisor

Figure 9:15 The shapes of the permanent teeth.

Structure of a tooth (Fig. 9:16)

Although the shape of the different teeth vary the structure is the same and consists of:

The crown—the part which protrudes from the gum
The root—the part embedded in the bone
The neck—the slightly constricted part where the crown merges with the root

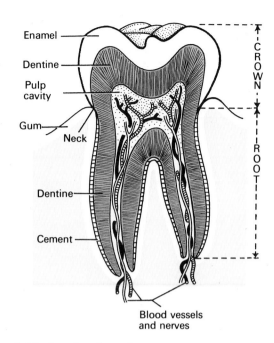

Figure 9:16 A section of a tooth.

In the centre of the tooth there is the *pulp cavity* containing blood vessels, lymph vessels and nerves, and surrounding this there is a hard ivory-like substance called *dentine*. Outside the dentine of the crown of the tooth there is a thin layer of very hard substance called *enamel*. The root of the tooth, on the other hand, is covered with a substance resembling bone, called *cement*, which fixes the tooth in its socket. There is a small foramen at the apex of the root of the tooth which allows for the passage of blood vessels and nerves to and from the tooth.

Blood supply

Most of the arterial blood supply to the teeth is by branches of the *maxillary arteries*. The venous drainage is by a number of veins which empty into the *internal jugular veins*.

Nerve supply

The nerve supply to the upper teeth is by branches of the *maxillary nerves* and to the lower teeth by branches of the *mandibular nerves*. These are both branches of the *trigeminal nerves* (5th cranial nerves) (see p. 191).

THE PHARYNX

As has already been described (see p. 94) the pharynx is divided for descriptive purposes into three parts, the *nasal*, the *pharyngeal* and the *laryngeal* parts. Of these, only the pharyngeal and laryngeal parts are associated with the alimentary tract. Food passes from the oral cavity to the pharynx then to the oesophagus below, with which it is continuous.

The lining membrane is stratified squamous epithelium and is continuous with the lining of the mouth at one end and with the oesophagus at the other.

The **middle layer** consists of fibrous tissue which becomes thinner towards the lower end.

The **outer layer** consists of a number of unstriped (involuntary) muscles called *constrictor* muscles which are involved in swallowing. When food reaches the pharynx swallowing is no longer under voluntary control.

Blood supply
The blood supply to the pharynx is by several branches of the *facial arteries*. Venous drainage is into the *facial veins* and the *internal jugular veins*.

Nerve supply
This is from the *pharyngeal plexus* which is formed by parasympathetic and sympathetic nerves. Parasympathetic supply is mainly by the *glossopharyngeal* and *vagus nerves* and sympathetic from the *cervical ganglia* (see p. 194).

THE SALIVARY GLANDS (Fig. 9:17)
There are three pairs of *compound racemose glands* which pour their secretions into the mouth. They are:

2 parotid
2 submandibular
2 sublingual

PAROTID GLANDS
These are situated one on each side of the face just below the external acoustic meatus. Each gland has a *parotid duct* opening into the mouth at the level of the second upper molar tooth.

SUBMANDIBULAR GLANDS
These lie one on each side of the face under the angle of the jaw. The two *submandibular ducts* open on to the floor of the mouth, one on each side of the frenulum of the tongue.

SUBLINGUAL GLANDS
These glands lie under the mucous membrane of the floor of the mouth in front of the submandibular glands. They have numerous small ducts which pierce the mucous membrane of the floor of the mouth.

Structure of the salivary glands
These glands are all surrounded by a *fibrous capsule*.

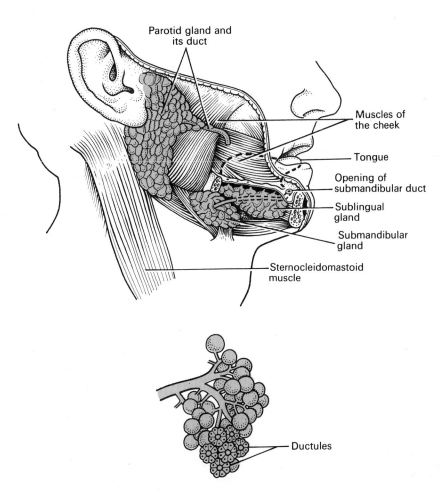

Figure 9:17   The positions of the salivary glands. *Bottom:* an enlargement of part of a gland.

They consist of a number of *lobules* made up of small alveoli lined with *secretory cells*. The secretions are poured into small ducts which join up to form larger ducts leading into the mouth.

Nerve supply

All the glands are supplied by parasympathetic and sympathetic nerve fibres.

Parasympathetic supply—stimulates secretion
Sympathetic supply—depresses secretion

Blood supply

Arterial supply is by various branches from the *external carotid arteries* and venous drainage is into the *external jugular veins*.

Saliva

This is the combined secretions from the salivary glands and the small mucus secreting glands of the lining of the oral cavity. It consists of:

Water
Mineral salts
Enzyme; ptyalin or salivary amylase
Mucus added by the glands in the mouth

THE OESOPHAGUS (Fig. 9:18)

The oesophagus or gullet is the first part of the alimentary tract to which the *general plan* described previously applies (see p. 117). It is about 25 cm (10 inches) long and is the narrowest part of the alimentary tract. It lies in the median plane in the thorax in front of the vertebral column and behind the trachea and the heart. It is continuous with the pharynx above and just below the diaphragm it joins the stomach. It passes through the central tendon of the diaphragm at the level of the 10th thoracic vertebra. Immediately the oesophagus passes through the diaphragm it curves upwards before becoming the stomach. This sharp angle is believed to be one of the factors which prevents the regurgitation (backward flow) of gastric contents into the oesophagus.

STRUCTURE

There are four layers of tissue as described in the general plan. As the oesophagus is almost entirely in the thorax the outer covering consists of *elastic fibrous tissue*. There is a slight thickening of the circular muscle layer at the gastric end which may contribute to preventing gastric contents from regurgitating into the oesophagus. The thoracic part of the oesophagus is lined with stratified squamous epithelium. Near its distal end this changes to columnar epithelium.

Blood supply

Arterial. The thoracic part of the oesophagus is supplied mainly by the oesophageal arteries. The abdominal

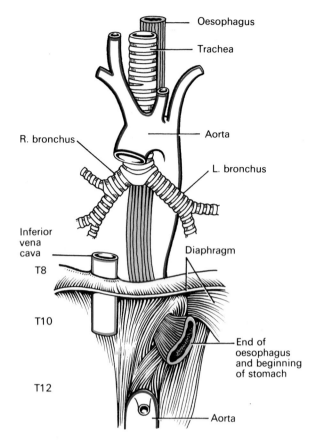

Figure 9:18 Some structures associated with the oesophagus.

part is supplied by branches from the inferior phrenic arteries and the left gastric branch of the coeliac artery.

Venous drainage. From the thoracic part venous drainage is into the azygos and hemiazygos veins. The abdominal part drains into the left gastric vein.

Nerve supply

Sympathetic and parasympathetic nerves terminate in the myenteric and submucosal plexuses. Parasympathetic fibres are branches of the vagus nerves (see Fig. 9:5).

FUNCTIONS OF THE MOUTH, LARYNX, OESOPHAGUS AND SALIVARY GLANDS

Digestion in the mouth

When food is taken into the mouth it is masticated or chewed by the teeth and moved round the mouth by the tongue and by the muscles of the cheeks (Fig. 9:19). It is mixed with saliva and formed into a soft mass or *bolus* ready for *deglutition* or swallowing. The length of time that food remains in the mouth depends, to a large extent, on the consistency of the food. Some foods need to be chewed longer than others before the individual feels that the mass is ready for swallowing.

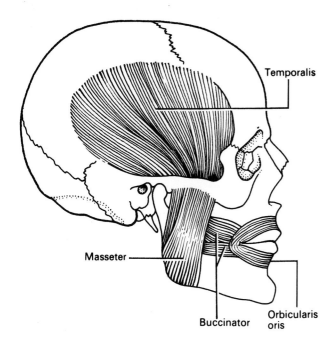

Figure 9:19   The muscles used when chewing.

Functions of saliva

Digestion. The enzyme *ptyalin* acts on cooked starches (polysaccharides) changing them to the disaccharide *maltose*. The optimum pH for this reaction is 6·7—slightly acid—and the action of ptyalin continues after the bolus has been swallowed. It is finally inhibited by the strongly acid reaction of gastric juices—pH 1·5 to 1·8 (see p. 130).

Lubrication of food. Dry food entering the mouth is moistened and lubricated by saliva before it can be made into a bolus ready for swallowing.

Cleansing and lubricating. An adequate flow of saliva is necessary to cleanse the mouth and keep the structures within the mouth soft and pliable.

Taste. The taste buds are stimulated by particles present in the food which are dissolved in water. Dry foods stimulate the sense of taste only after thorough mixing with saliva.

Secretion of saliva

The flow of saliva is controlled by sympathetic and parasympathetic nerve supply. Parasympathetic stimulation causes an increase in the secretion of saliva and sympathetic stimulation has an inhibitory effect on the glands. There are two types of reflex action involved in salivation.

Unconditioned reflex. This is the automatic response to the presence of an object other than food in the mouth. It is best demonstrated by placing something in the mouth of a child before he or she has learned what substances satisfy hunger.

Conditioned reflex. This too is an automatic response but it is a response which the individual has *learned* from previous experience. The sight, smell and even the thought of appetising food results in salivation sometimes called 'mouth watering'. This type of salivation occurs on the anticipation of food.

Deglutition or swallowing (Fig. 9:20)

This occurs in three stages after mastication is complete and the bolus has been formed. After the process has

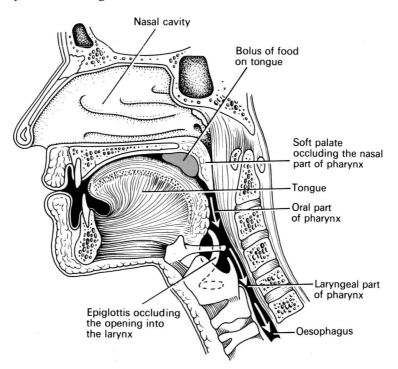

Figure 9:20   Section of the face and neck showing the positions of structures during swallowing.

been voluntarily initiated it is under autonomic nerve control.

1. The bolus is pushed backwards into the pharynx by the upward movement of the tongue.

2. The muscles of the pharynx propel the bolus down into the oesophagus. All other routes which the bolus could possibly take are closed. The soft palate rises up and occludes the nasal part of the pharynx; the tongue and the pharyngeal folds close the way back into the mouth; and the larynx is lifted up and forward so that its opening is occluded by the overhanging epiglottis coming into contact with the base of the tongue.

3. The presence of the bolus in the pharynx stimulates a wave of peristalsis which propels the bolus through the oesophagus to the stomach.

When the bolus reaches the pharynx it is no longer within voluntary control. The *peristaltic action* which takes over occurs throughout the remainder of the alimentary tract. It is the means by which the contents are moved on and is characterised by a wave of relaxation in front of the bolus which is pushed on by a following wave of contraction (Fig. 9:21).

The walls of the oesophagus are lubricated by mucus which assists the passage of the bolus during the peristaltic contraction of the muscular wall.

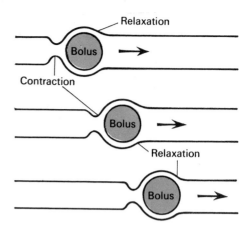

Figure 9:21 Illustration of the movement of the bolus through the oesophagus by peristalsis.

The Stomach and Gastric Juice

The stomach is a J-shaped dilated portion of the alimentary tract situated in the epigastric, umbilical and left hypochondriac regions of the abdominal cavity.

Organs in association with the stomach (Fig. 9:22)

Anteriorly— the left lobe of the liver and the anterior abdominal wall

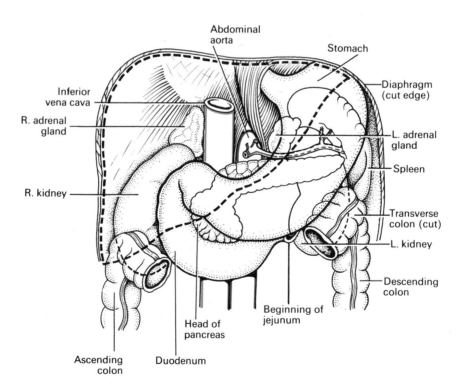

Figure 9:22 Stomach and its associated structures.

Posteriorly—	the abdominal aorta, the pancreas, the spleen, the left kidney and adrenal gland
Superiorly—	the diaphragm, the oesophagus and left lobe of the liver
Inferiorly—	the transverse colon and the small intestine
To the left—	the diaphragm and the spleen
To the right—	the liver and the duodenum

STRUCTURE (Fig. 9:23)

The oesophagus opens into the stomach at the *cardiac orifice*, and the duodenum is continuous with the stomach at the *pyloric orifice*.

The stomach is described as having two curvatures. The *lesser curvature* is short, lies on the posterior surface of the stomach and is a continuation downwards of the posterior wall of the oesophagus. Just before the pyloric sphincter it curves upwards to complete the J shape. The *greater curvature* is on the anterior surface of the stomach. Where the oesophagus enters the stomach the anterior part angles acutely upwards, curves downwards then slightly upwards towards the pyloric orifice.

The part of the stomach above the cardiac orifice is called the *fundus*, the main part is the *body* and the lower part is the *pyloric antrum*. At the distal end of the pyloric antrum there is a sphincter, the *pyloric sphincter*, which guards the opening between the stomach and the duodenum.

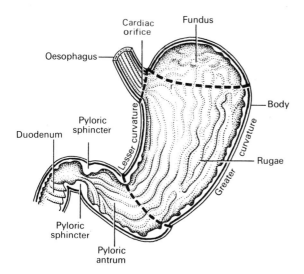

Figure 9:23 Longitudinal section of the stomach.

THE WALLS OF THE STOMACH

The four layers of tissue described in the *general plan* of the alimentary canal are to be found in the stomach but with some modifications.

The peritoneum

From Figure 9:24 it will be seen that the fold of peritoneum which attaches the stomach to the posterior abdominal wall extends beyond the greater curvature of the stomach. This is called the *greater omentum*. It is free at its distal end and hangs down in front of the abdominal organs like an apron.

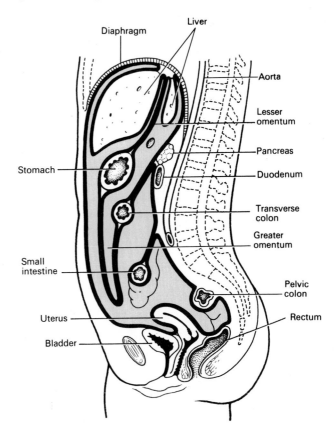

Figure 9:24A The peritoneum viewed from the side.

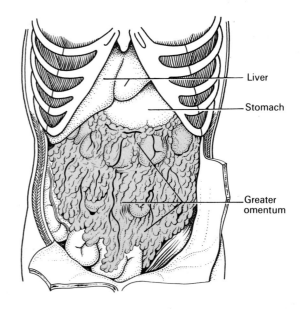

Figure 9:24B The omentum viewed from the front.

The greater omentum stores fat, is richly supplied with blood and lymph vessels and contains a considerable number of lymph nodes. It has the ability to isolate an area of slowly developing inflammation, such as chronic appendicitis, preventing the spread of infection to the peritoneal cavity as a whole.

The muscle layer (Fig. 9:25)

This consists of *three layers* of smooth muscle fibres. The outer layer has *longitudinal fibres*, the middle layer has *circular fibres* and the inner layer, *oblique fibres*. This arrangement allows for the churning motion characteristic of gastric activity, as well as the peristaltic movement.

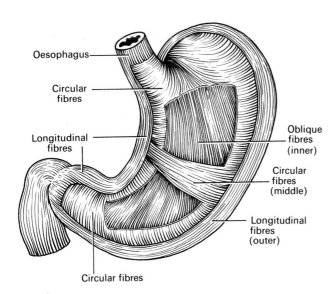

Oesophagus

Circular fibres

Longitudinal fibres

Oblique fibres (inner)

Circular fibres (middle)

Longitudinal fibres (outer)

Circular fibres

Figure 9:25   The muscle fibres of the stomach wall. Sections have been removed to show the three layers.

The mucous membrane lining

When the stomach is empty the mucous membrane lining is thrown into longitudinal folds or *rugae*, and when it is full the rugae are 'ironed out' and the surface has a smooth, velvety appearance. Within the mucous membrane there are numerous *gastric glands* which secrete *gastric juice*. These glands consist of specialised cells; they are situated below the surface and pour gastric juice into the stomach through tiny ducts.

BLOOD SUPPLY

Arterial blood is supplied to the stomach by branches of the coeliac artery and venous drainage is into the portal vein. Figures 9:7 and 9:8 give details of the names of these vessels.

NERVE SUPPLY

The sympathetic supply to the stomach is mainly from the *coeliac plexus* and the parasympathetic supply is from the *vagus nerves*. Sympathetic stimulation reduces the motility of the stomach and the secretion of gastric juice; vagal stimulation has the opposite effect (see Fig. 9:5, p. 119).

COMPOSITION OF GASTRIC JUICE AND THE FUNCTIONS OF THE STOMACH

The size of the stomach varies with the amount of food it contains. When a sizeable meal has been eaten the food accumulates in the stomach in layers, the last part of the meal remaining in the fundus for some time. Mixing with the gastric juice takes place gradually and it may be some time before the food is sufficiently acidified to stop the action of ptyalin.

In addition to the onward movement of gastric contents peristaltic action in the stomach consists of a *churning* movement brought about by contraction of the three layers of muscle tissue. This churning movement causes further mechanical breakdown of the food, the mixing of the food with gastric juice and its onward movement into the duodenum.

GASTRIC JUICE

This is secreted by special secretory glands in the walls of the stomach and consists of:

Water
Mineral salts
Mucus
Hydrochloric acid
Enzymes: pepsinogen and rennin
Intrinsic factor

Functions of gastric juice

1. *The water* further liquefies the food swallowed.
2. *The hydrochloric acid*:
    a. acidifies the food and stops the action of ptyalin
    b. converts *pepsinogen* to the active enzyme *pepsin*
    c. kills many micro-organisms which may be harmful to the body.
3. *Enzyme action*:

Pepsin begins the chemical digestion of proteins by converting them to *peptones*. It acts most effectively within the pH range of 1·6 to 3·2 produced by the hydrochloric acid.

Rennin curdles milk by changing soluble *caseinogen* (milk protein) into insoluble *casein*, which in turn is converted by pepsin into peptones. Rennin is present in the gastric juice of infants but not in adults.

4. *The intrinsic factor* (a protein compound) is necessary for the absorption of vitamin B_{12} (cyanocobalamin). Vitamin B_{12} is also called the *anti-anaemic factor*; it is present in food and is absorbed through the walls of the small intestine and stored in the liver until required in the red bone marrow for the normal development of erythrocytes.

5. *The mucus* prevents mechanical injury to the stomach wall by lubricating the contents. It prevents chemical

injury by acting as a barrier between the stomach wall and the other constituents of gastric juice. Hydrochloric acid is present in potentially damaging concentrations and pepsin digests protein.

Secretion of gastric juice

There is always a small quantity of gastric juice present in the stomach, even when it contains no food. This is known as *fasting juice*.

There are three phases of secretion of gastric juice (Fig. 9:26).

FUNCTIONS OF THE STOMACH

1. The stomach acts as a temporary reservoir for food thus allowing the digestive juices time to act on the different food substances.

2. It produces *gastric juice* which begins the chemical digestion of proteins.

3. Muscular action mixes the food with gastric juice then moves it on to the small intestine. When the contents of the pyloric end of the stomach have reached a suitable degree of acidity and liquefaction the pyloric sphincter relaxes and the muscular walls of the stomach

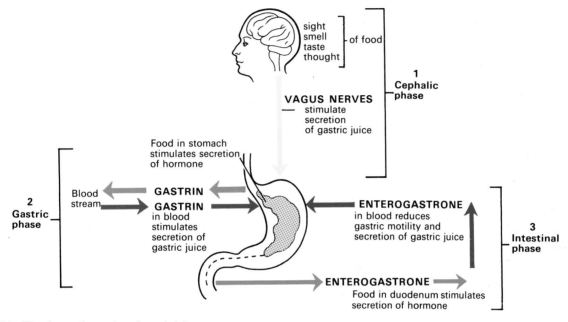

Figure 9:26   The phases of secretion of gastric juices.

Cephalic phase. This flow of juice occurs *before* food reaches the stomach and is due to reflex stimulation of the vagus nerves initiated by the sight, smell or taste of food. When the vagus nerves have been cut this phase of gastric secretion stops.

Gastric phase. When stimulated by the presence of food the stomach produces a hormone called *gastrin* which passes directly into the circulating blood. This hormone circulating in the blood which supplies the stomach stimulates the glands in the stomach wall to produce more gastric juice. In this way the secretion of digestive juice is continued after the completion of the meal and the end of the cephalic phase.

Intestinal phase. When the partially digested contents of the stomach reach the small intestine a hormone called *enterogastrone* is produced which slows down the secretion of gastric juice and reduces gastric motility. By slowing the emptying rate of the stomach, the contents of the duodenum become more thoroughly mixed with bile and pancreatic juice. This phase of gastric secretion is most marked when the meal has had a high fat content.

contract forcing small jets of gastric juice into the duodenum. The rate at which the stomach empties depends to a large extent on the type of food eaten. A carbohydrate meal leaves the stomach in 2 to 3 hours, a protein meal remains longer and a fatty meal remains in the stomach longest. The stomach contents entering the duodenum are called *chyme*.

4. Absorption takes place in the stomach to a limited extent. Water, glucose, alcohol and some drugs are absorbed through the walls of the stomach into the venous circulation.

5. Although iron absorption takes place in the small intestine it is dissolved out of foods most effectively in the presence of hydrochloric acid in the stomach.

The Small Intestine, Pancreas, Liver, Biliary Tract

THE SMALL INTESTINE (Figs. 9:27 and 9:28)
The small intestine is continuous with the stomach at the

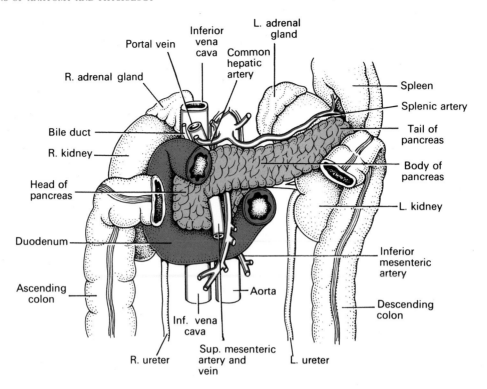

Figure 9:27   The duodenum and its associated structures.

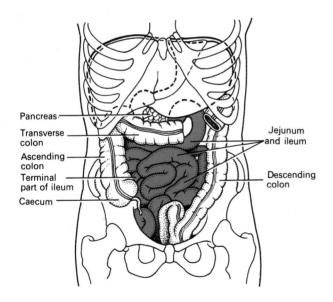

Figure 9:28   The jejunum and ileum and their associated structures.

pyloric sphincter and leads into the large intestine at the *ileocoecal valve*. It is a little over 5 metres (16 feet) in length and lies in the abdominal cavity surrounded by the large intestine. In the small intestine the chemical digestion of food is completed and most of the absorption of nutrient materials takes place.

The small intestine is described in three parts which are continuous with each other.

The duodenum is about 25 cm (10 inches) in length and curves in a C shape around the head of the pancreas. At its midpoint there is the opening which is common to the pancreatic duct and the bile duct and it is guarded by the *sphincter of Oddi*.

The jejunum is the middle part of the small intestine and is about 2 metres (6½ feet) long.

The ileum is the terminal part, is about 3 metres (10 feet) long and ends at the *ileocaecal valve* which controls the flow of material from the ileum to the large intestine and vice versa.

STRUCTURE

The walls of the small intestine are composed of the four layers of tissue described in the *general plan* (see p. 117). Modifications of the peritoneum and the mucous membrane lining are as follows:

The peritoneum

A double layer of peritoneum called *the mesentery* attaches the jejunum and the ileum to the posterior abdominal wall (Fig. 9:24A). The attachment is quite short in comparison with the length of the small intestine, therefore it is fan-shaped. The large blood vessels and nerves lie on the posterior abdominal wall and the branches from them to the small intestine pass between the two layers of the mesentery.

The mucous membrane

The surface area of the small intestine is greatly increased by two peculiarities in the arrangement of the mucous membrane.

The circular folds, unlike the rugae of the stomach, are not smoothed out when the small intestine is distended (Fig. 9:29).

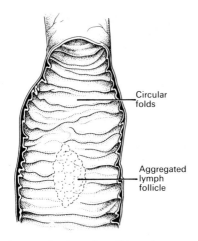

Figure 9:29 Section of a piece of small intestine.

The villi are tiny finger-like projections into the lumen of the organ about 0·5 mm to 1 mm long (Fig. 9:30). Their walls consist of columnar epithelial cells with tiny microvilli (1 μm long) on their free border. These epithelial cells enclose a network of blood and lymph capillaries. The lymph capillaries are called *lacteals*. Absorption of nutrient materials takes place across the wall of the villus into the blood and lymph capillaries.

Intestinal glands. These are simple tubular glands situated below the surface between the villi. The cells of

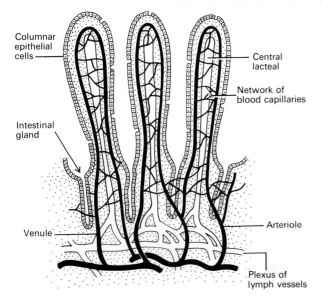

Figure 9:30 Highly magnified view of villi in the small intestine.

the glands migrate upwards to form the walls of the villi replacing those at the tips which are rubbed off by the intestinal contents. During migration the cells form digestive enzymes that lodge in the microvilli and, together with intestinal juice, complete the chemical digestion of carbohydrates, proteins and fats.

Lymph nodes. There are numerous lymph nodes in the mucous membrane at irregular intervals throughout the length of the small intestine. The smaller ones are known as *solitary lymphatic follicles*, and about 20 or 30 larger nodules situated towards the distal end of the ileum are called *aggregated lymphatic follicles* (Peyer's patches).

BLOOD SUPPLY
The *superior mesenteric artery* supplies the whole of the small intestine, and venous drainage is by the superior mesenteric vein which joins other veins to form the portal vein (see Fig. 5:39, p. 79).

NERVE SUPPLY
Sympathetic and parasympathetic.

FUNCTION
The functions of the small intestine will be described in conjunction with those of the liver and the pancreas as they are essentially interdependent.

THE PANCREAS (Fig. 9:31)
The pancreas is a pale grey gland which weighs about 60 grams (2 ounces). It is about 12 to 15 cm (5 to 6 inches) long and is situated in the *epigastric* and *left hypochondriac* regions of the abdominal cavity. It consists of a broad head, a body and a narrow tail. The head lies in the curve of the duodenum, the body behind the stomach and the tail, which just reaches the spleen, lies in front of the left kidney. The abdominal aorta and the inferior vena cava lie behind the gland.

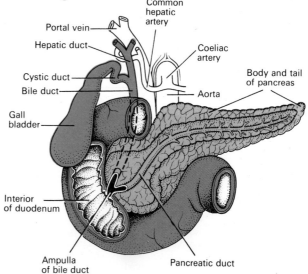

Figure 9:31 The pancreas in relation to the duodenum and biliary tract. Part of the anterior wall of the duodenum removed.

STRUCTURE

The pancreas consists of a large number of *lobules* made up of small *alveoli* the walls of which consist of *secretory cells*. Each lobule is drained by a tiny duct and these unite to form larger and larger ducts until they form the *pancreatic duct*. This duct extends the whole length of the gland and opens into the duodenum at its midpoint. Just before entering the duodenum the pancreatic duct joins the *bile duct* to form the *ampulla of the bile duct*. The duodenal opening of the ampulla is controlled by the *sphincter of Oddi*.

Islets of Langerhans

Distributed throughout the substances of the pancreas there are little collections of a different type of cell. These cells form what are known as the *islets of Langerhans*. Their secretion passes directly into the circulating blood and consists of the hormones *insulin* and *glucagon* (Ch. 14, p. 224).

FUNCTION

The function of the alveoli of the pancreas is to produce *pancreatic juice*, which plays an important part in the chemical digestion of food (see p. 137).

BLOOD SUPPLY

The splenic and mesenteric arteries supply arterial blood to the pancreas and the venous drainage is by the veins of the same names which join other veins to form the portal vein.

NERVE SUPPLY

Sympathetic and parasympathetic. As in the alimentary tract parasympathetic stimulation results in an increase in the secretion of pancreatic juice and sympathetic stimulation depresses secretion.

THE LIVER

The liver is the largest gland in the body. It weighs between 1 and 2·3 kg (2 to 5 pounds) and is heavier in males than females. It is situated in the upper part of the abdominal cavity occupying the greater part of the *right hypochondriac region*, part of the *epigastric region* and extending into the *left hypochondriac region*. Its upper and anterior surfaces are smooth and curved to fit the under surface of the diaphragm (Fig. 9:32); its posterior surface is irregular in outline (Fig. 9:33).

Organs in association with the liver

Superiorly and anteriorly—	diaphragm and anterior abdominal wall
Inferiorly—	stomach, bile ducts, duodenum, right colic flexure of the colon, right kidney and adrenal gland

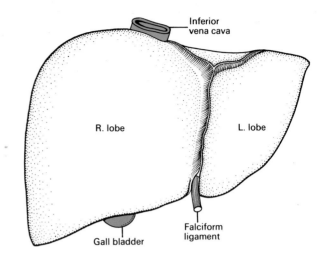

Figure 9:32 The liver—anterior view.

Posteriorly—	oesophagus, inferior vena cava, aorta, gall bladder, vertebral column and diaphragm
Laterally—	lower ribs and diaphragm

The liver is enclosed in a thin capsule and incompletely covered by a layer of peritoneum. Folds of peritoneum form supporting ligaments attaching the liver to the inferior surface of the diaphragm. It is held in position partly by these ligaments and partly by the pressure of the organs in the abdominal cavity.

The liver is described as having four lobes. The two most obvious are the large *right lobe* and the smaller, wedge-shaped, *left lobe*. The other two, the *caudate* and *quadrate* lobes, are areas on the posterior surface (Fig. 9:33).

THE PORTAL FISSURE

This is the name given to the part on the posterior surface of the liver where various structures enter and leave the gland.

The portal vein enters, carrying blood from the stomach, spleen, pancreas and the small and large intestines.

The hepatic artery enters, carrying arterial blood. It is a branch from the coeliac artery which is a branch from the abdominal aorta.

Nerve fibres, sympathetic and parasympathetic.

The right and left hepatic ducts leave, carrying bile from the liver to the gall bladder.

Lymph vessels leave the liver at this point.

BLOOD SUPPLY

The hepatic artery and the portal vein take blood to the liver. Hepatic veins, varying in number, leave the posterior surface of the liver. They immediately enter the inferior vena cava just below the diaphragm.

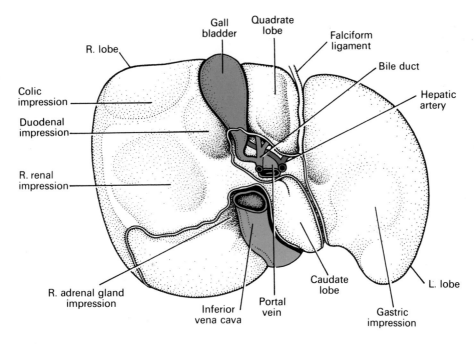

Figure 9:33   The liver—posterior view.

STRUCTURE

The lobes of the liver are made up of tiny lobules just visible to the naked eye. These lobules are *hexagonal* in outline and are formed by cubical shaped cells arranged in *pairs of columns* radiating from a *central vein*. Between two pairs of columns of cells there are *sinusoids* (blood vessels with incomplete walls) containing a mixture of blood from the tiny branches of the portal vein and hepatic artery (Fig. 9:34). This arrangement allows the arterial blood and venous blood (with a high concentration of nutritional materials) to mix and come into direct contact with the liver cells.

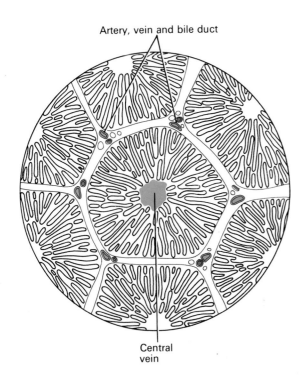

Figure 9:34A   Diagram of a magnified transverse section of liver lobules.

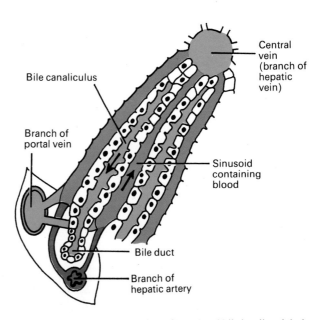

Figure 9:34B   Diagram of the flow of blood and bile in a liver lobule.

After the blood has been in contact with the liver cells it drains into the *central* or *intralobular vein*. Central veins from all the lobules join up, gradually becoming larger until eventually they unite to become the *hepatic veins* which drain blood from the liver as a whole and empty it into the inferior vena cava. Lymphoid tissue and a system of lymph vessels are also present in the lobules. Figure 9:35 shows the scheme of blood flow through the liver.

One of the functions of the liver cells is to form *bile*. In Figure 9:34 it will be seen that bile *canaliculi* run between the columns of liver cells which make up the pairs of columns separating the sinusoids. These canaliculi collect bile secreted by the liver cells and they join together to form larger bile canals. Eventually the *right and left hepatic ducts* are formed and drain all the bile from the liver.

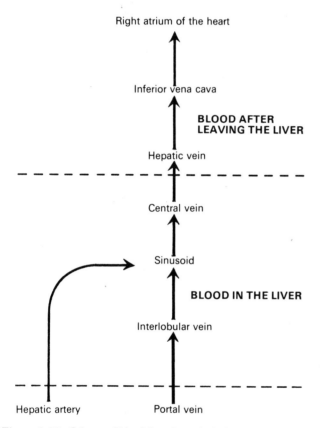

Figure 9:35   Scheme of blood flow through the liver.

FUNCTIONS

The liver is an extremely active organ. Some of its functions have already been described, therefore they will only be mentioned here.

1. *Deaminates amino acids*
   a. Removes the nitrogenous portion from the amino acids which are not required for the formation of new protein, and forms *urea* from this nitrogenous portion.
   b. Breaks down the nucleoprotein of worn-out cells of the body to form *uric acid* which is excreted in the urine.

2. *Converts glucose to glycogen* in the presence of *insulin*, and changes liver glycogen back to glucose in the presence of *glucagon*.

3. *Desaturates fat*, that is, converts stored fat to a form in which it can be used by the tissues to provide energy.

4. *Produces heat*. The liver uses a considerable amount of energy, has a high metabolic rate and produces a great deal of heat. It is the main heat producing organ of the body.

5. *Secretes bile*. The liver cells synthesise the constituents of bile from the mixed arterial and venous blood in the sinusoids. These include bile salts, bile pigments, cholesterol.

6. *Stores vitamin B_{12} (antianaemic factor)*. This vitamin is absorbed through the wall of the small intestine and is transported in the portal circulation to the liver where it is stored until it is needed in the bone marrow for the development of erythrocytes.

7. *Stores iron* derived from:
   a. the diet
   b. the break down of worn out erythrocytes in the spleen.

8. *Stores vitamins A, D, E and K* which have been taken into the body in the diet.

9. *Synthesises vitamin A*. This vitamin can be formed in the liver from carotene, the provitamin found in some plants, for example, in carrots and the green leaves of vegetables.

10. *Forms the plasma proteins* from the available amino acids. These include albumin, globulin, prothrombin and fibrinogen.

11. *Detoxicates drugs and other noxious substances*, such as toxins produced by micro-organisms.

THE BILIARY TRACT

THE BILE DUCTS
The right and left hepatic ducts join to form the *common hepatic duct* just outside the portal fissure. The hepatic duct passes downwards for about 3 cm ($1\frac{1}{4}$ inches) where it is joined at an acute angle by the *cystic duct* from the gall bladder. The cystic and hepatic ducts together form the *bile duct* which passes downwards posterior to the head of the pancreas to be joined by the main pancreatic duct at the *ampulla of the bile duct*. The opening of the combined ducts into the duodenum is controlled by the *sphincter of Oddi*. The bile duct is about 7·5 cm (3 inches) in length and has a diameter of about 6 mm ($\frac{1}{4}$ inch).

Structure
The walls of the bile ducts have the same layers of tissue as those described in the *general plan* of the alimentary canal (see p. 117). In the cystic duct the mucous

membrane lining is arranged in irregularly situated circular folds which have the effect of a *spiral valve*. Bile passes through the cystic duct twice—on its way into the gall bladder and again when it is expelled from the gall bladder to the bile duct and thence to the duodenum.

THE GALL BLADDER

The gall bladder is a pear-shaped sac attached to the posterior surface of the liver by connective tissue. It is described as having a *fundus* or expanded end, a *body* or main part and a *neck* which is continuous with the cystic duct.

Structure

The gall bladder has the same layers of tissue as those described in the *general plan* of the alimentary canal with some modifications.

Peritoneum covers only the inferior surface. The gall bladder is in contact with the posterior surface of the right lobe of the liver and is held in place by the visceral peritoneum of the liver (Fig. 9:33).

Muscle layer. There is an additional layer of oblique muscle fibres.

Mucous membrane displays small rugae when the gall bladder is empty which disappear when it is distended with bile.

Blood supply

The *cystic artery*, a branch of the hepatic artery, supplies blood to the gall bladder. Blood is drained away by the *cystic vein* which joins the portal vein.

Nerve supply

Nerve impulses are conveyed by sympathetic and para-sympathetic nerve fibres. It has the same autonomic plexuses as those described in the *general plan*.

Functions

1. It acts as a reservoir for bile.
2. The lining membrane adds mucus to the bile.
3. It absorbs water which concentrates the bile.
4. By the contraction of the muscular walls bile is expelled from the gall bladder and passed via the bile ducts into the duodenum. This occurs when a meal containing fat has been eaten. The sphincter of Oddi must relax before bile can pass into the duodenum. *Cholecystokinin-pancreozymin* (CCK-PZ), a complex hormone secreted by the duodenum, stimulates the gall bladder to contract. Figure 9:36 shows the flow of bile from the liver to the duodenum.

DIGESTION IN THE SMALL INTESTINE

When acid chyme passes into the small intestine it is mixed first with the *pancreatic juice* and *bile* then with *intestinal juice*. Although intestinal juice is secreted throughout the length of the small intestine its action is

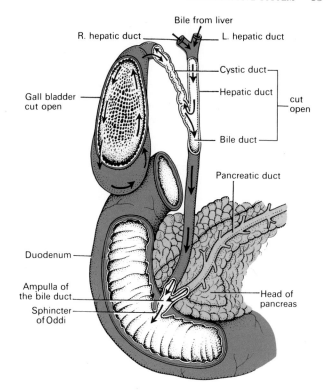

Figure 9:36 Diagram of the flow of bile from the liver and the duodenum.

very limited in the duodenum. It is the juice which *completes* the digestion of carbohydrates to monosaccharides, proteins to amino acids and fats to fatty acids and glycerol.

PANCREATIC JUICE

Pancreatic juice enters the duodenum at the ampulla of the bile duct and consists of:

Water
Mineral salts
Enzymes; trypsinogen, chymotrypsinogen, amylase, lipase

Pancreatic juice is strongly alkaline (pH 8). When acid stomach contents (pH 1 to 3) enter the duodenum they are mixed with pancreatic juice and bile and the pH is raised to between 6 and 7. This is the pH at which the pancreatic enzymes act most effectively.

Functions

Trypsinogen and *chymotrypsinogen* are inactive enzymes until they come in contact with *enteropeptidase* (enterokinase), an enzyme in intestinal juice, which converts them into *trypsin* and *chymotrypsin*. These enzymes convert peptones into peptides and polypeptides. It is important that they are produced as inactive precursors of protein splitting enzymes otherwise they would digest the pancreas.

Amylase converts *all digestible* polysaccharides (starches) not affected by ptyalin to disaccharides (sugars).

Lipase converts fats to fatty acids and glycerol. To aid the action of lipase *bile salts* emulsify fats, i.e., reduce the size of the globules.

Secretion

The secretion of pancreatic juice is stimulated by two hormones *secretin* and *cholecystokinin-pancreozymin* (CCK-PZ) produced by cells in the walls of the small intestine. The presence in the duodenum of acid material from the stomach stimulates the production of these hormones.

BILE

Bile, secreted by the liver, is unable to enter the duodenum when the sphincter of Oddi is closed, therefore it passes from the *hepatic duct* along the *cystic duct* to the gall bladder where it is stored. When a meal has been taken the gall bladder contracts, the sphincter of Oddi relaxes and bile passes through the cystic duct and the bile duct into the duodenum together with pancreatic juice. The hormone *cholecystokinin-pancreozymin* (CCK-PZ), produced by the walls of the duodenum, stimulates this activity. A more marked activity is noted if chyme entering the duodenum contains a high proportion of fat.

There is a slight difference between the composition of bile produced by the liver and that entering the duodenum. In the gall bladder water is absorbed and mucus is added by the goblet cells in the mucous membrane lining, therefore the bile is concentrated and becomes more viscid.

The constituents of bile from the gall bladder are:

Water
Mineral salts
Mucus
Bile salts—sodium taurocholate
              sodium glycocholate
Bile pigment—bilirubin
Cholesterol

Functions

1. The bile salts, *sodium taurocholate* and *sodium glycocholate*, emulsify fats in the small intestine.

2. Bile pigment, *bilirubin*, is a waste product of the breakdown of erythrocytes, and bile is the route by which it is excreted into the small intestine. Eventually it passes out of the body in the faeces.

3. The presence of bile in the small intestine is necessary for the absorption of vitamin K and digested fats.

4. It colours and deodorises the faeces.

5. It has an aperient effect.

INTESTINAL SECRETIONS

Intestinal juice is secreted by the glands of the small intestine. It consists of:

Water
Mucus
Enzymes: enteropeptidase (enterokinase)
           amylase

Most of the digestive enzymes in the small intestine are contained in the microvilli of the cells of the walls of the villi. The digestion of carbohydrate, protein and fat is completed by direct contact between these nutrients and the microvilli and within the cells of the walls of the villi.

The enzymes involved in completing the digestion of food in the small intestine are:

Peptidases
Lipase
Sucrase, maltase and lactase

Functions

Alkaline intestinal juice assists in raising the pH of the intestinal contents to between 6·5 and 7·5.

Enteropeptidase changes inactive pancreatic trypsinogen and chymotrypsinogen to active trypsin and chymotrypsin which convert peptones to peptides and polypeptides.

Amylase acts with pancreatic amylase in converting polysaccharides to disaccharides.

Peptidases complete the digestion of protein by converting peptides and polypeptides to *amino acids*.

Lipase completes the digestion of fats to *fatty acids and glycerol*.

Sucrase, maltase and *lactase* complete the digestion of carbohydrates by converting dissacharides to *monosaccharides*, of which glucose is the most commonly occurring.

Secretion

Mechanical stimulation of the intestinal glands by chyme is believed to be the main stimulus to the secretion of intestinal juice, although the hormone secretin may be involved.

FUNCTIONS OF THE SMALL INTESTINE

1. Onward movement of its contents which is produced by peristaltic, segmental and pendular movements.

2. Secretion of intestinal juice.

3. Completion of digestion of carbohydrates, proteins and fats.

4. Protection against infection by micro-organisms, which have survived the bactericidal action of the hydrochloric acid in the stomach, by the solitary lymph follicles and aggregated lymph follicles.

5. Secretion of the hormones cholecystokinin—pancreozymin and secretin.

6. Absorption of nutrient materials.

ABSORPTION OF NUTRITIONAL MATERIALS

(Fig. 9:37)

Carbohydrates, proteins and fats in their undigested or partly digested form cannot pass *through* the mucous membrane of the small intestine into the blood and lymph. However monosaccharides, amino acids, fatty acids and glycerol can permeate through the columnar epithelial walls of the villi. *Glucose* and *amino acids* are absorbed into the blood capillaries. *Fatty acids* and *glycerol* are absorbed into the lacteals giving the lymph a milky appearance and the name *chyle*. Other nutritional materials such as vitamins, mineral salts and water are absorbed from the small intestine into the blood capillaries.

The surface area through which absorption takes place in the small intestine is greatly increased by the *circular folds* of mucous membrane and by the very large number of *villi* present. It has been calculated that the surface area of the small intestine is about five times that of the whole body.

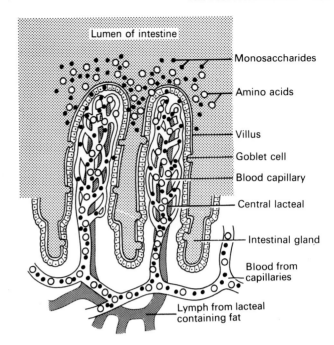

Figure 9:37 Diagram of the absorption of nutrient materials.

Summary of chemical digestion of food

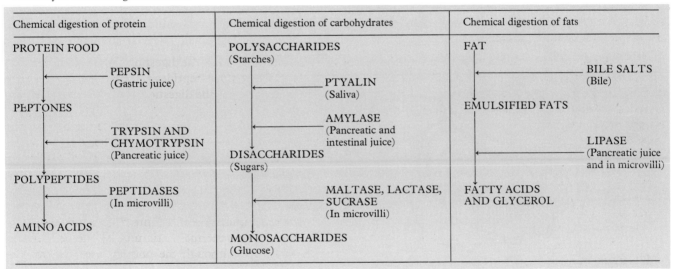

Chemical digestion of protein	Chemical digestion of carbohydrates	Chemical digestion of fats
PROTEIN FOOD	POLYSACCHARIDES (Starches)	FAT
PEPSIN (Gastric juice)	PTYALIN (Saliva)	BILE SALTS (Bile)
PEPTONES	AMYLASE (Pancreatic and intestinal juice)	EMULSIFIED FATS
TRYPSIN AND CHYMOTRYPSIN (Pancreatic juice)	DISACCHARIDES (Sugars)	LIPASE (Pancreatic juice and in microvilli)
POLYPEPTIDES		FATTY ACIDS AND GLYCEROL
PEPTIDASES (In microvilli)	MALTASE, LACTASE, SUCRASE (In microvilli)	
AMINO ACIDS	MONOSACCHARIDES (Glucose)	

The Large Intestine or Colon, Rectum and Anal Canal

The large intestine is about 1·5 metres (5 feet) long, beginning at the *caecum* in the right iliac fossa and terminating at the *rectum* and *anal canal* deep in the pelvis. Its lumen is larger than that of the small intestine. It forms an arch round the coiled up small intestine (Fig. 9:38).

For descriptive purposes the colon is divided into the caecum, ascending colon, transverse colon, descending colon, sigmoid or pelvic colon, rectum and anal canal.

The caecum is the first part of the colon. It is a

dilated portion which has a blind end inferiorly and is continuous with the *ascending colon* superiorly. Just below the junction of the two the *ileocaecal valve* opens from the ileum on the medial aspect of the caecum. The *vermiform appendix* is a fine tube, closed at one end, which leads from the caecum. It is usually about 13 cm (5 inches) long and has the same structure as the walls of the colon but contains more lymphoid tissue (Fig. 9:39).

The ascending colon passes upwards from the caecum to the level of the liver where it bends acutely to the left at the *right colic flexure* (hepatic flexure) to become the *transverse colon*.

The transverse colon is a loop of colon which extends

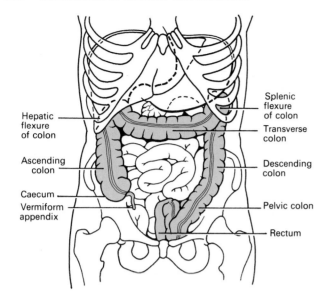

Figure 9:38 Diagram showing the parts of the large intestine (colon) and their positions.

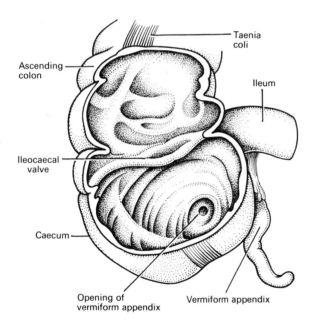

Figure 9:39 Interior of the caecum.

across the abdominal cavity in front of the duodenum and the stomach to the area of the spleen where it forms the *left colic flexure* (splenic flexure) by bending acutely downwards to become the *descending colon*.

The descending colon passes down the left side of the abdominal cavity then curves towards the midline. After it enters the true pelvis it is known as the *pelvic colon*.

The pelvic colon describes an S-shaped curve in the pelvis then continues downwards to become the *rectum*.

Rectum

This is a slightly dilated part of the colon which is about 13 cm (5 inches) long. It leads from the pelvic colon and terminates in the *anal canal*.

Anal canal

This is a short canal about 3·8 cm (1½ inches) long in the adult and leads from the rectum to the exterior. There are two sphincter muscles which control the anus; the internal sphincter which consists of smooth muscle fibres is under the control of the *autonomic nervous system* and the external sphincter, formed by striated muscle, is under *voluntary nerve* control.

STRUCTURE

The four layers of tissue described in the *general plan* are present in the colon, the rectum and the anal canal. The arrangement of the *longitudinal muscle fibres* is modified in the colon. They do not form a smooth continuous layer of tissue but are collected into three bands called *taeniae coli*, situated at regular intervals round the colon. As these bands of muscle tissue are slightly shorter than the total length of the colon they give a sacculated or puckered appearance to the organ (Fig. 9:40).

The longitudinal muscle fibres completely surround the rectum and the anal canal. The anal sphincters are formed by thickening of the circular muscle layer.

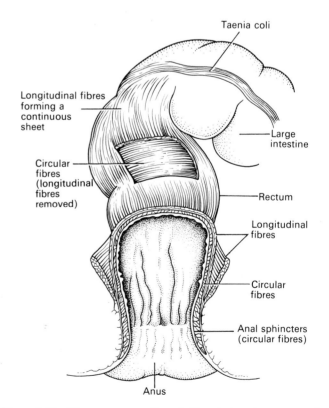

Figure 9:40 The arrangement of muscle fibres in the colon, recum and anus. Sections have been removed to show the layers.

In the submucous layer there is more lymphoid tissue than in any other part of the alimentary tract.

In the mucous membrane lining of the colon and the upper part of the rectum there are large numbers of goblet cells forming simple tubular glands which secrete mucus. They are not present beyond the junction between the rectum and the anus. The lining membrane of the *anus* consists of stratified squamous epithelium which is continuous with the mucous membrane lining of the rectum above and merges with the skin beyond the external anal sphincter.

BLOOD SUPPLY

Arterial supply is mainly by the superior and inferior mesenteric arteries (Fig. 9:6).

The superior mesenteric artery supplies the caecum, ascending and most of the transverse colon.

The inferior mesenteric artery supplies the remainder of the colon and the rectum.

The distal part of the rectum and the anus are supplied by branches from the *internal iliac arteries*.

Venous drainage is mainly by *the superior and inferior mesenteric veins* which drain blood from the parts supplied by arteries of the same names. These veins join the splenic and gastric veins to form the portal vein (Fig. 9:8). Veins draining the distal part of the rectum and the anus join the *internal iliac veins*.

NERVE SUPPLY

This is via sympathetic and parasympathetic nerves to the myenteric and submucous plexuses. The *external anal sphincter* is under *voluntary control* and is supplied by *motor nerves* from the spinal cord.

FUNCTIONS OF THE LARGE INTESTINE, RECTUM AND ANAL CANAL

Absorption

The contents of the ileum which pass through the ileocaecal valve into the caecum are fluid, even although some water has been absorbed in the small intestine. In the large intestine absorption of water continues until the familiar semisolid consistence of faeces is achieved. Mineral salts and some drugs are also absorbed into the blood capillaries from the large intestine.

Micro-organism activity

There are large numbers of micro-organisms in the colon. They include *Escherichia coli, Enterobacter aerogenes, Streptococcus faecalis, Clostridium welchii*. These micro-organisms can cause disease in man but they are harmless and may have some useful functions when they remain in the colon. They are known to synthesise a number of vitamins, but only *folic acid*, required for erythrocyte development, is absorbed in significant amounts.

Large numbers of micro-organisms are present in the faeces.

Defaecation

The large intestine does not exhibit peristaltic movement as it is seen in other parts of the digestive tract. Only at fairly long intervals does a wave of strong peristalsis sweep along the transverse colon forcing its contents into the descending and pelvic colons. This is known as *mass movement* and it is often precipitated by the entry of food into the stomach. This combination of stimulus and response is called the *gastro-colic reflex*. Usually the rectum is empty, but when a mass movement forces the contents of the pelvic colon into the rectum the nerve endings in its walls are stimulated by stretch. In the infant defaecation occurs by reflex action which is in no way controlled. However, after the nervous system has fully developed, nerve impulses are conveyed to consciousness when the stretch receptors in the rectum are stimulated and the brain can inhibit the reflex until such time as it is convenient to defaecate. The external anal sphincter is under conscious control. Thus defaecation involves involuntary contraction of the muscle of the rectum and relaxation of the internal anal sphincter and voluntary relaxation of the external anal sphincter. Contraction of the abdominal muscles and lowering of the diaphragm increases the intra-abdominal pressure and so assists the process of defaecation.

Constituents of faeces. The faeces consist of a semi-solid brown mass. The brown colour is due to the presence of bilirubin.

Even although absorption of water takes place in the large intestine it still makes up about 60 to 70 per cent of the weight of the faeces. The remainder consists of undigestible cellular material (roughage), dead and live micro-organisms, epithelial cells from the walls of the tract, some fatty acids and mucus secreted by the lining mucosa of the large intestine. Mucus helps to lubricate the faeces and an adequate amount of roughage in the diet ensures that the contents of the colon are sufficiently bulky to stimulate defaecation.

Metabolism

When nutritional materials are oxidised in the cells of the body energy is released, some in the form of heat. Energy may be used immediately to do work, for example to synthesise new muscle cells from amino acids, or it may be stored in chemical form as adenosine triphosphate (ATP). Heat is used to maintain the body temperature at the optimum level for chemical activity in the body (37°C or 98·6°F). Excess heat is disposed of through the skin and the body excreta.

The energy produced in the body may be measured and expressed in units of work (*joules*) or units of heat (*Calories*).

A Calorie (capital C) is the amount of heat required to

raise the temperature of 1 litre of water through 1 centi-grade degree.

$$1 \text{ Calorie} = 4184 \text{ joules (J)} = 4{\cdot}184 \text{ kilojoules (kJ)}$$

The nutritional value of carbohydrates, proteins and fats eaten in the diet may be expressed in *kilojoules per gram* or *Calories per gram*.

1 gram of carbohydrate provides 17 kilojoules (4 Calories)
1 gram of protein provides 17 kilojoules (4 Calories)
1 gram of fat provides 38 kilojoules (9 Calories).

The minimum energy requirement per day is described as the *basal metabolic rate* (BMR). This is the rate at which the sum of the metabolic processes takes place when the individual is at rest and has eaten no food for a period of at least 12 hours, i.e., after the last meal has been digested and absorbed. The individual is then described as being at *rest and in the post-absorptive state*.

An indirect measure of the BMR is made by measur-ing *either* the *amount of oxygen* taken into the body or the amount of *carbon dioxide excreted* in a given number of minutes. These are reliable measures because when energy is released oxygen is utilised and carbon dioxide is produced as a waste product. The *surface area* of the body in *square metres* (M^2) is also taken into account when estimating the BMR and it is calculated from measure-ments of the height and weight of the individual (Fig. 9:41).

The basal metabolic rate in men is higher than in women, therefore they require more energy foods.

Men: BMR is about 170 kJ (40 Calories) per M^2 per hour
Women: BMR is about 155 kJ (37 Calories) per M^2 per hour

Example
A woman weighing 57 kg (9 stones), 161 cm (5 ft 3 inches) in height has a surface area of 1·6 M^2. Her basal meta-bolic requirements, if she remains at rest and in the post-absorptive state throughout a 24 hour period, are:

$$155 \times 1{\cdot}6 \times 24 = 5952 \text{ kJ}$$

or

$$37 \times 1{\cdot}6 \times 24 = 1420 \text{ Calories}$$

In normal life most of the basal energy requirement is used during the 8 hours of sleep because even during sleep metabolism is above the basal level. This means that the normal diet must provide considerably more kilojoules to supply the individual's energy needs for 24 hours. The amount of extra energy needed depends on the amount of physical work being done (mental work requires little if any additional energy).

Most people would need at least an additional 4200 kJ (1000 Calories) but if they are involved in strenuous

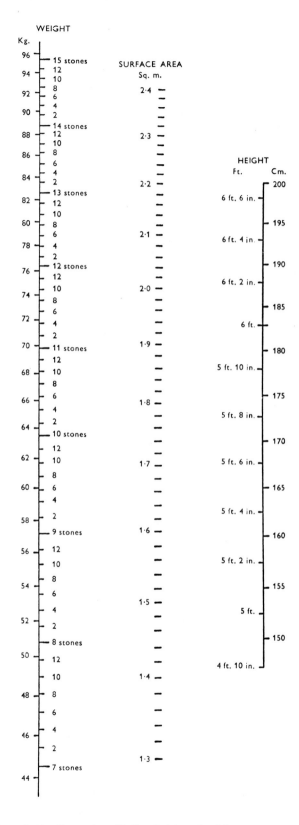

Figure 9:41 Conversion table from height and weight to square metres of surface area of the body. (*From* Green J H 1976 An introduction to human physiology, 4th SI edn. Oxford University Press, London. Reproduced with permission.)

exercise or hard physical work their needs would be greater.

Using the above example a busy woman's energy output would be:

$$5952 + 4200 = 10\,152 \text{ kJ}$$

or

$$1420 + 1000 = 2420 \text{ Calories}$$

The three main constituents of the diet should be present in approximately the following proportions (Fig. 9:42):

Carbohydrates 55%
Proteins 15% } from all food sources
Fats 30%

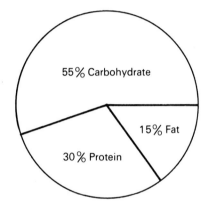

Figure 9:42   Percentages of food substances in the normal diet.

The following table shows the application of these proportions to a 10 152 kJ (2421 Calorie) diet:

Foods	% of whole diet	kJ	Calories	Grams of food
Carbohydrate	55	5584	1332	330
Proteins	15	1522	363	90
Fats	30	3046	726	80
Total	100	10152	2421	500

METABOLISM OF CARBOHYDRATE

When digested, carbohydrate in the form of monosaccharides, mainly glucose, is absorbed into the blood capillaries of the villi of the small intestine. It is transported by the portal circulation to the liver, where it is dealt with in several ways (see Fig. 9:43).

1. Glucose may be used to provide the energy necessary for the considerable metabolic activity which takes place in the liver.

2. Some of the glucose may remain in the circulating blood to maintain the normal blood glucose of about 3·5 to 5·5 millimoles per litre (mmol/l) [60 to 100 mg %].

3. Some of the glucose may be converted to the insoluble polysaccharide, *glycogen*, in the liver and in the

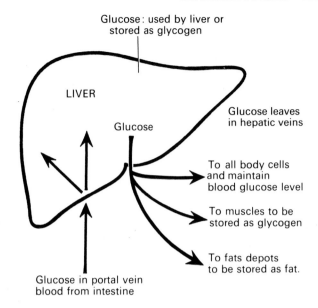

Figure 9:43   Summary of the sources, distribution and utilisation of glucose.

muscles. *Insulin* is the hormone necessary for this change to take place. The formation of glycogen inside cells is a means of storing carbohydrate without upsetting the osmotic equilibrium (see p. 53). Before it can be used it must be broken down again into its constituent monosaccharides. Liver glycogen constitutes a store for the supply of glucose for liver activity and to maintain the blood glucose level. Muscle glycogen provides the glucose requirement of muscle activity. *Adrenalin, thyroxin,* and *glucagon* are three hormones which are associated with the conversion of glycogen to glucose.

4. Carbohydrate, which is in excess of that required to maintain the blood glucose level and glycogen level in the tissues, is converted to fat and stored in the fat depots.

All the cells of the body require energy to carry out their metabolic processes which include: multiplication of cells for replacement of worn out cells; contraction of muscle fibres; synthesis of secretions produced by the cells of glands. The oxidation of carbohydrate and fat provides most of the energy required by the body.

OXIDATION OF CARBOHYDRATE

Complete oxidation of glucose requires an adequate supply of oxygen. This is the process by which energy is released during prolonged physical activity, for example, the man who runs 1500 metres in 4 minutes depends upon *aerobic oxidation*. The energy release takes place slowly and is balanced by oxygen intake. Complete oxidation of carbohydrate in the body results in the production of energy, carbon dioxide and water.

Some energy can be provided by glucose in the absence of oxygen. This *anaerobic process* does not release all the energy from the glucose molecule and the process can be maintained for only a limited period of time. This is the energy used in a sudden spurt of activity over a very short period of time, for example, the man who runs 100 metres in 10 seconds could not take in enough oxygen in that time to provide energy by the complete oxidation of glucose, so he has to depend on the anaerobic process. One of the end products of this process is lactic acid, and if it accumulates in excess in the muscles it causes the pain which is associated with unaccustomed exercise.

The fate of the end products of carbohydrate metabolism

1. *Lactic acid.* Some of the lactic acid produced by anaerobic catabolism of glucose may be oxidised in the tissues to carbon dioxide and water but first it must be changed to pyruvic acid. If complete oxidation does not take place lactic acid passes to the liver in the circulating blood where it is converted to glucose and may then take any of the pathways open to glucose (see Fig. 9:43).

2. *Carbon dioxide* is excreted from the body as a gas by the lungs. This has already been described (see p. 104).

3. *Water.* The water of metabolism is added to the considerable amount of water already present in the body.

METABOLISM OF PROTEIN

Protein foods which are taken as part of the diet consist of a number of amino acids (see p. 109). About *20 amino acids* have been named and about 8 of these are described as *essential* because they can not be synthesised in the body, the remainder are described as *non-essential* amino acids because they can be synthesised by many tissues. The enzymes involved in this process are called *trans-*

aminases. Digestion breaks down the protein of the diet to its constituent amino acids in preparation for absorption into the blood capillaries of the villi in the wall of the small intestine. In the portal circulation amino acids are transported to the liver then into the general circulation, thus making them available to all the cells and tissues of the body. Different cells choose from those available the particular amino acids required for building or repairing their specific type of tissue.

Amino acids which are not required for building and repairing body tissues are broken down into two parts in the liver:

1. The *nitrogenous part* is converted to *urea* and excreted in the urine. This process is called the *deamination of amino acids* and takes place in the liver.

2. The remaining part is used to provide energy. Like carbohydrates and fats, the residue after deamination may be stored as fat if in excess of immediate energy requirements.

THE AMINO ACID POOL

A pool of amino acids is maintained within the body. This is the source from which the different cells of the body draw the amino acids they need to synthesise their own materials, for example, new cells, secretions such as enzymes and hormones, blood proteins.

Sources of amino acids (Fig. 9:44)

1. *Exogenous.* These are derived from the protein eaten in the diet, digested and absorbed through the mucosa of the small intestine.

2. *Endogenous.* These are obtained from the breakdown

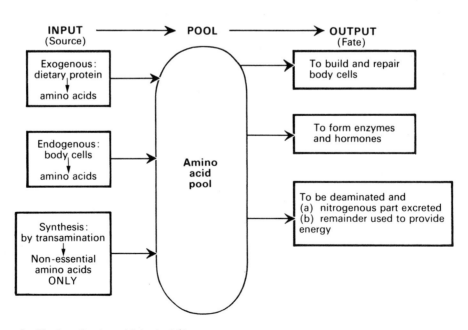

Figure 9:44   Sources and utilisation of amino acids in the body.

of body protein. In an adult about 80 to 100 g of protein are broken down and replaced each day.

Loss of amino acids

1. *Deamination*. Amino acids not needed by the body are deaminated. The nitrogenous part is excreted as urea by the kidneys, and the remainder is used to provide energy and heat.

2. *Excretion*. The faeces contain a considerable amount of protein consisting of desquamated cells from the lining of the alimentary tract.

Endogenous and exogenous amino acids are mixed in the 'pool' and the body is said to be in *nitrogen balance* when the rate of removal from the pool is equal to the additions to it. Unlike carbohydrates the body has no capacity for the storage of amino acids except for this relatively small pool. Figure 9:45 depicts what happens to amino acids in the body.

METABOLISM OF FAT (Fig. 9:46)

Fats which have been digested and absorbed into the *lacteals* are transported via the receptaculum chyli and the thoracic duct to the bloodstream and so, by a circuitous route, to the liver. Fatty acids and glycerol circulating in the blood are used by organs and glands to provide energy and in the synthesis of some of their secretions. In the liver some fatty acids and glycerol are used to provide energy and heat, and some are reorganised and recombined to form a variety of fatty compounds and human fat which is stored in the fat depots of the body. These include subcutaneous fat, fat supporting some organs such as the kidneys, and between the layers of the omentum. Before this fat can be used for metabolic purposes it has to be transported back to the liver, undergo further change called *desaturation*, then it is passed into the circulation in a form which can gain entry to all cells and be oxidised. The end products of fat metabolism are energy, heat, carbon dioxide and water.

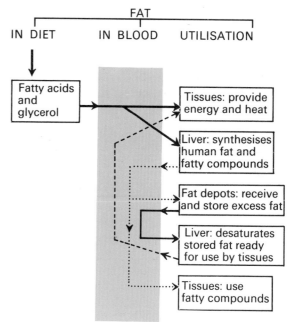

Figure 9:46   Sources, distribution and utilisation of fats in the body.

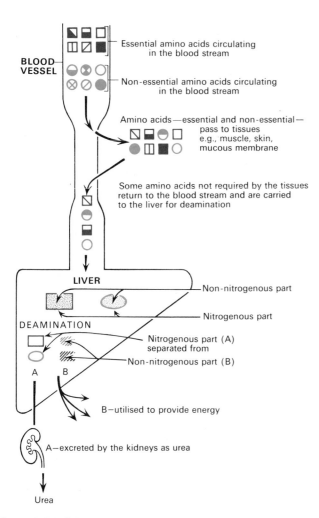

Figure 9:45   Scheme showing the fate of amino acids in the body.

Ketone bodies are the *ketoacids* which are produced during the process of oxidation of fats and are always present in the blood in very small amounts. They are excreted in the urine and in the expired air as *acetone*. When there has been an insufficient intake of carbohydrate foods in the diet, fat is used up in excessive quantities to provide energy and heat, thus a state of ketosis arises due to the increase in the amount of ketone acids in the blood.

Fat is synthesised from carbohydrates and proteins which are taken into the body in excess of its needs.

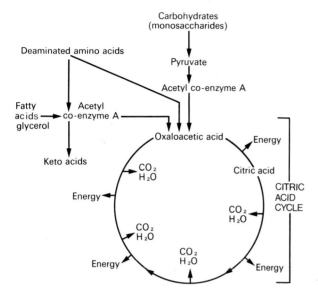

Figure 9:47 Diagram showing the relationship between carbohydrates, deaminated amino acids and fats as energy-releasing substances.

RELATIONSHIPS BETWEEN CARBOHYDRATES, FATTY ACIDS, GLYCEROL AND DEAMINATED AMINO ACIDS AS ENERGY-RELEASING SUBSTANCES

The degradation of carbohydrates, fatty acids, glycerol and deaminated amino acids occurs inside the cells releasing *energy* and forming the waste products *carbon dioxide* and *water*. The catabolism of these molecules occurs in a series of steps, a little energy being released at each stage. Up to a certain point each nutrient passes through a series of separate and distinct stages but thereafter, they all follow a common pathway of degradation. This final common pathway is called the *citric acid cycle* or *Krebs cycle*.

Figure 9:47 provides a diagrammatic representation of the processes involved and only a few of the many steps are shown.

Carbohydrates go through a series of stages to *pyruvate* and *acetyl co-enzyme A*. It is in this form that it joins *oxaloacetic acid* in the citric acid cycle.

Fatty acids pass through a series of oxidative stages to *acetyl coenzyme A* and, under normal circumstances, progress to oxaloacetic acid and the citric acid cycle. If, however, an excessive amount of acetyl coenzyme A is produced some of it develops into *keto acids*.

Deaminated amino acids are of two types: those which go through a series of stages to *oxaloacetic acid* and so to the citric acid cycle and those which follow a different series of changes to become *acetyl coenzyme A* and thereafter take the pathway either to oxaloacetic acid or to keto acids.

The formation of abnormal amounts of keto acids

occurs in starvation and in diabetes mellitus when excessive amounts of fat and amino acids are used to provide energy, that is, when acetyl coenzyme A is produced more rapidly than it can be used in the citric acid cycle. In both these examples there is an insufficiency of carbohydrate inside the cells. In diabetes this is due to a shortage in the supply of the hormone *insulin* which facilitates the transportation of carbohydrate from the extracellular fluid across the cell membrane. Excess keto acids are excreted in the urine and in the expired air as acetone.

Summary of Digestion, Absorption and Utilisation of Carbohydrates, Proteins and Fats

CARBOHYDRATES

Digestion

Organ	Digestive juice	Enzyme and action
Mouth	Saliva	Ptyalin converts cooked starches to *maltose*
Stomach	Gastric juice	Hydrochloric acid stops the action of salivary ptyalin
Small intestine	Pancreatic juice	Amylase converts all starches to *disaccharides* (sugars)
Small intestine	In microvilli	Sucrase ⎫ Convert all sugars to Maltase ⎬ *monosaccharides*, mainly Lactase ⎭ *glucose*

Absorption

Glucose is absorbed into the capillaries of the villi and transported in the portal circulation to the liver.

Utilisation

1. For liver metabolism.
2. To maintain a constant blood glucose level so that all body tissues have a constant supply.
3. Some of the excess is converted to glycogen in the presence of insulin and stored in the liver and in the muscles.
4. Any remaining glucose is converted into fat and stored in the fat depots.
5. Glucose is used in the body to provide energy and heat. Oxygen is necessary to obtain all the energy it contains and the waste products left are carbon dioxide and water.

PROTEIN

Digestion

Organ	Digestive juice	Enzyme and action
Mouth	Saliva	No action
Stomach	Gastric juice	Hydrochloric acid converts pepsinogen to *pepsin*
		Pepsin converts all proteins to *peptone*
		Rennin converts soluble caseinogen to insoluble *casein* in children
Small intestine	Pancreatic juice	Enteropeptidase of intestinal juice converts trypsinogen and chymo-trypsinogen to *trypsin* and *chymotrypsin* which convert peptones to *polypeptides*
Small intestine	In microvilli	Peptidases convert peptides and polypeptides to *amino acid*

Absorption

Amino acids are absorbed into the capillaries of the villi and transported in the portal circulation to the liver.

Utilisation

1. In the liver to form albumin, globulin, prothrombin and fibrinogen.

2. In various combinations by cells of the body for cell replacement, cell repair, the production of secretions, e.g. hormones and enzymes.

3. To maintain the amino acid pool.

4. Amino acids not required are deaminated in the liver. The nitrogenous part is converted into urea and excreted in the urine. The remaining part is used to provide energy and heat or deposited as fat in the fat depots.

FATS

Digestion

Organ	Digestive juice	Enzyme and action
Mouth	Saliva	No action
Stomach	Gastric juice	No action
Small intestine	Bile	Bile salts emulsify fats
Small intestine	Pancreatic juice	Lipase converts fats to *fatty acids* and *glycerol*
Small intestine	In microvilli	Lipase completes the digestion of fats to *fatty acids* and *glycerol*

Absorption

Fatty acids and glycerol are absorbed into the lacteals of the villi and are transported via the receptaculum chyli and the thoracic duct to the left subclavian vein. In this way they are transported by the circulating blood to the liver where fatty acids and glycerol are reorganised and recombined.

Utilisation

1. Utilised in the presence of oxygen to provide energy and heat, the waste products carbon dioxide and water being produced.

2. Stored in the fat depots.

3. When depot fat is required for oxidation it must first be desaturated by the liver.

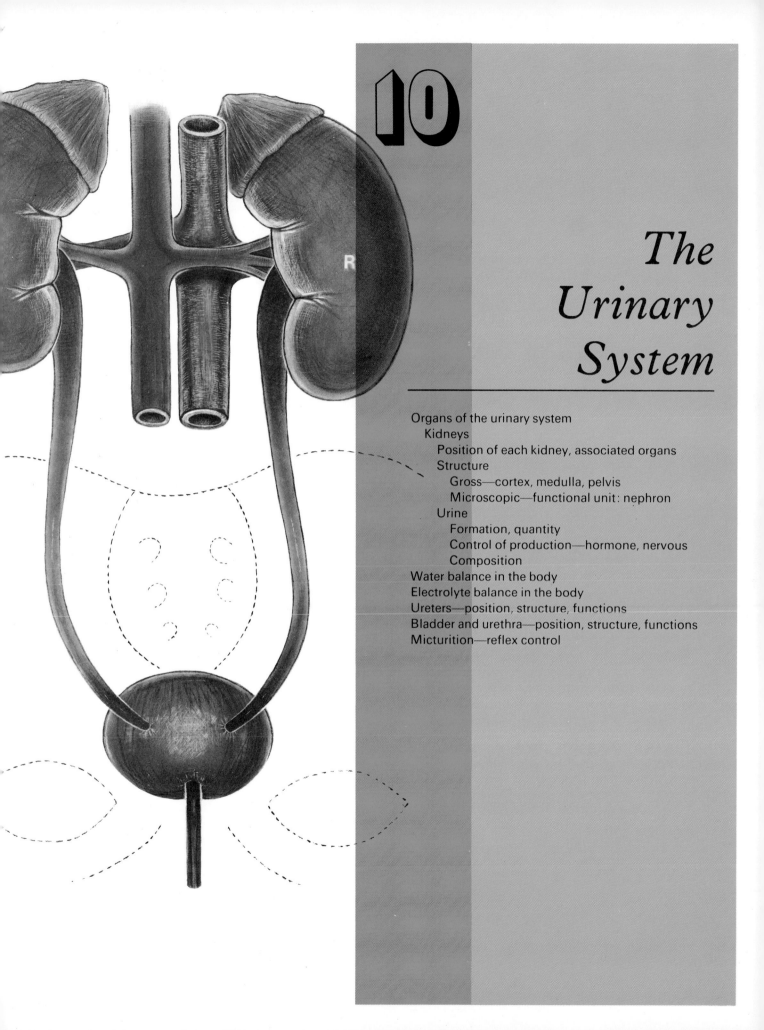

10

The Urinary System

10. The Urinary System

The urinary system is one of the excretory systems of the body. It consists of the following structures:

2 *kidneys* which secrete urine

2 *ureters* which convey the urine from the kidneys to the urinary bladder

1 *urinary bladder* where urine collects and is temporarily stored

1 *urethra* through which the urine is discharged from the urinary bladder to the exterior

Figure 10:3 shows an overview of the urinary system.

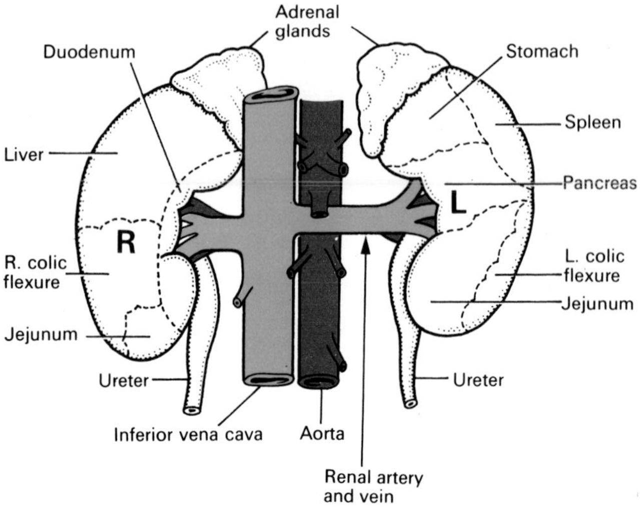

The Kidneys

The kidneys lie on the posterior wall of the abdomen, one on each side of the vertebral column, behind the peritoneum.

They extend from the level of the twelfth thoracic vertebra to the third lumbar vertebra. The right kidney is usually slightly lower than the left, probably due to the considerable amount of space occupied by the liver. The left kidney is slightly longer and narrower than the right and lies nearer the median plane.

The kidneys are described as bean-shaped organs and are approximately 11 cm (4¼ inches) long, 6 cm (2¼ inches) wide and 3 cm (1¼ inches) thick. They are embedded in, and held in position by, a mass of adipose tissue termed the *renal fat*. A sheath of fibroelastic tissue known as the *renal fascia* encloses the kidney and the renal fat.

ORGANS IN ASSOCIATION WITH THE KIDNEYS (Figs. 10:1 and 10:2)

As the kidneys lie on either side of the vertebral column each is associated with a different group of structures.

The right kidney

Superiorly— the right *adrenal gland* lies on the upper pole of the kidney.

Anteriorly— the *right lobe of the liver*, the *duodenum* and the *right colic flexure* of the large intestine.

Posteriorly— the *diaphragm*, and the *muscles of the posterior abdominal wall*.

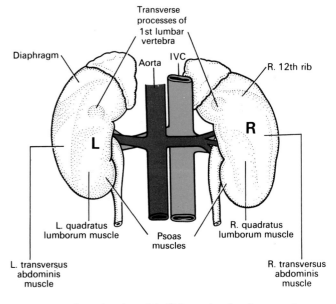

Figure 10:1 Anterior view of the kidneys showing the areas of contact with associated structures.

Figure 10:2 Posterior view of the kidneys showing the areas of contact with associated structures.

150

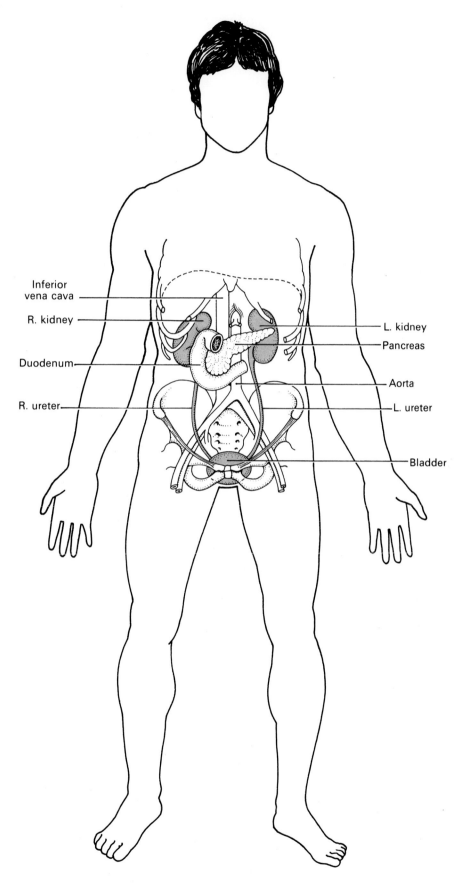

Inferior
vena cava

R. kidney

Duodenum

R. ureter

L. kidney

Pancreas

Aorta

L. ureter

Bladder

Figure 10:3   The parts of the urinary system and the structures associated with them viewed from the front.

The left kidney

Superiorly— the left *adrenal gland* lies on its upper pole.

Anteriorly— the *spleen*, the *stomach*, the *pancreas*, the *jejunum* and the *left colic flexure* of the large intestine.

Posteriorly— the *diaphragm* and the *muscles of the posterior abdominal wall*.

THE GROSS STRUCTURE OF THE KIDNEY

There are three areas of tissue which can be distinguished when a longitudinal section of the kidney is viewed with the naked eye (Fig. 10:4):

1. *The fibrous capsule* surrounds the kidney.
2. *The cortex* is a reddish-brown layer of tissue immediately below the capsule and between the pyramids.
3. *The medulla* is the innermost layer which consists of pale conical-shaped striations called the *renal pyramids*.

The hilus is the name given to the concave medical border of the kidney where the renal blood and lymph vessels and nerves enter and leave.

The renal pelvis is a funnel-shaped structure which acts as a receptacle for the urine formed by the kidney. The pelvis has a number of branches called *calyces* at its upper end, each of which surrounds the apex of a renal pyramid. Urine which is formed in the kidney passes through a *papilla* at the apex of a pyramid into a lesser calyx, then into a greater calyx before passing through the pelvis into the ureter.

MICROSCOPIC STRUCTURE OF THE KIDNEYS

The kidney substance is composed of a large number of microscopic structures known as *nephrons* and *collecting tubules* which together are called *uriniferous tubules*. The nephron is described as the *functional unit* of the kidney and it is estimated that there are approximately one million in each kidney. The uriniferous tubules are supported by a small amount of connective tissue which contains blood vessels, nerves and lymph vessels.

The nephron

The nephron consists of a tubule which is closed at one end. The other end opens into a collecting tubule. The closed or blind end is indented to form a cup-shaped structure called the *glomerular capsule* which almost completely encloses a network of arterial capillaries called the *glomerulus*. Continuing from the glomerular capsule the remainder of the nephron is described in three parts: the *proximal convoluted tubule*, the *loop of Henle* and the *distal convoluted tubule* which leads into a *collecting tubule* (Fig. 10:5).

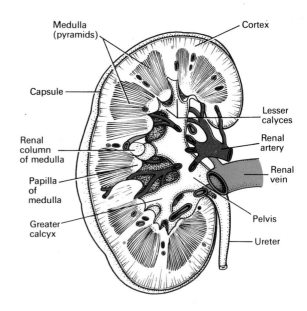

Figure 10:4 A longitudinal section of the right kidney.

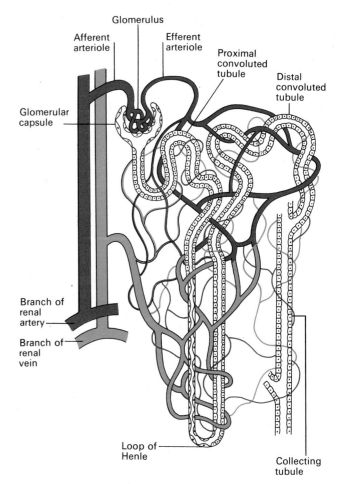

Figure 10:5 Diagram of a nephron including the arrangement of the blood vessels.

The renal artery, after entering the kidney at the hilus, divides into small arteries and arterioles. In the cortex one arteriole, the *afferent arteriole*, enters each glomerular capsule then subdivides to form the glomerulus. The blood vessel leading away from the glomerulus is called the *efferent arteriole*; it breaks up into a *second* capillary network to supply oxygen and nutritional materials to the remainder of the nephron. Venous blood drained away from this capillary bed eventually leaves the kidney in the renal vein which empties into the inferior vena cava (Fig. 10:6). The blood pressure in the glomerulus is higher than in other capillaries because the calibre of the afferent arteriole is greater than that of the efferent arteriole.

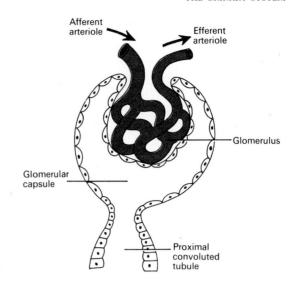

Figure 10:7   Diagram of the glomerulus and glomerular capsule.

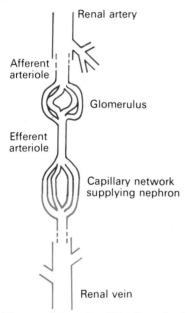

Figure 10:6   Diagram of the series of blood vessels of the kidney.

The walls of the glomerulus and the glomerular capsule consist of a single layer of *flattened epithelial cells* (Fig. 10:7). The remainder of the nephron and the collecting tubule are formed by a single layer of highly specialised cells.

The nerve supply consists of sympathetic and parasympathetic nerves.

FUNCTIONS OF THE KIDNEY

The function of the kidneys is to form urine which passes through the ureters to the bladder for excretion. In doing this, vital functions in relation to the maintenance of fluid and electrolyte balance and the disposal of waste material from the body are carried out.

FORMATION OF URINE

Urine is formed by the nephrons of the kidney and the process occurs in three phases:

   Simple filtration
   Selective reabsorption
   Secretion

Simple filtration (Fig. 10:8)

Filtration takes place through the *semipermeable* walls of the glomerulus and the glomerular capsule. The blood cells are too large to pass through this membrane so they leave the glomerulus in the efferent arteriole. The size of the molecules of chemical substances in the blood determines whether they remain in the capillaries or pass through the filter into the glomerular capsule.

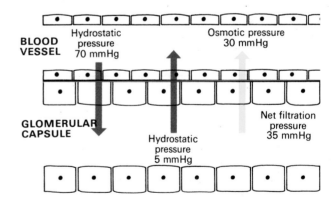

Figure 10:8   Diagram of filtration in the nephron.

The main factor which assists filtration is the difference between the blood pressure in the glomerulus and the pressure of the filtrate in the glomerular capsule. Because the calibre of the efferent arteriole is less than that of the afferent arteriole, a capillary hydrostatic pressure of about 70 mmHg builds up in the glomerulus. This pressure is opposed by the osmotic pressure of the blood, about 30 mmHg, and by filtrate hydrostatic pressure of about 5 mmHg in the glomerular capsule. The net filtration pressure is, therefore:

$$70 - (30 + 5) = 35 \text{ mmHg}$$

Blood constituents which pass into the glomerular capsule	Blood constituents which remain in the glomerulus
water	leucocytes
mineral salts	erythrocytes
amino acids	platelets
ketoacids	blood proteins
glucose	globulin
hormones	albumin
urea	prothrombin
uric acid	fibrinogen
toxins	
drugs	

Selective reabsorption (Fig. 10:9)

Selective reabsorption is the process by which the composition and volume of the glomerular filtrate is altered during its passage through the convoluted tubules, the loop of Henle and the collecting tubule. The general purpose of this process is to reabsorb those constituents of the filtrate which are essential to the body, maintain the fluid and electrolyte balance of the body and the alkalinity of the blood.

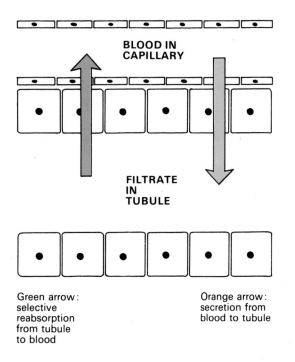

BLOOD IN CAPILLARY

FILTRATE IN TUBULE

Green arrow: selective reabsorption from tubule to blood

Orange arrow: secretion from blood to tubule

Figure 10:9   Diagram of selective reabsorption and secretion in the nephron.

Some constituents of the glomerular filtrate do not normally appear in the urine, for example, glucose, amino acids, vitamins. Such substances are described as *high threshold substances*. However, the capacity of the tubules to reabsorb these substances is limited, for example, when the blood glucose level rises to about 10 mmol/l (160 mg %), all of it cannot be reabsorbed and some appears in the urine.

Some substances are absorbed in varying amounts according to conditions within the body. This is the type of reabsorption associated with the maintenance of the alkalinity of the blood and the fluid and electrolyte balance in the body. This group of substances includes water and mineral salts.

Waste products, such as urea and uric acid, are absorbed only to a slight extent. These are called *low threshold substances*

Secretion (Fig. 10:9)

Filtration occurs as the blood flows through the glomerulus. Non-threshold substances and foreign materials, such as drugs, may not be cleared from the blood by filtration because the blood does not remain for a sufficient length of time in the glomerulus. Such substances are cleared by *secretion into the convoluted tubules* and passed from the body in the urine.

COMPOSITION OF URINE

water	96 per cent
urea	2 per cent
uric acid	
creatinine	
ammonia	
sodium	
potassium	2 per cent
chlorides	
phosphates	
sulphates	
oxalates	

Urine is amber in colour due to the presence of bile pigment, has a specific gravity of between 1·020 and 1·030, and is acid in reaction. A healthy adult passes 1000 to 1500 ml (1¾ to 2½ pints) per day. The amount of urine secreted and the specific gravity vary according to the amount of the fluid intake. During sleep and muscular exercise the amount of urine produced is decreased.

WATER BALANCE AND URINE OUTPUT

Water is *taken into* the body through the alimentary tract and a small amount is formed by the metabolic processes. The *excretion* of water occurs in saturated expired air, as a constituent of the faeces, through the skin as sweat and in the urine. The amount lost in expired air and in the faeces is fairly constant and the amount of sweat produced is associated with the maintenance of normal body temperature (see p. 164).

The balance between fluid intake and output is controlled by the kidneys. The minimum urinary output, consistent with the essential removal of waste material, is about 500 ml per day. The amount produced in excess of this is controlled mainly by the *antidiuretic hormone* (ADH) which is released into the blood by the *posterior lobe of the pituitary gland*. There is a close link between the posterior pituitary and the *hypothalamus* in the brain.

It has been shown that there are cells in the hypothalamus which are sensitive to changes in the osmotic pressure of the blood. These cells are called *osmoreceptors*. The link between the hypothalamus and the posterior pituitary is provided by circulating blood and by nerve fibres.

If the osmotic pressure of the blood is raised the osmoreceptors in the hypothalamus are stimulated and there is an increase in the output of ADH by the pituitary gland. Conversely, if the osmotic pressure of the blood is reduced the amount of ADH released is reduced.
As a result, the amount of water reabsorbed and returned to the blood is increased. This dilutes the blood circulating to the hypothalamus, and consequently, reduces the amount of ADH released. This cyclic effect maintains the concentration of the blood within normal limits (see Fig. 10:10).

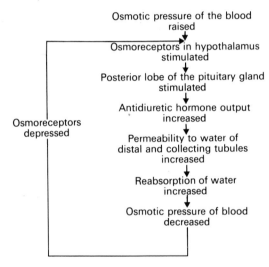

Figure 10:10   Feed-back mechanism for the control of antidiuretic hormone (ADH).

Although the antidiuretic hormone is the most important means by which the water balance of the body is maintained it is not the only one. If there is an excessive amount of any dissolved substance in the blood which *must be excreted through the kidneys* extra water is excreted with it. If excessive amounts of electrolytes have to be excreted from the body the amount of water required for this purpose is increased. This may lead to dehydration in spite of increased production of ADH but it is usually accompanied by acute thirst and increased water intake.

ELECTROLYTE BALANCE

Changes in the *concentration of electrolytes* in the body may be due to changes in the amounts of water or to the amounts of electrolytes. There are several methods of maintaining the balance between water and electrolyte concentration.

SODIUM AND POTASSIUM CONCENTRATION

Sodium is the most common of the cations (positively charged ions) to exist in the extracellular fluid.

Intake

Sodium is a constituent of almost all foods that are taken in the diet and it is often added to food during cooking. This means that the intake is usually in excess of the body's needs.

Output

There are two main routes by which sodium is excreted from the body.

1. *In urine*. Sodium is a normal constituent of urine and the amount excreted is controlled by the hormone *aldosterone* produced by the cortex of the *adrenal gland* (suprarenal gland). Aldosterone influences the absorption of sodium from the filtrate in the nephron. When the amount of sodium in the blood is decreased aldosterone production is increased and an increased amount of sodium is reabsorbed. This system influences the level of potassium in the blood and, indirectly, the intracellular potassium level. When the amount of sodium reabsorbed is increased the amount of potassium excreted is increased and vice versa.

The secretion of aldosterone by the adrenal cortex is stimulated by the *adrenocorticotrophic hormone* (ACTH) from the *anterior lobe of the pituitary gland*; by a fall in the concentration of sodium in the blood and by *renin* produced in the kidneys.

When the blood pressure in the afferent arteriole falls renin is produced by the cells surrounding the arteriole. Renin acts on globulin in the blood to produce *angiotensin* which stimulates the adrenal cortex to produce aldosterone. The resultant reabsorption of sodium, accompanied by water, increases the blood volume which raises the blood pressure and reduces the renin output. (See Fig. 10:11.) Angiotensin has a powerful vasoconstrictor effect on the blood vessels generally.

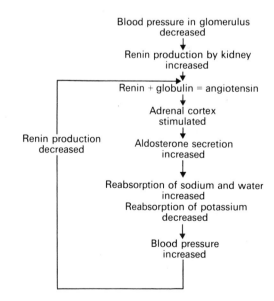

Figure 10:11   Summary of the relationship between blood pressure and selective reabsorption by the nephron.

Sodium and potassium are in high concentration in digestive juices, sodium in gastric juice and potassium in pancreatic and intestinal juice. Normally these ions are reabsorbed by the colon but in acute and prolonged diarrhoea they may be excreted in large quantities with resultant electrolyte imbalance.

2. *In sweat*. The amount of sodium excreted in sweat is insignificant except when sweating is excessive. This may occur when there is a high environmental temperature or during sustained physical exercise. Normally the renal mechanism described above maintains the cation concentration within physiological limits.

The Ureters (Fig. 10:12)

The ureters are the two tubes which convey the urine from the kidneys to the urinary bladder. Each tube measures approximately 25 to 30 cm (10 to 12 inches) in length, and its diameter is approximately 3 mm.

The ureter is continuous with the funnel-shaped *pelvis of the kidney*. It passes downwards through the abdominal cavity, behind the peritoneum and in front of the psoas

muscle into the pelvic cavity, and opens into the posterior aspect of the base of the urinary bladder. The ureter passes obliquely through the bladder wall (Fig. 10:13). Because of this arrangement the ureters are compressed and the opening occluded when the pressure rises in the bladder. This occurs when it fills with urine and when its muscular walls contract during micturition.

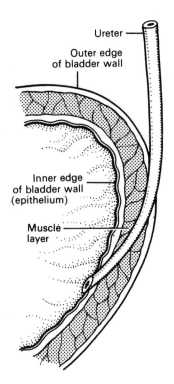

Figure 10:13 Diagram of the position of the ureter in relation to the bladder wall.

STRUCTURE

The ureters consist of three layers of tissue:

An outer coat of *fibrous tissue* which is continuous with the fibrous capsule of the kidney

A middle *muscular layer* consisting of interlacing muscle fibres forming a syncytium which spirals round the ureter in clockwise and anticlockwise directions

An inner lining of *mucous membrane* consisting of *transitional epithelium*

FUNCTION

The ureters propel the urine from the kidneys into the bladder by peristaltic contraction of their muscular walls. This is an intrinsic function not under nervous control. Peristalsis is stimulated by the presence of urine in the ureter and the stretch effect stimulates the syncytium of muscle.

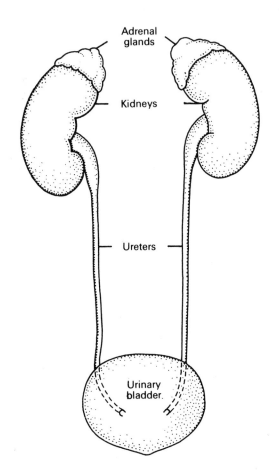

Figure 10:12 The ureters and their relationships to the kidneys and bladder.

The Urinary Bladder

The urinary bladder is described as a sac which acts as a reservoir for urine. It lies in the pelvic cavity, and its size and position vary depending on the amount of urine it contains. When distended the bladder rises into the abdominal cavity.

ORGANS IN ASSOCIATION WITH THE BLADDER

The organs in association with the bladder differ in the male and the female.

In the female (Fig. 10:14A)

Anteriorly—	the symphysis pubis
Posteriorly—	the uterus
Superiorly—	the small intestine
Inferiorly—	the urethra and the muscles forming the pelvic floor

In the male (Fig. 10:14B)

Anteriorly—	the symphysis pubis
Posteriorly—	the rectum and seminal vesicles
Superiorly—	the small intestine
Inferiorly—	the urethra and prostate gland

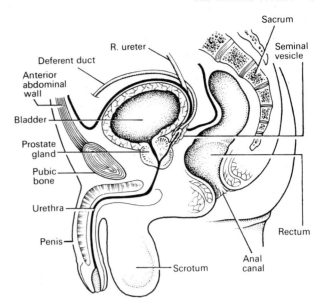

Figure 10:14B The pelvic organs associated with the bladder and the urethra—in the male.

STRUCTURE (Fig. 10:15)

The bladder is roughly pear-shaped, but becomes more oval in shape as it fills with urine. It is described as having anterior, superior and posterior surfaces. The posterior surface is known as the *base*. The bladder opens into the urethra at its lowest point, *the neck*.

The bladder is composed of four layers of tissue.

1. *The peritoneum* covers only the superior surface of the bladder from which it is reflected upwards to become the parietal peritoneum lining the anterior abdominal wall. Posteriorly it is reflected on to the uterus in the female and the rectum in the male (see Fig. 9:3).

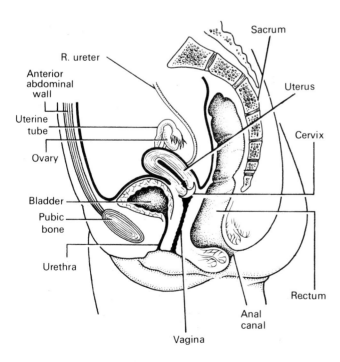

Figure 10:14A The pelvic organs associated with the bladder and the urethra—in the female.

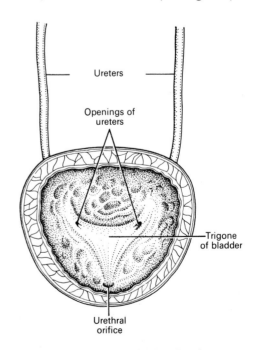

Figure 10:15 Section of the bladder showing the trigone.

2. *The muscle layer* consists of a mass of interlacing smooth muscle fibres.

3. *The submucous* coat joins the inner lining and the muscular layer and is made up of areolar tissue containing blood vessels, lymph vessels and sympathetic and parasympathetic nerves.

4. *The mucous membrane* forms the inner lining and is composed of transitional epithelium. When the bladder is empty or contracted the inner lining is arranged in folds and these gradually disappear as the bladder fills.

The three orifices in the bladder wall form a triangle called the *trigone*. The upper two orifices on the posterior wall are the openings of the two ureters. The inferior orifice is the point of origin of the urethra. Where the urethra commences there is a thickening of the smooth muscle layer which acts as a sphincter and controls the passage of urine from the bladder into the urethra.

The Urethra

The urethra is a canal which extends from the neck of the bladder to the exterior and its length differs in the male and in the female. The male urethra is associated with the urinary and the reproductive systems, and is described in Chapter 15.

The female urethra is approximately 4 cm (1½ inches) in length. It runs downwards and forwards behind the symphysis pubis and opens at the *external urethral orifice* just in front of the vagina. The external urethral orifice is guarded by a sphincter muscle which is under the control of the will. Except during the passage of urine the walls of the urethra are in close apposition.

The urethra is composed of three layers of tissue.

1. *A muscular coat* which is continuous with that of the bladder. At its origin there is an *internal sphincter*, composed mainly of elastic tissue and some smooth muscle fibres controlled by the autonomic part of the nervous system. Near the external urethral orifice the smooth muscle is replaced by striated muscle which forms the *external sphincter* and is under voluntary control.

2. *A thin spongy coat* containing large numbers of blood vessels.

3. *A lining of mucous membrane* which is continuous with that of the bladder in the upper part of the urethra. The lower part consists of stratified squamous epithelium and is continuous externally with the skin of the vulva.

Functions of the Bladder and Micturition

The urinary bladder acts as a reservoir for urine. As the urine gradually collects there is little change in pressure for some time. The bladder as a whole adapts itself to the increased volume. When approximately 200 to 300 ml

(7 to 10 fl oz) of urine have accumulated the pressure will have risen sufficiently to stimulate the autonomic nerve endings within the bladder wall. In the infant micturition occurs by *reflex action* (see p. 183) which is in no way controlled (Fig. 10:16). However, after the nervous system has fully developed, nerve impulses are conveyed to consciousness, and the brain can inhibit the reflex for a limited period of time or until it is convenient to micturate (Fig. 10:17).

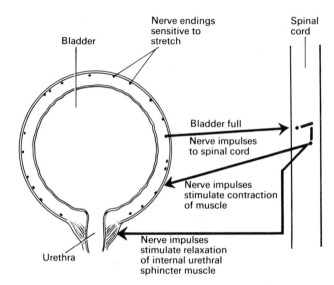

Figure 10:16   Diagram of simple reflex control of micturition in the infant.

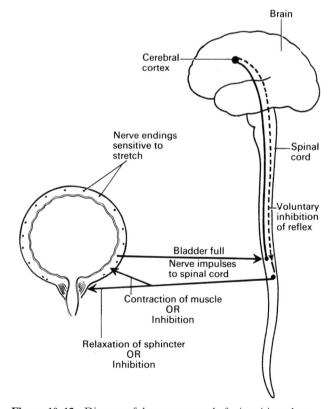

Figure 10:17   Diagram of the nerve control of micturition when inhibition of the reflex action is possible.

Micturition occurs when the muscular wall of the bladder contracts, the internal sphincter dilates and the external sphincter relaxes. It can be assisted by lowering the diaphragm and contracting the abdominal muscles, thus increasing the pressure within the pelvic cavity. Inhibition of reflex contraction of the bladder and relaxation of the internal sphincter is possible for only a limited period of time. Over distension of the bladder is extremely painful, and when a painful degree of distension is reached there is a tendency for involuntary relaxation of the external sphincter to occur and a small amount of urine to escape, provided there is no mechanical obstruction present.

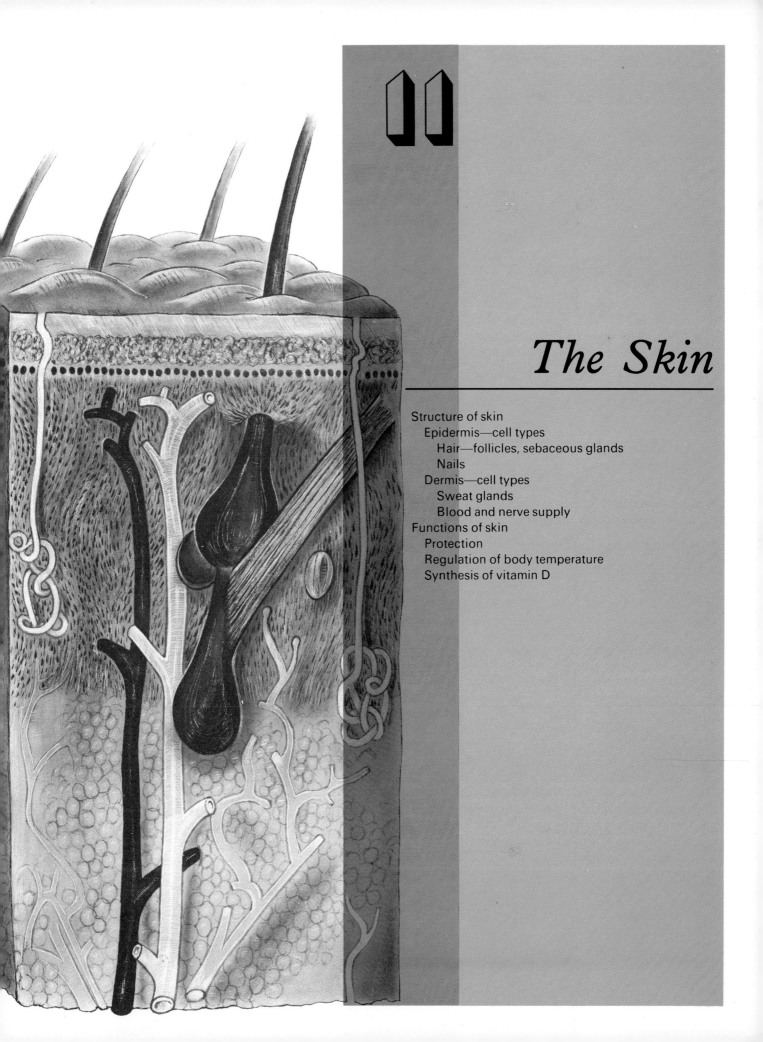

The Skin

11. The Skin

The skin completely covers the body, is continuous with the membranes lining the orifices of the body and is one of the most active organs.

It contains the nerve endings of many of the sensory nerves.

It is one of the main excretory organs.

It plays an important part in the regulation of the body temperature.

It protects the deeper organs from injury and the invasion of micro-organisms.

Structure of the Skin

The skin is composed of two main parts:

The epidermis

The dermis or corium

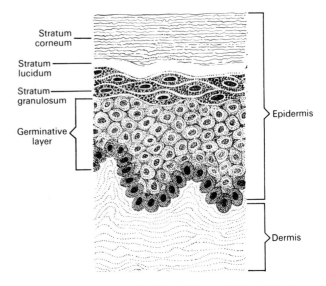

Figure 11:1 The skin showing the main layers of the epidermis.

THE EPIDERMIS (Fig. 11:1)

The epidermis is the most superficial part of the skin and is composed of *keratinised stratified squamous epithelium* which varies in thickness in different parts of the body. It is thickest on the palms of the hands and soles of the feet. There are no blood vessels or nerve endings in the epidermis, but its deeper layers are bathed in interstitial fluid which is drained away as lymph.

There are several layers of cells in the epidermis which extend from the superficial *stratum corneum* (horny layer) to the deepest *germinative layer*. The cells on the surface are flat, thin, non-nucleated, dead cells in which the protoplasm has been replaced by *keratin*.

Cells on the surface are constantly being rubbed off and they are replaced by cells which originated in the germinative layer and have undergone gradual change as they progressed towards the surface.

The maintenance of healthy epidermis depends upon three processes being synchronised:

1. Desquamation of the keratinised cells from the surface
2. Effective keratinisation of the cells approaching the surface
3. Continual cell division in the deeper layers with cells being pushed to the surface

Passing through the epidermis are the hairs, secretion from the sebaceous glands and the ducts of the sweat glands.

The surface of the epidermis is ridged by projections of cells in the dermis called the *papillae*. The pattern of ridges formed in this way is different in every individual and the impression made by them is called the 'finger-print'.

THE DERMIS (Fig. 11:2)

The dermis is tough and elastic. It is composed of *white fibrous tissue* interlaced with *yellow elastic fibres*. In the deeper layer forming the subcutaneous tissue there is areolar and adipose tissue.

The structures in the dermis are:

Blood vessels

Lymph vessels

Sensory nerve endings

Sweat glands and their ducts

Hair roots, hair follicles and hairs

Sebaceous glands

The arrectores pilorum—involuntary muscles attached to the hair follicles

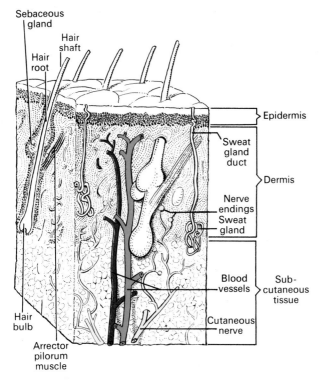

Figure 11:2   The skin showing the main structures of the dermis.

1. *Blood vessels.* Arterioles form a fine network with capillary branches supplying sweat glands, sebaceous glands, hair follicles and the deeper layers of the epidermis.

2. *Lymph vessels.* These form a network throughout the dermis and the deeper layers of the epidermis.

3. *Sensory nerve endings.* Nerve endings which are sensitive to *touch*, *change in temperature* and *pressure* are widely distributed in the dermis. There are no nerve endings in the epidermis.

The skin is an important sensory organ. It is one of the organs through which the individual is aware of his environment. Nerve impulses which originate in these nerve endings are conveyed to the spinal cord by sensory, or cutaneous, nerves. From there they are conveyed to the sensory area of the cerebrum where the sensations of touch, temperature and pain are perceived. (See Ch. 12)

4. *Sweat glands.* These are found widely distributed throughout the skin and are most numerous in the palms of the hands, soles of the feet, axillae and groins.

The glands are composed of *epithelial cells* and the *body* of the gland has a coiled appearance. The *duct of the gland* traverses both the dermis and epidermis to open on to the surface of the skin at a minute depression known as the *pore.* Each gland is supplied by a network of blood capillaries.

The most important function of sweat is in relation to the maintenance of the normal body temperature. It has a lesser function as a route for the excretion of waste

materials. Electrolytes may be lost in abnormally large amounts if there is excessive sweating.

The composition of sweat is as follows:

Water	99·4 per cent
Potassium	
Sodium	
Chloride	0·2 per cent
Sulphate	
Waste substances	0·4 per cent

5. *Hair follicles.* These consist of a downward growth of epidermal cells into the dermis or even the subcutaneous tissue. At the base of the follicle there is a cluster of cells, called *the bulb*, from which the hair grows. The hair is formed by the multiplication of cells of the bulb and, as they are pushed upwards and away from their source of nutrition, the cells die and are converted to keratin.

The hair consists of *the shaft* which protrudes from the surface of the skin and the remainder is called *the root* (Fig. 11:3).

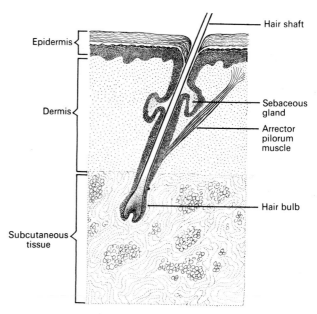

Figure 11:3   A hair in the skin.

The colour of the hair depends on the amount of melanin present. White hair is the result of the replacement of melanin by tiny air bubbles.

6. *The sebaceous glands* (Fig. 11:3). These consist of secretory epithelial cells derived from the same tissue as the hair follicles. They pour their secretion, *sebum*, into the hair follicles and therefore they are present in the skin of all parts of the body except the palms of the hands and the soles of the feet. They are most numerous in the skin of the scalp, the face, the axillae and the groins.

Sebum keeps the hair soft and pliable and gives it a shiny appearance. On the skin it provides some waterproofing, acts as a bactericidal agent preventing the successful invasion of micro-organisms and it prevents drying especially on exposure to heat and sunshine.

7. *The arrectores pilorum* (Fig. 11:3). These are little bundles of involuntary muscle fibres connected with the hair follicles. When these muscles contract they make the hair stand erect. This also causes the skin around the hair to become elevated giving the appearance of 'goose flesh'. The muscles are stimulated by sympathetic nerve fibres in fear and in response to cold. Although each muscle is very small the contraction of a large number generates an appreciable amount of heat.

PIGMENTATION OF THE SKIN

When no pigment is present the skin looks pinkish white in colour due to the blood in the capillaries of the dermis. In most individuals this colour is modified by varying amounts and proportions of several pigments. The three most important are:

1. Melanin—a brown pigment found in the germinative layer of the epidermis.
2. Melanoid—a brownish pigment found distributed widely in the cells of the epidermis.
3. Carotene—a yellow pigment found in the stratum corneum or horny layer of the epidermis.

THE NAILS (Fig. 11:4)

The nails in human beings are equivalent to the claws, horns and hoofs of animals. They are derived from the same cells as epidermis and hair and consist of a hard, horny type of keratinised dead cell. They protect the tips of the fingers and toes.

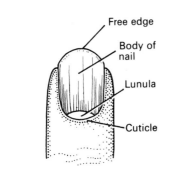

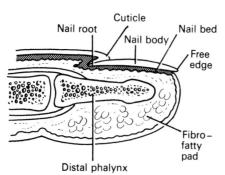

Figure 11:4   The nail and related structures.

The root of the nail is embedded in the skin, is covered by the *cuticle* and forms the hemispherical pale area called the *lunula*.

The body of the nail is the exposed part and grown out from the germinative zone of the epidermis called the *nail bed*.

The finger nails grow more quickly than the toe nails. Growth is quicker when the environmental temperature is higher and vice versa.

Functions of the Skin

PROTECTION

The skin is one of the main protective organs of the body. It protects the deeper and more delicate organs, and acts as the main barrier against the invasion of micro-organisms and other harmful agents.

Due to the presence of the sensory nerve endings the body reacts by reflex action to unpleasant or painful stimuli, and thus is protected from further injury (see p. 183).

FORMATION OF VITAMIN D$_3$

There is a fatty substance called *7-dehydrocholesterol* in the skin and ultraviolet light from the sun converts it to vitamin D. The vitamin D circulates in the blood and is used, with calcium and phosphorus, in the formation and maintenance of bone. Excess of immediate requirements is stored in the liver.

REGULATION OF BODY TEMPERATURE

Human beings are warm-blooded animals and the body temperature is maintained at an average of 36·8°C (98·4°F). In health, variations are usually limited to between 0·5 and 0·75C°, although it may be found that the temperature in the evening is a little higher than in the morning. This is the optimum temperature for the many complex chemical processes to occur. If the temperature is raised the metabolic rate is increased and if it is lowered the rate of metabolism is reduced.

To ensure this constant temperature a fine balance is maintained between heat produced in the body and heat lost to the environment.

Heat production

Some of the energy released in the cells when carbohydrates, fats and deaminated amino acids are metabolised is in the form of heat. Because of this the organs which are the most active, chemically and physically, produce the most heat.

The principal organs involved are:

1. *The muscles.* Contraction of voluntary muscles produces a large amount of heat. The more strenuous the muscular exercise the greater the heat produced. Shiver-

ing involves muscle contraction and produces heat when there is the risk of the body temperature falling below normal.

2. *The liver.* It will be remembered that the liver performs many chemical activities each involving the production of heat.

3. *The digestive organs.* Heat is produced by the contraction of the muscle of the alimentary tract and by the chemical reactions involved in digestion.

Heat loss
Heat is lost from the body in several ways:

97 per cent by the skin
2 per cent in expired air
1 per cent in urine and faeces

Only the heat lost by the skin can be regulated to maintain a constant body temperature. The heat lost by the other routes is obligatory.

Heat loss from the body is affected by the difference between body and environmental temperature, the amount of the body surface exposed to the air and the type of clothes worn. Air is a poor conductor of heat and when layers of air are trapped in the clothing and between the skin and the clothing they act as effective insulators against excessive heat loss. For this reason several layers of light weight clothes provide more effective insulation against a low environmental temperature than one heavy garment.

Nervous control
The centre controlling temperature is situated in the *cerebrum* and involves a group of nerve cells in the *hypothalamus* called the *heat regulating centre.* There is also a group of nerve cells in the *medulla oblongata* known as the *vasomotor centre* which controls the calibre of the blood vessels, especially the small arteries and the arterioles, and they control the amount of blood which circulates in the capillaries in the dermis.

The heat regulating centre and vasomotor centre are thought to be extremely sensitive to the temperature of the blood and any significant *change* stimulates them to activity. From these centres sympathetic nerves convey impulses to the *sweat glands, arterioles* and the *arrector muscles* of the hairs in the skin.

Activity of the sweat glands
If the temperature of the body is increased by 0·25 to 0·5C° the sweat glands are stimulated to secrete sweat which is conveyed to the surface of the body by ducts. This moisture *evaporates* into the atmospheric air *cooling the body* because the heat which evaporates the water is taken from the skin. When sweat droplets can be seen on the skin the rate of production of sweat exceeds the rate of evaporation. This is most likely to happen when the environmental air is humid and the temperature high.

Loss of heat from the body by *evaporation* is described as occurring by:

Insensible water loss
Sweating

In *insensible water loss* heat is being continuously lost by evaporation, even although the sweat glands are not active. Water diffuses upwards from the deeper layers of the skin to the surface of the body and evaporates into the air.

In *sweating* the sweat glands are active and secrete sweat on to the surface of the body which evaporates and, in the process, cools the skin.

Effects of vasodilation
The amount of heat lost from the skin depends to a great extent on the amount of blood in the vessels in the dermis. As the amount of heat produced in the body increases the arterioles become dilated and more blood pours into the capillary network in the skin. In addition to increasing the amount of sweat produced the *temperature of the skin is raised.* When this happens there is an increase in the amount of heat lost by:

Radiation
Conduction
Convection

In *radiation* the exposed parts of the body radiate heat away from the body.

In *conduction* the clothes in contact with the skin conduct heat away from the body.

In *convection* the air passing over the exposed parts of the body is heated and rises, cool air replaces it and convection currents are set up. Heat is also lost from the clothes by convection.

If the external environmental temperature is low or if heat production is decreased, the blood vessels, under the influence of the sympathetic nerves, constrict thus decreasing the blood supply to the skin and so preventing heat loss.

In man, therefore, this fine balance of heat production and heat loss must continuously be maintained to ensure no drastic change in body temperature.

12

The Nervous System

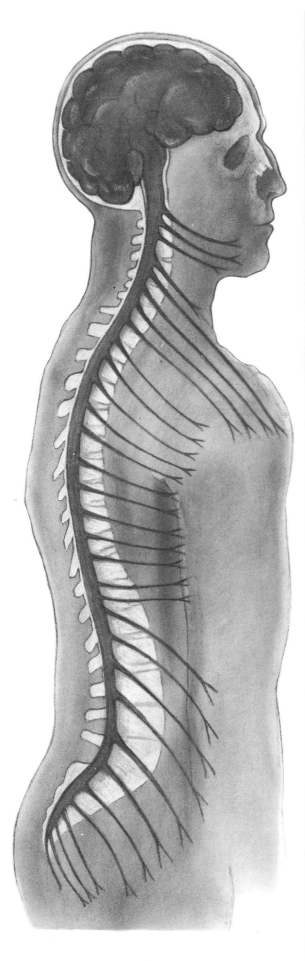

12. The Nervous System

In a systematic study of anatomy and physiology the systems of the body are described separately, but it must be appreciated that they are dependent upon each other. In previous chapters the nerve supply to the organs has been mentioned but not described in detail.

For descriptive purposes the parts of the nervous system are grouped as follows:
1. The central nervous system consisting of the brain and the spinal cord
2. The peripheral nervous system consisting of
    31 pairs of spinal nerves
    12 pairs of cranial nerves
    the autonomic nervous system

A general view of the nervous system is given in Figure 12:1.

The Neurone

The nervous system consists of a vast number of units called *neurones* (Fig. 12:2) which are supported by a special type of connective tissue called *neuroglia*. Each neurone consists of a *nerve cell* and its processes, called *axons and dendrites*. Neurones are commonly referred to simply as nerves.

The physiological 'units' of the nervous system are *nerve impulses* which are akin to tiny electrical charges. However, unlike ordinary electrical wires, the neurones are actively involved in conducting nerve impulses. In effect the strength of the impulse is maintained through the length of the neurone.

The cells of some neurones initiate nerve impulses while others act as 'relay stations' where impulses are passed on and sometimes redirected.

NERVE CELLS

The nerve cells vary considerably in size and shape but they are all too small to be seen by the naked eye. They form the grey matter of the nervous system and are found at the *periphery* of the brain, in the *centre* of the spinal cord, in groups called *ganglia* outside the brain and spinal cord and as *single cells* in the walls of organs.

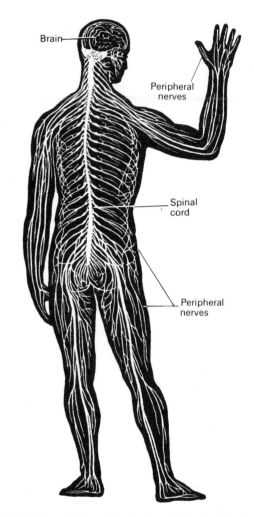

Brain

Peripheral nerves

Spinal cord

Peripheral nerves

Figure 12:1   General view of the brain, spinal cord and spinal nerves.

AXONS AND DENDRITES

Axons and dendrites are the processes of the nerve cells and form the *white matter* of the nervous system. They are found *deep* in the brain, at the *periphery* of the spinal cord and they are described as *nerves* or *nerve fibres* outside the brain and spinal cord.

AXONS

Each nerve cell has only one axon or process which

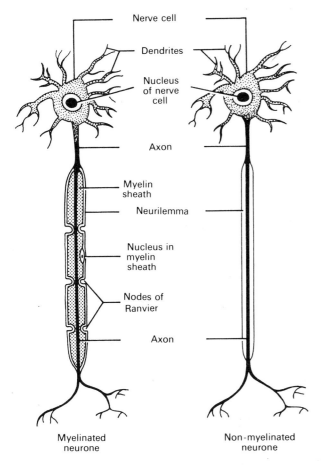

Myelinated neurone   Non-myelinated neurone

Figure 12:2 The neurone.

carries nerve impulses away from the cell. They are usually longer than the dendrites, and may be as long as 100 cm (approximately 40 inches).

The structure of an axon

1. *Axolemma* is the name given to the membrane of the axon and the internal cytoplasm is called *axoplasm*.

2. *Myelin* is a sheath of fatty material which surrounds most axons and gives them a white appearance, thus the term *white matter* used to describe collections of axons in the spinal cord and the brain. *Non-myelinated fibres* consist of post-ganglionic autonomic fibres (see p. 192), some small fibres in the central nervous system and some fine peripheral sensory fibres.

The functions of the myelin sheath are:

To act as an insulator
To protect the axon from pressure or injury
To speed up the flow of nerve impulses through the axon

The myelin sheath is absent at intervals along the length of the axon and near its branching end. The breaks in the myelin sheath are called the *nodes of Ranvier* and their presence contributes to the rapid transmission of nerve impulses along myelinated fibres.

3. *The neurilemma* is a very fine, delicate membrane which surrounds the axons of all peripheral nerves. It consists of a series of *Schwann cells* which surround the axon and the myelin sheath when it is present. In myelinated fibres there is one Schwann cell between adjacent nodes of Ranvier. Each axon is encased in Schwann cells while *a group* of non-myelinated fibres is enclosed within one series of Schwann cells.

DENDRITES

The dendrites are the processes or nerve fibres which carry impulses *towards* nerve cells. They have the same structure as axons but they are usually shorter and branching. Each neurone has many dendrites.

TYPES OF NERVES

SENSORY OR AFFERENT NERVES

These are the nerves which transmit impulses from the periphery of the body to the spinal cord and then to the brain where they are interpreted and perceived as, for example, touch, heat and cold, taste, sight.

MOTOR OR EFFERENT NERVES

These are the nerves which convey impulses from the brain and the spinal cord to other parts of the body stimulating glandular secretion and contraction of all types of muscle.

MIXED NERVES

In the spinal cord sensory nerves are arranged in groups called *tracts* and are separated from tracts of motor nerves. Outside the spinal cord when sensory and motor nerves are enclosed within the same tube of connective tissue they are known as *mixed nerves*.

THE SYNAPSE AND CHEMICAL TRANSMITTERS

There is always more than one neurone involved in the transmission of a nerve impulse from its origin to its effector organ, whether it is sensory or motor. There is no *anatomical continuity* between these neurones and the point at which the nerve impulse passes from one to another is called the *synapse* (Fig. 12:3). At its free end the axon of one neurone breaks up into minute branches which terminate in small swellings called *end feet, presynaptic knobs* or *boutons* which are in close proximity to the dendrites and the cell body of the next neurone. When nerve impulses reach the end feet a chemical substance is released which stimulates the next neurone. There are a number of different substances known to function in this way which are called *chemical transmitters*. Their action is very short lived as immediately they have stimulated the next neurone they are neutralised by an enzyme and rendered inactive. A knowledge of the action of different chemical transmitters has become more important because of the drugs available to neutralise them or prolong their effect.

The chemical transmitters and their modes of action in the brain and spinal cord are not yet fully understood but it is believed they include *noradrenalin*, *5-hydroxytryptamine (serotonin)*, *gamma aminobutyric acid (GABA)* and *dopamine*.

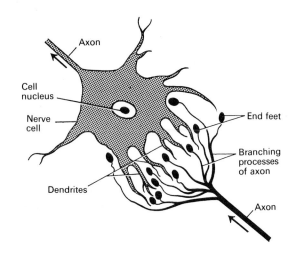

Figure 12:3 Diagram of a synapse.

Figure 12:4 summarises the chemical transmitters which are known to function outside the brain and spinal cord.

TERMINATION OF NERVES

The sensory nerves, for example in the skin, lose their myelin sheath and neurilemma and divide into fine branching filaments known as the *sensory nerve endings* (Fig. 12:5).

It is the *sensory nerve endings* which are stimulated in the skin by touch, pain, heat and cold. The impulse is then transmitted to the brain where the sensation is perceived.

The *motor nerves* conveying impulses to skeletal muscle to produce contraction divide into fine filaments which terminate in minute pads called the *motor end plates* (Fig. 12:6). At the point where the nerve reaches the muscle the myelin sheath and neurilemma are absent and the fine filament passes to a sensitive area on the surface of the muscle fibre. Each muscle fibre is stimulated through a single motor end plate, and one motor nerve can have as many as a hundred motor end plates. There is a tiny space between the motor end plate and the muscle fibre into which the chemical transmitter, *acetylcholine*, is secreted which stimulates the muscle fibre to contract.

The endings of *autonomic nerves* supplying smooth muscle and glands branch near their effector structure and secrete a transmitter substance which stimulates or depresses the activity of the structure.

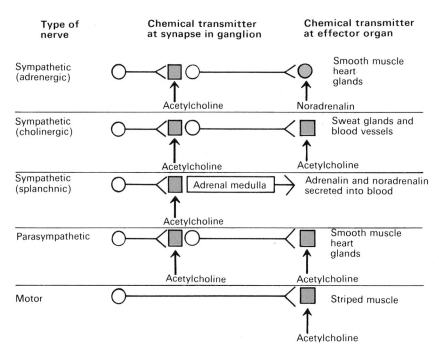

Figure 12:4 Chemical transmitters of nerve impulses at synapses outside the brain and spinal cord.

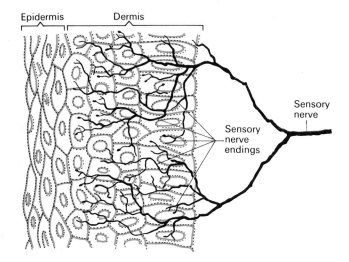

Figure 12:5   Sensory nerve endings in the skin.

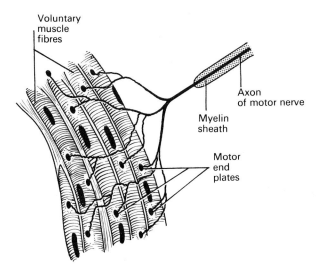

Figure 12:6   Termination of a motor nerve in striped muscle.

Neuroglia

Neuroglia is specialised connective tissue found in the brain and spinal cord supporting the nerve cells and their fibres. There are three types of neuroglial cells called *oligodendrocytes*, *astrocytes* and *microglia*. Some of these cells are part of the reticuloendothelial system.

THE PROPERTIES OF NERVE TISSUE

Nerve tissue has the characteristics of *irritability* and *conductivity*.

Irritability is the ability to initiate nerve impulses in response to stimuli from:

1. Outside the body, for example, touch, light waves
2. Inside the body, for example, an increase in the concentration of carbon dioxide in the blood, a thought which results in voluntary movement

In the body this stimulation may be described as partly electrical and partly chemical.

Conductivity means the ability to transmit an impulse from:

1. One part of the brain to another
2. The brain to striated muscle resulting in voluntary muscle contraction
3. Muscles and joints to the brain, contributing to the maintenance of balance
4. The brain to organs of the body resulting in the contraction of smooth muscle or the secretion of glands
5. Organs of the body to the brain in association with the regulation of body functions
6. The outside world to the brain through sensory nerve endings in the skin which are stimulated by temperature and touch
7. The outside world to the brain through the special sense organs, i.e., eyes, ears, nose, tongue, skin.

The Central Nervous System

The central nervous system consists of the brain and the spinal cord.

THE MEMBRANES COVERING THE BRAIN AND SPINAL CORD

The brain and spinal cord are completely surrounded by three membranes known as the *meninges* which lie between the skull and the brain and between the vertebrae and the spinal cord (Fig. 12:7). Named from without inwards they are:

The dura mater
The arachnoid mater
The pia mater

The dura and arachnoid maters are separated by a potential space called the *subdural space*. The arachnoid and pia maters are separated by the *subarachnoid space* which contains *cerebrospinal fluid*.

THE DURA MATER

The cerebral dura mater consists of two layers of dense fibrous tissue. The outer layer takes the place of the periosteum on the inner surface of the skull bones and the inner layer provides a protective covering for the brain.

There is only a potential space between the two layers except where the inner layer sweeps inwards between the:

Cerebral hemispheres to form the *falx cerebi*
Cerebellar hemispheres to form the *falx cerebelli*

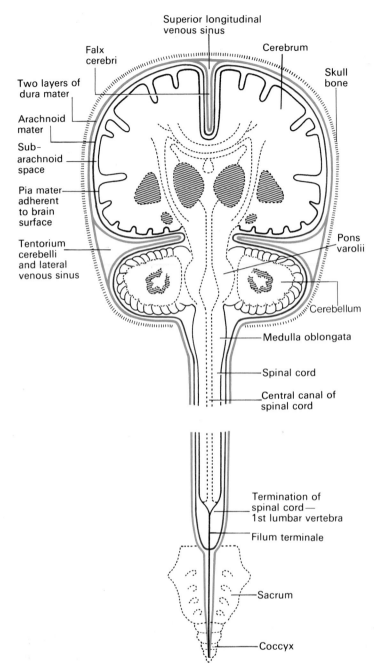

Figure 12:7 The meninges covering the brain and spinal cord.

Cerebrum and the cerebellum to form the *tentorium cerebelli*

It will be remembered that the venous blood from the brain is drained into venous sinuses. These are formed between the layers of dura mater where they are separated. The *superior saggital sinus* is formed by the falx cerebri and the tentorium cerebelli forms the *straight* and *transverse sinuses*.

The spinal dura mater corresponds to the inner layer of cerebral dura mater and forms a loose sleeve around the spinal cord. Between it and the periosteum and ligaments which line the vertebral canal, there is a space which contains blood vessels and areolar tissue called the *epidural* or *extradural space*.

The spinal dura mater begins at the foramen magnum, where the spinal cord leaves the skull, and extends to the level of the second sacral vertebra. Thereafter it invests the filum terminale and fuses with the periosteum of the coccyx.

THE ARACHNOID MATER

This is a delicate serous membrane situated between the

dura mater and the pia mater. It is separated from the dura mater by a potential space known as the *subdural space*, and from the pia mater by a definite space, the *subarachnoid space*. The subarachnoid space contains *cerebrospinal fluid*. The arachnoid mater passes over the convolutions of the brain and accompanies the inner layer of dura mater in the formation of the falx cerebri, tentorium cerebelli and falx cerebelli. It continues downwards to envelop the spinal cord and ends by merging with the dura mater at the level of the second sacral vertebra.

THE PIA MATER

This is a fine vascular membrane consisting mainly of minute blood vessels supported by fine connective tissue. It closely invests the brain completely covering the convolutions, and dips into each fissure. It continues downwards to invest the spinal cord. Beyond the end of the cord it continues as the *filum terminale*, pierces the arachnoid tube and goes on with the dura mater, to fuse with the periosteum of the coccyx.

THE VENTRICLES OF THE BRAIN AND THE CEREBROSPINAL FLUID

Within the brain there are four irregular-shaped cavities, or *ventricles*, containing cerebrospinal fluid (Fig. 12:8). They are called:

The right and left lateral ventricles
The third ventricle
The fourth ventricle

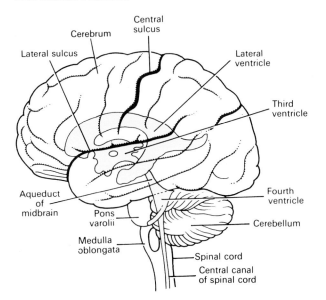

Figure 12:8   The positions of the ventricles of the brain superimposed on its surface. Viewed from the left side.

The lateral ventricles

These cavities lie within the cerebral hemispheres, one on either side of the median plane just below the corpus

callosum. They are separated from each other by a thin membrane known as the *septum lucidum*, and are lined with ciliated epithelium. They are approximately 6 cm (2½ inches) in length.

The third ventricle

The third ventricle is a cavity containing cerebrospinal fluid situated below the lateral ventricles and between the two parts of the thalamus. It communicates with the lateral ventricles by openings known as the *interventricular foramina*.

The fourth ventricle

The fourth ventricle is a lozenge-shaped cavity containing cerebrospinal fluid situated below and behind the third ventricle, between the cerebellum and the pons varolii. It communicates with the third ventricle above by a canal known as the *aqueduct of the midbrain* and is continuous below with the *central canal* of the spinal cord.

The fourth ventricle communicates with the subarachnoid space through openings in its roof. Thus cerebrospinal fluid enters the subarachnoid space through these openings and through the open distal end of the central canal of the spinal cord.

THE CEREBROSPINAL FLUID (Fig. 12:9)

The cerebrospinal fluid is formed and secreted into each ventricle of the brain by its *choroid plexus*. The choroid plexuses consist of areas in the walls of the ventricles where the lining membrane is thinner and where there is a profusion of blood capillaries.

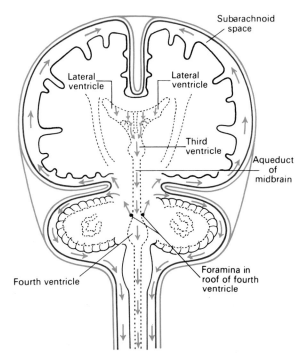

Figure 12:9   Arrows showing the flow of cerebrospinal fluid.

The walls of the capillaries in the brain are less permeable than other capillaries. This means that some substances are unable to cross their walls unless they are actively transported. Because of this the composition of the tissue fluid in the brain and cerebrospinal fluid is different from tissue fluid elsewhere in the body. The difference in permeability is called *the blood-brain barrier*.

From the roof of the fourth ventricle the cerebrospinal fluid flows through two foramina, the *median and lateral foramina*, into the *subarachnoid space* and subsequently completely surrounds the brain and spinal cord. It also flows from the floor of the fourth ventricle downwards through the *central canal* of the spinal cord.

The cerebrospinal fluid is reabsorbed into blood capillaries in the arachnoid mater, and in this way it is returned to the circulating blood.

Cerebrospinal fluid is a clear, slightly alkaline fluid with a specific gravity of 1·005 and it consists of:

Water
Mineral salts
Glucose
Protein: small amounts of albumin and globulin derived from plasma
Creatinine ⎫
Urea ⎭ small amounts

The normal hydrostatic pressure of cerebrospinal fluid is about 130 mm water (10 mmHg).

Functions of the cerebrospinal fluid

1. It supports and protects the delicate structures of the brain and spinal cord.

2. It maintains a uniform pressure around these delicate structures.

3. It acts as a cushion and shock absorber for the brain and spinal cord.

4. It keeps the brain and spinal cord moist and there may be interchange of substances between the fluid and nerve cells.

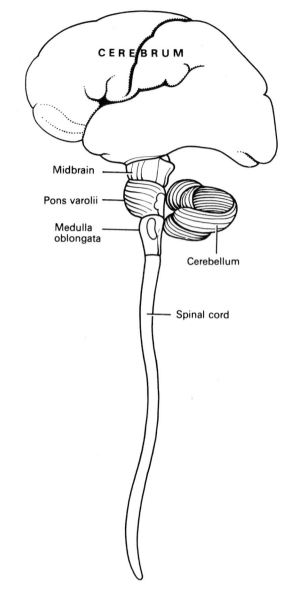

Figure 12:10 The parts of the central nervous system.

The Brain

The brain constitutes about one-fiftieth of the body weight and lies within the cranial cavity.

The structures forming the brain are (Fig. 12:10):

The cerebrum or fore brain
The midbrain ⎫
The pons varolli ⎬ the brain stem
The medulla oblongata ⎭
The cerebellum or hind brain

THE CEREBRUM

The cerebrum constitutes the largest part of the brain and it occupies the anterior and middle cranial fossae (see Fig. 16:4, p. 246). It is divided by a deep cleft called the *longitudinal cerebral fissure* into two distinct parts, *the right and left cerebral hemispheres*, each of which contains one of the lateral ventricles.

Deep within the brain these two hemispheres are connected by a mass of white matter (nerve fibres) known as the *corpus callosum*. The falx cerebri separates the two hemispheres and penetrates to the depth of the corpus callosum.

The superficial or peripheral part of the cerebrum is composed of nerve cells or grey matter forming *the cerebral cortex* and the deeper layers consist of white matter or nerve fibres.

The cerebral cortex shows many infoldings or furrows which vary in depth. The exposed areas of the folds are termed *gyri or convolutions* and they are separated by *sulci or fissures*. These convolutions greatly increase the surface area of the cerebrum.

Each hemisphere of the cerebrum is divided for descriptive purposes into *lobes* which have the names of the bones of the cranium under which they lie:

Frontal
Parietal
Temporal
Occipital

In each hemisphere there are deep fissures or sulci some of which form the boundaries of the lobes (Fig. 12:11).

The central sulcus separates the frontal lobe from the parietal lobe.

The lateral sulcus separates the frontal and parietal lobes from the temporal lobe.

The parieto-occipital sulcus separates the parietal and temporal lobes from the occipital lobe.

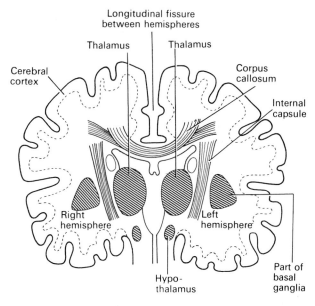

Figure 12:12 A section of the cerebrum showing some connecting nerve fibres.

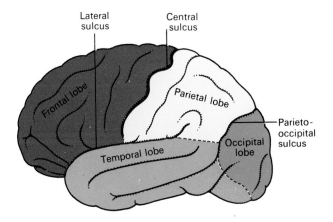

Figure 12:11 The lobes and sulci of the cerebrum.

Interior of the cerebrum (Fig. 12:12)
The cerebral cortex is composed mainly of nerve cells. Within the cerebrum the lobes are connected by masses of nerve fibres, or tracts, which make up the white matter of the brain. The fibres which link the different parts of the brain and spinal cord, consisting of afferent and efferent fibres, are:

Arcuate (association) fibres which connect different parts of the cerebral cortex by extending from one gyrus to the next or between adjacent lobes

Commisural fibres which connect the two cerebral hemispheres (corpus callosum)

Projection fibres which connect the cerebral cortex with grey matter of the brain and the spinal cord e.g. internal capsule

The *internal capsule* is an important area consisting of projection fibres. It lies deep within the brain between the basal ganglia and the thalamus. All nerve impulses ascending to, and descending from, the cerebral cortex are carried by fibres which form the internal capsule.

Functions of the cerebrum
There are three main varieties of activity associated with the cerebral cortex.

1. The mental activities involved in memory, intelligence, sense of responsibility, thinking, reasoning, moral sense and learning are attributed to the *higher centres*.
2. Sensory perception, which includes the perception of pain, temperature, touch and the special senses of sight, hearing, taste and smell.
3. The initiation and control of the contraction of voluntary muscle.

Functional areas of the cerebrum (Fig. 12:13)
The main areas of the cerebrum associated with sensory perception and voluntary motor activity are known but it should be appreciated that it is unlikely that any area is associated exclusively with only one function. Except where specially mentioned the different areas are active in both hemispheres.

The precentral (motor) area lies in the frontal lobe immediately anterior to the *central sulcus*. The nerve cells are called pyramidal cells and they initiate the contraction of voluntary muscles. A nerve fibre from a pyramidal cell passes downwards through the internal capsule to the medulla oblongata where it crosses to the opposite side then descends in the spinal cord. At the appropriate level in the spinal cord the nerve impulse crosses a synapse to stimulate a second neurone which terminates at the motor end plate of a muscle fibre.

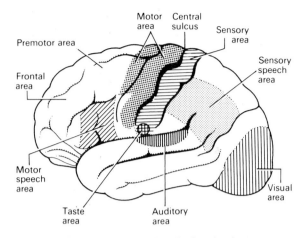

Figure 12:13   The cerebrum showing the functional areas.

This means that the motor area of the *right hemisphere* of the cerebrum controls voluntary muscle movement on the *left side of the body* and vice versa. The neurone with its cell in the cerebrum is called the *upper motor neurone* and the other, with its cell in the spinal cord, is the *lower motor neurone* (Fig. 12:14). Damage to either of these neurones may result in paralysis.

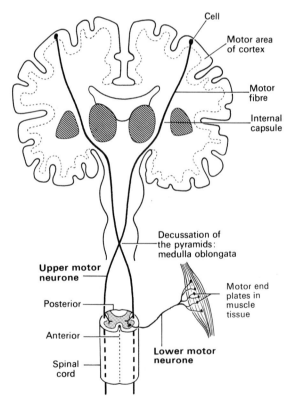

Figure 12:14   The motor nerve pathways: upper and lower motor neurones.

In the motor area of the cerebrum the body is represented upside down, the cells in the upper part controlling the feet and the cells in the deepest part controlling the head, neck, face and fingers (Fig. 12:15A).

The premotor area lies in the frontal lobe immediately anterior to the motor area. The cells are thought to exert a controlling influence over the motor area, ensuring an orderly series of movements. For example, in tying a shoe lace or writing, many muscles contract but the movements must be co-ordinated and carried out in a particular sequence. Such a pattern of movement, when established, is described as manual dexterity.

In the lower part of this area just above the lateral sulcus there is a group of nerve cells known as the *motor speech (Broca's) area* which controls the movements necessary for speech. It is dominant in the *left hemisphere* in *right-handed people* and vice versa.

The frontal area or pole extends anteriorly from the premotor area to include the remainder of the frontal lobe. It is a large area and is more highly developed in man than in other animals. It is thought that communications between this and the other regions in the cerebrum are responsible for the behaviour, character and emotional state of the individual. No particular behaviour, character or intellectual trait has, so far, been attributed to the activity of any one group of cells.

The postcentral (sensory) area is the area of the cerebrum which lies behind the central sulcus (Fig. 12:15B). Here sensations of pain, temperature, pressure and touch, knowledge of muscular movement and the position of joints are perceived. The sensory area of the *right hemisphere* receives impulses from the *left side of the body* and vice versa.

The parietal area lies behind the post-central area and includes the greater part of the parietal lobe of the cerebrum. Its functions are believed to be associated with obtaining and retaining accurate knowledge of objects. It has been suggested that objects can be recognised by touch alone because of the knowledge from past experience retained in this area.

The sensory speech area is situated in the lower part of the parietal lobe and extends into the temporal lobe. It is here that the spoken word is perceived. There is a dominant area in the *left hemisphere* if the individual is *right-handed* and vice versa.

The auditory (hearing) area lies immediately below the lateral sulcus within the temporal lobe. The cells receive and interpret impulses transmitted from the inner ear by the vestibulocochlear nerve.

The olfactory (smell) area lies deep within the temporal lobe where impulses received from the nose via the olfactory nerve are received and interpreted.

The taste area is thought to lie just above the lateral sulcus in the deep layers of the sensory area, and it is here that the nerve impulses from the tongue are interpreted.

The visual area lies behind the parieto-occipital sulcus and includes the greater part of the occipital lobe. The optic nerves, or nerves of the sense of sight, pass from the eye to this area which receives and interprets the impulses as visual impressions.

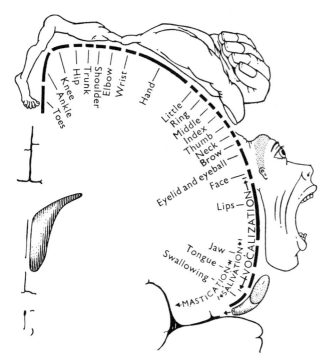

Figure 12:15A *The motor homonculus* showing how the body is represented in the motor area of the cerebrum.

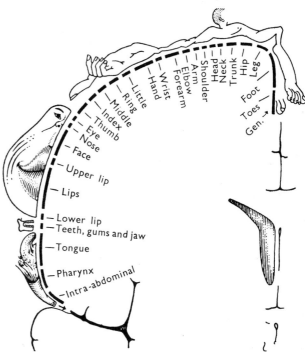

Figure 12:15B *The sensory homonculus* showing how the body is represented in the sensory area of the cerebrum. (Both A and B *from* Penfield W, Rasmussen T 1950 The cerebral cortex of man. Macmillan, New York. Reproduced with permission.)

Deep within the cerebral hemispheres there are groups of nerve cells known as *ganglia or nuclei*. These ganglia act as *relay stations* where synapses occur between neurones. Important masses of grey matter include:

The basal ganglia
The thalamus
The hypothalamus

The basal ganglia

This area of grey matter, lying deep within the cerebral hemispheres, is thought to influence skeletal muscle tone. If this control is inadequate or absent movements are jerky, clumsy and unco-ordinated.

The thalamus

The thalamus consists of a mass of nerve cells situated within the cerebral hemispheres just below the corpus callosum.

All sensory nerves from the periphery of the body associated with impulses of pain, temperature, pressure and touch are conveyed to the thalamus. It is here that very crude uncritical sensations reach consciousness. The thalamus is unable to distinguish finer sensations. The discrimination and highly critical interpretation of these sensations occur in the sensory area of the cerebral cortex. This means that the individual is aware of, for example, pain, but is unable to identify the part of the body involved.

The hypothalamus

The hypothalamus is composed of a number of groups of nerve cells. It is situated below and in front of the thalamus and immediately above the pituitary gland. The hypothalamus is linked to the posterior lobe of the pituitary gland by nerve fibres and to the anterior lobe by a complex system of blood vessels. Through these connections the cells of the hypothalamus control the output of hormones from both lobes of the gland (see p. 216).

Other functions with which the hypothalamus is concerned include centres for the control of the autonomic nervous system, for example, control of hunger, thirst, body temperature, heart and blood vessels, and defensive reactions, such as those associated with fear and rage.

THE MIDBRAIN

The midbrain is the area of the brain between the cerebrum above and the *pons varolii* below. It consists of groups of nerve cells and nerve fibres which connect the cerebrum with lower parts of the brain and the spinal cord.

The nerve cells act as relay stations for the ascending and descending nerve fibres. Two groups of cells of particular note are the *medial and lateral geniculate bodies* which provide relay stations for the transmission of nerve impulses from the optic nerves and the vestibular portion of the vestibulocochlear nerves to the *cerebellum*. These nerve impulses play a major part in the maintenance of balance of the body.

THE PONS VAROLII

The pons varolii is situated in front of the cerebellum below the midbrain and above the medulla oblongata. It consists mainly of nerve fibres which form a bridge between the two hemispheres of the cerebellum and of fibres passing between the higher levels of the brain and the spinal cord. There are groups of cells within the pons which act as relay stations, some of these are associated with the cranial nerves (see p. 189).

The anatomical structure of the pons varolii differs from that of the cerebrum in that *the nerve cells lie deeply* and the *nerve fibres are on the surface.*

THE MEDULLA OBLONGATA

The medulla oblongata extends from the pons varolii above and is continuous with the spinal cord below. It is about 2·5 cm (1 inch) long, is shaped like a pyramid with its base upwards, and lies just within the cranium above the foramen magnum. Its anterior and posterior surfaces are marked by central fissures.

The outer aspect is composed of *white matter* or *nerve fibres* which pass between the brain and the spinal cord. *Grey matter* or *nerve cells lie centrally* within the medulla. Some of these constitute relay stations for sensory nerves passing from the spinal cord to the cerebrum. Several cranial nerves arise from groups of cells (nuclei) in the medulla.

The vital centres, associated with autonomic reflex activity, are present in its deeper structure. These are:

The cardiac centre

The respiratory centre

The vasomotor centre

The reflex centres of vomiting, coughing, sneezing and swallowing

Functions

1. *Decussation of the pyramids.* In the medulla the majority of *motor nerves* descending from the motor area in the cerebrum to the spinal cord cross from one side to the other. This means that the left hemisphere of the cerebrum controls the right half of the body, and vice versa (Fig. 12:16).

2. *Sensory decussation.* Some of the *sensory nerves* ascending to the cerebrum from the spinal cord cross from one side to the other in the medulla.

3. *The cardiac centre* controls the rate and force of cardiac contraction. Sympathetic and parasympathetic nerve fibres originating in the medulla pass to the heart. Sympathetic stimulation increases the rate and force of the heart beat and parasympathetic stimulation has the opposite effect.

4. *The respiratory centre* controls the rate and depth of respiration. From this centre nerve impulses pass to the phrenic and intercostal nerves which stimulate contraction of the diaphragm and intercostal muscles, thus initiating inspiration. The respiratory centre is stimulated

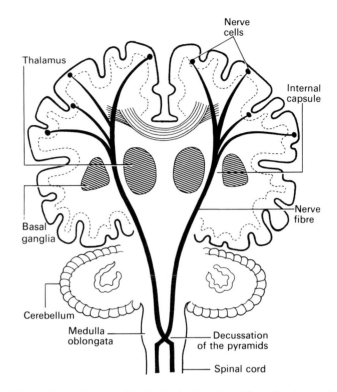

Figure 12:16 Section of the brain showing the origin and pathway of motor neurones.

by excess carbon dioxide and deficiency of oxygen in the blood.

5. *The vasomotor centre* controls the calibre of the blood vessels, especially the small arteries and arterioles which have a large proportion of smooth muscle fibres in the tunica media.

Vasomotor impulses reach the blood vessels through the autonomic nervous system. Stimulation may cause either constriction or dilatation of blood vessels depending on the site (See Figs. 12:40 and 12:41.)

The sources of stimulation of the vasomotor centre are the arterial baroreceptors, emotions, such as sexual excitement and anger. Pain usually causes vasoconstriction although severe pain may cause vasodilation, a fall in blood pressure and fainting.

6. *Reflex centres.* When irritating substances are present in the stomach or respiratory tract, nerve impulses pass to the medulla oblongata. These impulses stimulate the reflex centres which initiate the reflex actions of vomiting, coughing or sneezing.

THE CEREBELLUM

The cerebellum is situated behind the pons varolii and immediately below the posterior portion of the cerebrum occupying the posterior cranial fossa (Fig. 12:17). It is ovoid in shape and presents two hemispheres which are separated by a narrow median strip known as the *vermis.* Grey matter forms the surface of the cerebellum, and the white matter lies deeply.

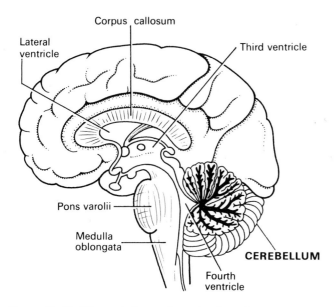

Lateral ventricle

Corpus callosum

Third ventricle

Pons varolii

Medulla oblongata

CEREBELLUM

Fourth ventricle

Figure 12:17 The cerebellum and associated structures.

The nerve fibres which enter and leave the cerebellum do so by three tracts known as the cerebellar peduncles.

The superior cerebellar peduncle connect the cerebellum with the midbrain and cerebrum.

The middle cerebellar peduncle connect the cerebellum with the pons varolii.

The inferior cerebellar peduncle connect the cerebellum with the medulla oblongata and the spinal cord.

Functions

The cerebellum is concerned with voluntary muscular movement and balance. Cerebellar activities are carried out *below the level of consciousness* and are, therefore, not under the control of the will.

It controls and co-ordinates the movements of various groups of muscles ensuring smooth, even, and precise actions.

It co-ordinates activities associated with the maintenance of the balance and equilibrium of the body. The sensory input is derived from the muscles and joints, the eyes and the ears. Impulses from the muscles and joints indicate their position in relation to the body as a whole and those from the eyes and the semicircular canals in the ears provide information about the position of the head in space. Impulses from the cerebellum influence the contraction of skeletal muscle so that the balance of the body is maintained.

Damage to the cerebellum results in clumsy unco-ordinated muscular movement, staggering gait and inability to carry out smooth, steady precise actions.

THE RETICULAR FORMATION

In the brain stem there are collections of neurones which are referred to as the *reticular formation*. These neurones

lie in the core of the brain stem and are surrounded by neural pathways which pass nerve impulses between the brain and the spinal cord, in both directions. It has a vast number of synaptic links with other parts of the brain and is therefore constantly receiving 'information' being transmitted in the ascending and descending tracts.

Functions of the reticular formation

1. It is involved in the co-ordination of skeletal muscle activity which involves voluntary motor movement and the maintenance of balance.

2. It is involved in the regulation of activity controlled by the autonomic nervous system, such as, cardiovascular, respiratory, gastrointestinal activity.

3. It is involved in the phenomena of *sleep and wakefulness*. Ascending impulses from the reticular formation 'arouse' the cerebral cortex, and the individual, from sleep. This is called the *reticular activating system* (RAS) which functions on a selective basis, for example, the slight sound made by a sick child moving in bed may awaken his mother but the noise of regularly passing trains may be suppressed and have no effect. The RAS selectively blocks or passes on sensory nerve impulses between the sensory organ and the cerebral cortex.

The Spinal Cord

The spinal cord is the elongated, almost cylindrical part of the central nervous system which lies within the vertebral canal of the vertebral column. It is continuous above with the medulla oblongata and extends from the *upper border of the atlas* to the lower border of the *first lumbar vertebra* (Fig. 12:18). It is approximately 45 cm (18 inches) long in an adult caucasian male, and is about the thickness of a little finger. It is surrounded by the dura, arachnoid and pia maters as described previously (Fig. 12:19). Cerebrospinal fluid is present in the central canal of the spinal cord and in the subarachnoid space. When a specimen of cerebrospinal fluid is required it is taken at a point beyond the end of the cord, i.e., below the level of the second lumbar vertebra.

The spinal cord is the nervous tissue link between the brain and the organs of the body (Fig. 12:20). Nerves conveying impulses from the brain to the various organs of the body descend through the spinal cord. At the appropriate level they leave the cord and pass to the organ which they supply. Similarly, sensory nerves from the skin and other organs enter and pass upwards in the spinal cord to the brain.

STRUCTURE

The spinal cord is incompletely divided into two equal parts, anteriorly by a short, shallow *median fissure* and

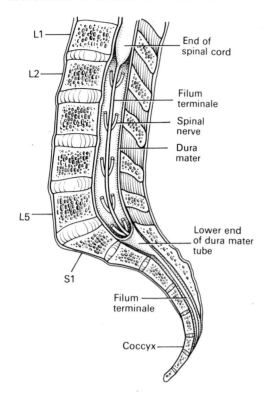

L1
L2
L5
S1

End of
spinal cord

Filum
terminale

Spinal
nerve

Dura
mater

Lower end
of dura mater
tube

Filum
terminale

Coccyx

Figure 12:18   Section of the distal end of the vertebral canal.

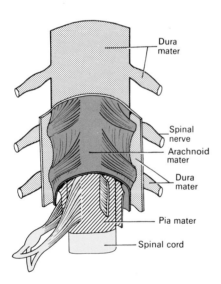

Dura
mater

Spinal
nerve

Arachnoid
mater

Dura
mater

Pia mater

Spinal cord

Figure 12:19   The membranes covering the spinal cord. Each cut
away to show the various layers.

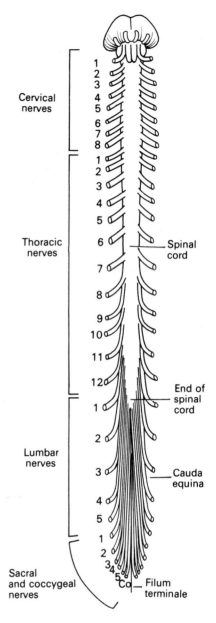

Cervical
nerves

1
2
3
4
5
6
7
8

Thoracic
nerves

1
2
3
4
5
6
7
8
9
10
11
12

Spinal
cord

Lumbar
nerves

1
2
3
4
5

End of
spinal
cord

Cauda
equina

Sacral
and coccygeal
nerves

1
2
3 4 5
Co

Filum
terminale

Figure 12:20   The spinal cord and the spinal nerves.

posteriorly by a deep narrow septum known as the
posterior median septum.

Examination of a cross section of the spinal cord shows
that it is composed of grey matter in the centre which is
surrounded by white matter. It should be noted that this
arrangement is the opposite to that in the cerebrum and
cerebellum (Fig. 12:21).

GREY MATTER

The grey matter is composed of nerve cells and the white
matter of nerve fibres and both are supported by
neuroglia. Within the centre of the spinal cord is a canal,
the *central canal*, which is continuous with the fourth
ventricle of the brain and contains cerebrospinal fluid.

The arrangement of grey matter in the spinal cord
bears a resemblance to the shape of the letter H and is
described as having *two posterior*, *two anterior* and *two
lateral columns*. The area of grey matter lying transversely
is known as the *transverse commissure* and it is pierced by
the central canal.

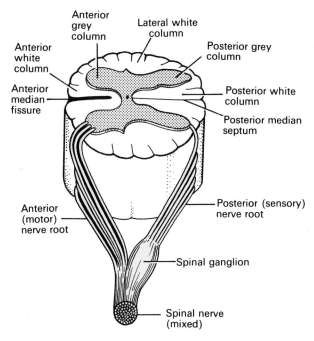

Figure 12:21 A section of the spinal cord showing nerve roots.

The grey matter of the spinal cord consists of nerve cells which may be:

1. The *sensory cells* which receive impulses from the periphery of the body
2. The cells of *lower motor neurones* which transmit impulses to the skeletal muscles
3. The cells of *connector neurones* which link the sensory and motor neurones in the formation of spinal reflex arcs

At each point where nerve impulses are passed from one neurone to another there is a synaptic gap and a chemical transmitter (see p. 169).

The posterior columns of grey matter

These are composed of nerve cells which are stimulated by *sensory impulses* from the periphery of the body. The nerve fibres of these cells, which contribute to the formation of the white matter of the cord, transmit the sensory impulses to the brain.

The anterior columns of grey matter

These are composed of the *cells of the lower motor neurones* which are stimulated by the axons of the upper motor neurones or by the *cells of connector neurones* linking the anterior and posterior columns of the spinal cord in the formation of reflex arcs.

The posterior root (spinal) ganglia. These ganglia are composed of nerve cells which lie just *outside* the spinal cord on the pathway of the sensory nerves. They are the cells of the sensory nerve fibres conveying impulses to

the cord from the periphery of the body. The fibres of these cells form the posterior roots of the spinal nerves.

WHITE MATTER

The white matter of the spinal cord is arranged in three columns or tracts:

The anterior columns
The posterior columns
The lateral columns

These columns are formed by *sensory nerve fibres ascending* to the brain from the periphery of the body, *motor nerve fibres descending* from the brain prior to reaching the organs or muscles which they stimulate and the *fibres of connector neurones*.

SENSORY NERVE TRACTS (AFFERENT OR ASCENDING) IN THE SPINAL CORD

There are two main sources of sensation which are transmitted to the brain via the spinal cord.

1. *The skin*. Sensory nerve endings in the skin (cutaneous receptors) are stimulated by *pain, temperature and touch*. The nerve impulses pass by three neurones to the sensory area in the *opposite hemisphere of the cerebrum* where the sensation and its location are perceived (Fig. 12:22).

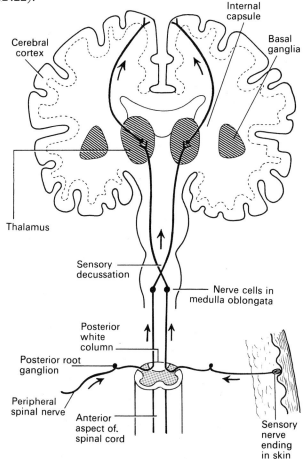

Figure 12:22 A sensory nerve pathway from the skin to the cerebrum.

2. *The tendons, muscles and joints.* Sensory nerve endings in these structures (proprioceptors) are stimulated by stretch. These nerve impulses have two destinations:

(a) By a three neurone system the impulses reach the sensory area of the *opposite hemisphere of the cerebrum.* The resultant perceptions are of the position of the body or parts of the body in space.

(b) By a two neurone system the nerve impulses reach the *cerebellar hemisphere on the same side.*

Together with impulses from the eyes and the ears proprioceptor impulses are associated with the maintenance of balance and posture.

Sensory nerve impulses from the left side of the body are carried to the right hemisphere of the brain. This means that one of the neurones must cross to the opposite side. The crossing is called *decussation* and it takes place either at the level in the spinal cord at which the nerve fibre enters or in the medulla.

Table 12:1 provides further information about the origins, routes of transmission and the destinations of sensory nerve impulses.

MOTOR NERVE TRACTS (EFFERENT OR DESCENDING) IN THE SPINAL CORD

Neurones which transmit nerve impulses away from the brain are called motor, efferent or descending neurones.

Motor neurone stimulation results in:

1. The contraction of voluntary, striated or skeletal muscle

2. The contraction of smooth or involuntary muscle and the secretion of glands controlled by nerves of the *autonomous part of the nervous system* (see p. 192).

Voluntary muscle movement

The contraction of the muscles which move the joints is, in the main, under the control of the will, which means that the stimulus to contraction occurs at the level of consciousness in the cerebrum. However, some nerve impulses which affect skeletal muscle contraction are initiated in the midbrain, brain stem and cerebellum. This is below the level of consciousness and it results in the co-ordination of muscle activity, such as, when very fine movement is required and in the maintenance of posture and balance.

Efferent nerve impulses are transmitted from the brain to the body via bundles of nerve fibres or *tracts* in the spinal cord. *All the motor pathways* from the brain to the muscles are made up of *two neurones.*

1. *The upper motor neurone* has its cell in the *precentral sulcus area of the cerebrum.* The axons pass through the internal capsule, the pons and the medulla. These bundles of fibres form the *lateral corticospinal tracts* of white matter in the spinal cord and the fibres terminate in close association with the dendrites and cells of the lower

Table 12:1 Sensory nerve impulses: origins, routes, destinations

Receptor	Route		Destination
Pain, touch, temperature	Neurone 1 ⟶	to spinal cord by posterior root	
	Neurone 2 ⟶	decussation on entering spinal cord then in lateral spinothalamic tract to thalamus	
			Neurone 3 ⟶ to parietal lobe of cerebrum
Touch, proprioceptors	Neurone 1 ⟶	to medulla in posterior spinothalamic tract	
	Neurone 2 ⟶	decussation in medulla, transmission to thalamus	
			Neurone 3 ⟶ to parietal lobe of cerebrum
Proprioceptors	Neurone 1 ⟶	to spinal cord	
			Neurone 2 ⟶ no decussation, to cerebellum in posterior spinocerebellar tract

motor neurones in the anterior columns of grey matter. The axons of most upper motor neurones decussate either at the level of the medulla or in the spinal cord just before they terminate.

2. *The lower motor neurone* has its cell in the *anterior column of grey matter* in the spinal cord. Its axon emerges from the spinal cord by the anterior root, joins with the in-coming sensory fibres and forms the mixed *spinal nerve* which passes through the intervertebral foramen. Near its termination the axon branches into a variable number of tiny fibres each of which ends on a sensitive area of a voluntary muscle fibre (see p. 171).

Involuntary muscle movement

1. *Upper motor neurones* which have their cells in the brain at a level *below* the cerebrum, i.e., midbrain, brain stem, cerebellum, spinal cord, influence muscle activity in relation to the maintenance of posture and balance, the co-ordination of muscle movement and the control of muscle tone.

Table 12:2 shows details of the area of origin of these neurones and the tracts which their axons form before reaching the cell of the lower motor neurone in the spinal cord.

2. *Spinal reflexes.* This is the type of reflex action in

Table 12:2   Upper motor neurones: origins and tracts

Origin	Name of tract	Situation in spinal cord	Functions
Midbrain and pons	Rubrospinal tract decussates in brain stem	Lateral column	Control of skilled muscle movement
Reticular formation	Reticulospinal tract does not decussate	Lateral column	Co-ordination of muscle movement, maintenance of posture and balance
Midbrain and pons	Tectospinal tract decussates in midbrain	Anterior column	
Midbrain and pons	Vestibulospinal tract some fibres decussate in the cord	Anterior column	

The lower motor neurone has been described as the *final common pathway* for the transmission of nerve impulses to striated muscles. The cell of this neurone is influenced by a number of upper motor neurones originating from various sites in the brain and by some neurones which begin and end in the spinal cord. Some of these neurones stimulate the cells of the lower motor neurone while others have an inhibiting effect. The outcome of these influences is smooth, co-ordinated muscle movement, some of which is voluntary and some involuntary.

which the lower motor neurone cell is stimulated by a neurone which originates at the same or a different level *in the spinal cord* (Fig. 12:23). This is called a *connector neurone* because it is the link between an afferent or sensory neurone and an efferent or lower motor neurone. These are the elements of most *reflex arcs* and a *reflex action* may be described as an immediate motor response to a sensory stimulus. Many connector and motor neurones may be stimulated by afferent impulses from a very small area of skin. For example, the pain impulses which are initiated by touching red hot metal with the

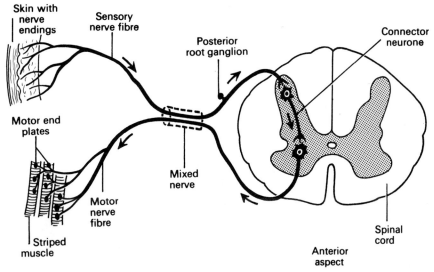

Figure 12:23   A simple reflex arc.

finger are transmitted to the spinal cord by an afferent nerve. This stimulates many connector and lower motor neurones in the cord which results in the removal of the finger by the contraction of many skeletal muscles of the hand, arm and shoulder. Reflex action takes place very quickly, in fact, the motor response may have occurred simultaneously with the perception of the pain in the cerebrum. Reflexes of this type are invariably protective but they can on occasion be inhibited. For example, if it is a precious plate which is very hot when lifted every effort will be made to relieve the pain but prevent dropping the plate!

3. *Stretch reflexes.* Only two neurones are involved. The cell of the lower motor neurone is stimulated by the afferent neurone. There is no connector neurone involved. The *knee jerk* is one example, but this type of reflex can be demonstrated at any point where a stretched tendon crosses a joint. By tapping the tendon just below the knee when the knee is bent the sensory nerve endings in the tendon and in the thigh muscles are stretched. This initiates a nerve impulse which passes into the spinal cord to the cell of the lower motor neurone in the anterior column of grey matter on the same side. As a result the thigh muscles suddenly contract and the foot kicks forward. This is used as a test of the integrity of the reflex arc. The physiological value of this type of reflex is protective as it may prevent excessive joint movement and damage to tendons, ligaments and muscles.

4. *Autonomic reflexes.* See page 178.

The Peripheral Nervous System

This part of the nervous system consists of:
  31 pairs of spinal nerves
  12 pairs of cranial nerves
  The autonomic part of the nervous system

Most of the nerves of the peripheral nervous system are composed of *sensory nerve fibres* conveying impulses from sensory end organs to the brain, and *motor nerve fibres* conveying impulses from the brain through the spinal cord to the effector organs, for example, skeletal muscles, smooth muscle and glands.

Each nerve consists of numerous nerve fibres collected into bundles. Each bundle has several coverings of protective connective tissue (see Fig. 12:24).

1. *The endoneurium* is a delicate connective tissue which surrounds each individual fibre.

2. *The perineurium* is a smooth connective tissue which surrounds each *bundle* of fibres.

3. *The epineurium* is the connective tissue which surrounds and encloses a number of bundles of nerve fibres. Most of the large nerves possess this outer protective covering.

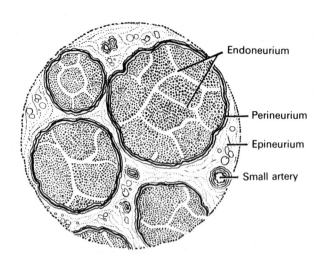

Figure 12:24 Transverse section of a peripheral nerve showing the protective coverings.

THE SPINAL NERVES

There are *31 pairs of spinal nerves* which leave the vertebral canal by passing through the intervertebral foramina formed by adjacent vertebrae. They are named and grouped according to the vertebrae with which they are associated (see Fig. 12:20):
  8 cervical nerves
  12 thoracic nerves
  5 lumbar nerves
  5 sacral nerves
  1 coccygeal nerve

Although there are only 7 cervical vertebrae there are 8 nerves because the 1st pair leave the vertebral canal between the occipital bone and the atlas and the 8th pair leave below the last cervical vertebra. Thereafter the nerves are given the name and number of the vertebra immediately above, for example, the 1st thoracic nerve emerges between thoracic vertebrae 1 and 2.

The lumbar, sacral and coccygeal nerves leave the *spinal cord* before its termination at the level of the first lumbar vertebra, and extend downwards inside the vertebral canal in the subarachnoid space below this level. In this way they form a sheaf of nerves which resembles a horse's tail, called the *cauda equina*. These nerves leave the vertebral canal at the appropriate lumbar, sacral or coccygeal level.

The spinal nerves arise from both sides of the spinal cord and emerge through the intervertebral foramina. Each nerve is formed by the union of *a motor and a sensory nerve root* and is, therefore, a mixed nerve. Each spinal nerve has a contribution from the sympathetic part of the autonomic nervous system in the form of a *pre-ganglionic fibre*. For details of the bones and muscles mentioned in the following section see Chapters 16, 17 and 18.

NERVE ROOTS (Fig. 12:25)

The anterior nerve root consists of *motor nerve fibres* which are the axons of the nerve cells in the anterior column of grey matter in the spinal cord and, in the thoracic and lumbar regions, *sympathetic nerve fibres* which are the axons of cells in the lateral columns of grey matter.

The posterior nerve root consists of *sensory nerve fibres* which enter at the posterior column of the spinal cord. They are characterised by the presence of the *spinal ganglion* (posterior root ganglion) consisting of a little cluster of nerve cells.

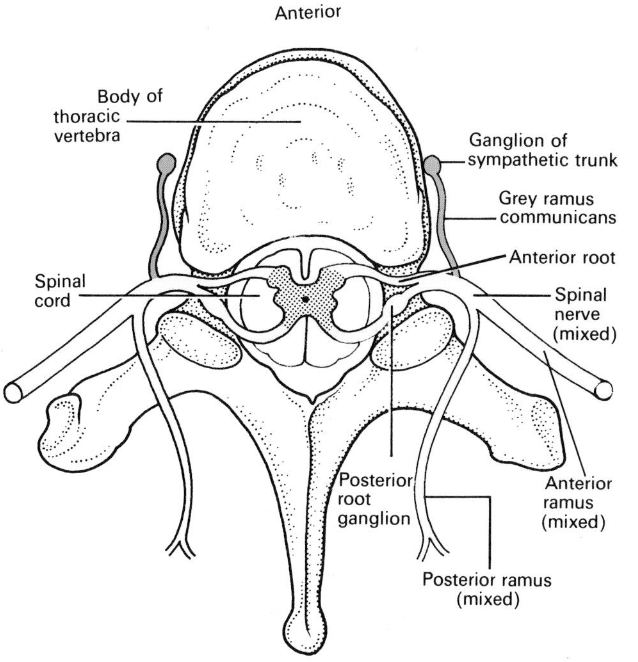

Immediately after emerging from the intervertebral foramen each spinal nerve divides into:

A grey ramus communicans
A posterior ramus
An anterior ramus

The rami communicans are part of the sympathetic portion of the autonomic nervous system (see p. 193).

The posterior rami pass backwards and divide into medial and lateral branches to supply the muscles and skin at the posterior aspect of the trunk.

The anterior rami supply the anterior and lateral aspects of the trunk and the upper and lower limbs.

In the cervical, lumbar and sacral regions the anterior rami unite near their origins to form large masses of nerves known as *plexuses* (Fig. 12:26) where nerves are regrouped and rearranged before proceeding to supply the skin, bones, muscles and joints of a particular area.

In the thoracic region the anterior rami do not form plexuses.

Figure 12:25 Diagram showing the relationship between sympathetic and mixed spinal nerves. Sympathetic part in green.

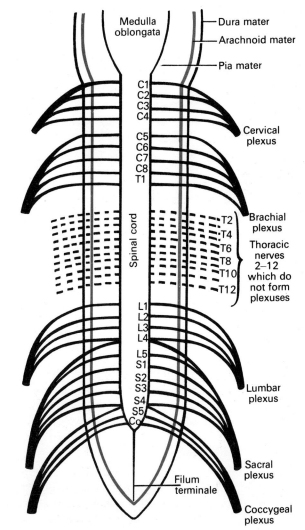

Figure 12:26 Diagram showing the membranes covering the spinal cord and the plexuses formed by the spinal nerves.

For a very short distance after leaving the spinal cord the nerve roots have a covering of *dura* and *arachnoid maters*. These coverings terminate before the two roots join to form the mixed spinal nerve. The nerve roots have no covering of pia mater.

There are five large plexuses formed *on each side* of the vertebral column. They are:

The cervical plexuses
The brachial plexuses
The lumbar plexuses
The sacral plexuses
The coccygeal plexuses

The cervical plexus (Fig. 12:27)

The anterior rami of the *first four cervical nerves* form the cervical plexus.

The plexus lies opposite the 1st, 2nd, 3rd and 4th cervical vertebrae under the protection of the sternocleidomastoid muscle.

The branches of the cervical plexus are divided into two groups: *superficial and deep*.

The superficial branches supply the structures at the back and side of the head and the skin in front of the neck to the level of the sternum.

The deep branches supply muscles of the neck, for example, the sternocleidomastoid and the trapezius.

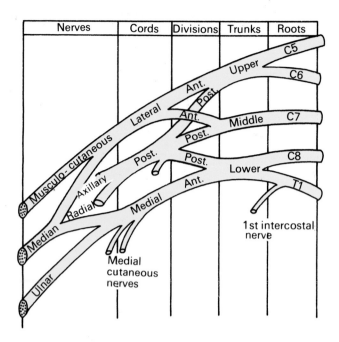

Figure 12:28   The brachial plexus.

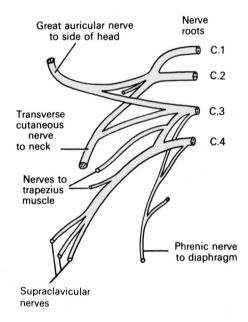

Figure 12:27   The cervical plexus.

The phrenic nerve originates from cervical roots 3, 4 and 5 and passes downwards through the thoracic cavity in front of the root of the lung to supply the muscle of the diaphragm with impulses which stimulate contraction.

The brachial plexus

The anterior rami of the *lower four cervical nerves* and a large part of the *first thoracic nerve* form the brachial plexus. Figure 12:28 shows the formation of this plexus and the nerves which emerge from it.

The plexus is situated above and behind the subclavian vessels and in the axilla.

The branches of the brachial plexus supply the skin and muscles of the upper limb and some of the chest muscles. Five large nerves and a number of smaller ones emerge from this plexus, each with a contribution from more than one nerve root:

The axillary (circumflex) nerve—C 5, 6
The radial nerve—C 5, 6, 7, 8, T 1
The musculocutaneous nerve—C 5, 6, 7
The median nerve—C 5, 6, 7, 8, T 1
The ulnar nerve—C 7, 8, T 1
The medial cutaneous nerve—C 8, T 1

The axillary or circumflex nerve winds round the humerus at the level of the surgical neck. It then breaks up into minute branches to supply the deltoid muscle and the shoulder joint.

The radial nerve is the largest branch of the brachial plexus. It winds round the posterior aspect of the humerus to supply the triceps muscle. It then crosses in front of the elbow joint and passes downwards for a short distance on the lateral aspect before winding round to the posterior aspect of the forearm to become the *posterior interosseous nerve*. It supplies the muscles lying on the posterior aspect of the forearm, i.e., the extensors of the wrist and finger joints. It continues into the back of the hand to supply the skin of the thumb, the first two fingers and the lateral half of the ring finger.

The musculocutaneous nerve passes downwards to the lateral aspect of the forearm. It supplies the muscles of the upper arm and the skin of the forearm.

The median nerve passes down the midline of the arm in close association with the brachial artery. It passes in front of the elbow joint then downwards to supply the muscles of the anterior aspect of the forearm. It continues into the hand where it supplies small muscles and the skin of the anterior aspect of the thumb, the first two fingers and the lateral half of the third finger. It gives off no branches above the elbow.

The ulnar nerve descends through the upper arm lying medial to the brachial artery. It passes behind the medial epicondyle of the humerus to supply the muscles on the ulnar aspect of the forearm. It continues downwards to supply the muscles in the palm of the hand and the skin of the whole of the little finger and both sides of the medial half of the ring finger. It gives off no branches above the elbow.

The main nerves of the arm are presented in Figure 12:29. The distribution and origins of the cutaneous nerves of the arm are shown in Figure 12:30.

Iliohypogastric nerve—L 1
Ilioinguinal nerve—L 1
Genitofemoral—L 1, 2
Lateral cutaneous nerve of thigh—L 2, 3
Femoral nerve—L 2, 3, 4
Obturator nerve—L 2, 3, 4
Lumbosacral trunk—L 4, (5)

The iliohypogastric, ilioinguinal and genitofemoral nerves supply muscles and the skin in the area of the lower abdomen, upper and medial aspects of the thigh and the inguinal region.

The lateral cutaneous nerve of thigh supplies the skin of the lateral aspect of the thigh including part of the anterior and posterior surfaces.

The femoral nerve is one of the larger branches of the

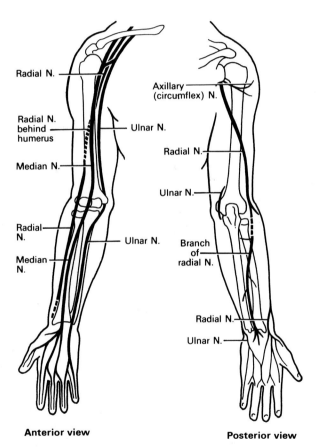

Radial N.
Radial N. behind humerus
Median N.
Radial N.
Median N.
Ulnar N.
Ulnar N.
Axillary (circumflex) N.
Radial N.
Ulnar N.
Branch of radial N.
Radial N.
Ulnar N.
Anterior view
Posterior view

Figure 12:29 The main nerves of the arm.

Supraclavicular N. C3, 4
Axillary (circumflex) N. C5, 6
Radial N. C5, 6
Musculo-cutaneous N. C5, 6, 7
Radial N. C 7, 8
Medial cutaneous N. C8, T1
Medial cutaneous N. C8, T1
Ulnar N. C8, T1
Median N. C6, 7, 8
Supraclavicular N. C3, 4
Axillary (circumflex) N. C5, 6
Radial N. C5, 6, 7, 8
Musculo-cutaneous N. C5, 6, 7
Radial N. C6, 7, 8
Ulnar N. C8, T1
Median N. C6, 7, 8
Anterior view
Posterior view

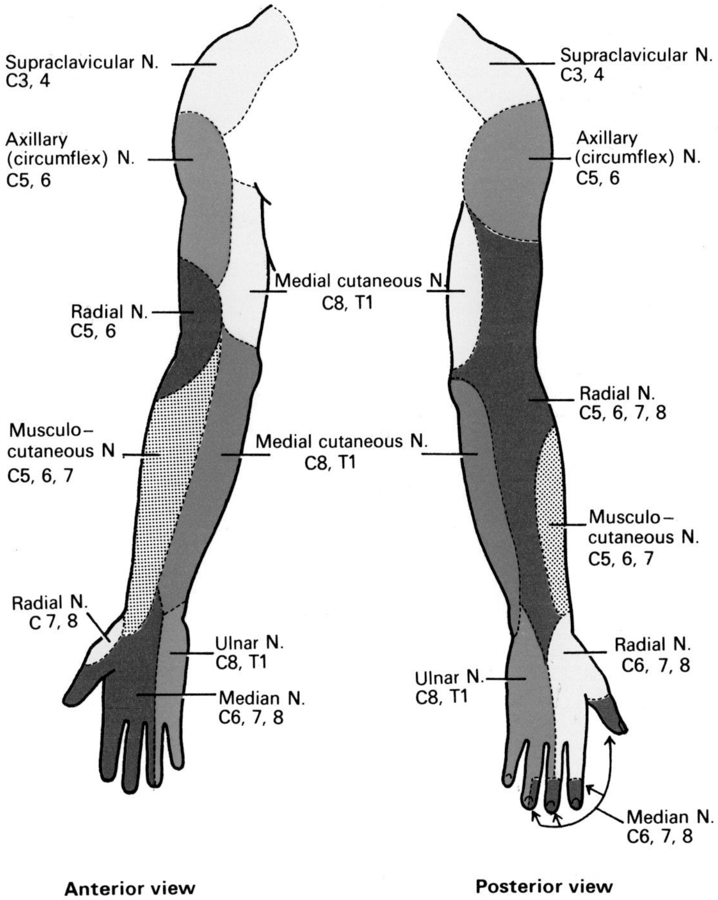

Figure 12:30 The distribution and origins of the cutaneous nerves of the arm.

The lumbar plexus (Fig. 12:31)

The lumbar plexus is formed by the anterior rami of the *first three* and *part of the fourth* lumbar nerves. The plexus is situated in front of the transverse processes of the lumbar vertebrae and behind the psoas muscle.

The main branches of the lumbar plexus and the nerve roots which contribute to them are:

lumbar plexus. It passes behind the inguinal ligament to enter the thigh in close association with the femoral artery. It divides into cutaneous and muscular branches to supply the skin and the muscles of the front of the

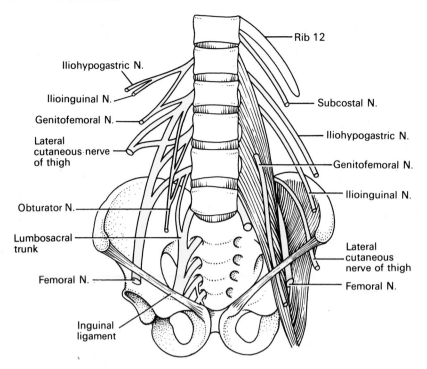

Figure 12:31 The lumbar plexus.

thigh. It has one branch, the *saphenous nerve*, which supplies the medial aspect of the leg.

The obturator nerve supplies the adductor muscles of the thigh and the skin of the medial aspect of the thigh. It ends just above the level of the knee joint.

The lumbosacral trunk descends into the pelvis and makes a contribution to the sacral plexus.

The sacral plexus (Fig. 12:32)
The sacral plexus is formed by the anterior rami of the *first, second and third sacral nerves* and by the *lumbosacral trunk* which is formed by the *fifth* and *part of the fourth lumbar nerves*. It lies in the posterior wall of the pelvic cavity.

The sacral plexus divides into a number of branches which supply the muscles and skin of the pelvic floor, muscles around the hip joint and the pelvic organs. In addition to these it provides the *sciatic nerve* which contains fibres from L 4, 5, S 1, 2, 3.

The sciatic nerve is the largest nerve in the body. It is about 2 cm ($\frac{3}{4}$ inch) wide at its origin. It passes through the greater sciatic foramen into the buttock then descends through the posterior aspect of the thigh supplying the hamstring muscles. At the level of the middle of the femur it divides to form the *tibial* and the *common peroneal nerves*.

The tibial nerve descends through the popliteal fossa to the posterior aspect of the leg where it supplies the muscles and skin. Branches supply muscles and skin of

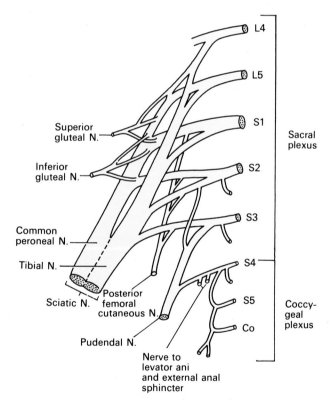

Figure 12:32 The sacral and coccygeal plexuses.

the sole of the foot and the toes. One of the main branches is the *sural nerve* which supplies the tissues in the area of the heel, the lateral aspect of the ankle and the dorsum of the foot.

The common peroneal nerve descends obliquely along the lateral aspect of the popliteal fossa, winds round the neck of the fibula into the front of the leg where it divides into the *deep peroneal* (anterior tibial) and the *superficial peroneal* (musculocutaneous) nerves. These nerves supply the skin and muscles of the anterior aspect of the leg and the dorsum of the foot and toes.

The main nerves of the leg are shown in Figure 12:33. The distribution and origins of the cutaneous nerves of the leg are shown in Figure 12:34.

The coccygeal plexus (Fig. 12:32)

The coccygeal plexus is a very small plexus formed by part of the fourth sacral nerve and by the fifth sacral and the coccygeal nerves. The nerves from this plexus supply the skin in the area of the coccyx and the muscles of the pelvic floor.

The thoracic nerves

The thoracic nerves do not intermingle to form plexuses.

There are 12 pairs and the first 11 pairs are called the *intercostal nerves*. They pass between the ribs supplying them, the intercostal muscles and the overlying skin. The 12th pair are called the subcostal nerves.

The 7th to the 12th thoracic nerves also supply the muscles and the skin of the posterior and anterior abdominal wall.

THE CRANIAL NERVES (Fig. 12:35)

There are 12 pairs of cranial nerves. These nerves have their cells in the brain. Some of the cranial nerves are sensory, some are motor, and those containing both sensory and motor fibres are described as mixed. They are given names and numbers.

1. Olfactory—sensory
2. Optic—sensory
3. Oculomotor—motor
4. Trochlear—motor
5. Trigeminal—mixed
6. Abducent—motor

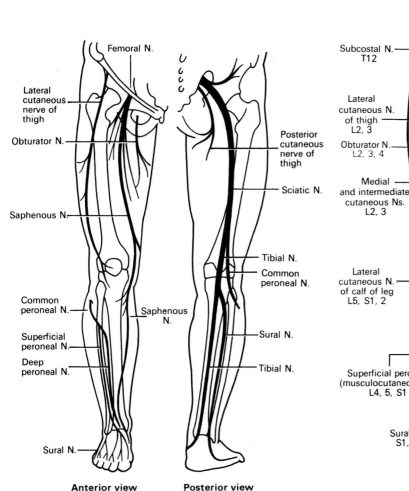

Figure 12:33   The main nerves of the leg.

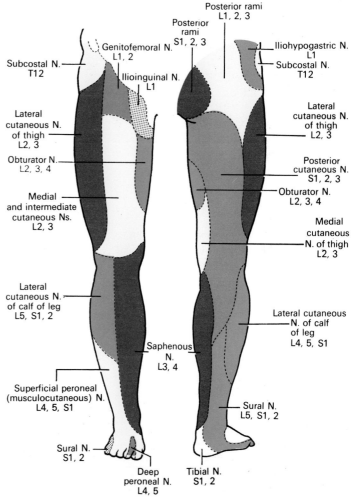

Figure 12:34   The distribution and origins of the cutaneous nerves of the leg.

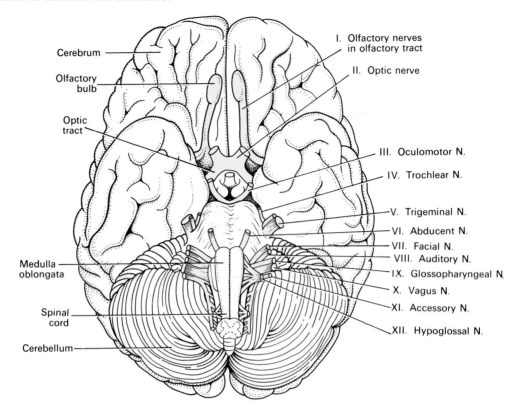

Cerebrum

Olfactory bulb

Optic tract

Medulla oblongata

Spinal cord

Cerebellum

I. Olfactory nerves in olfactory tract

II. Optic nerve

III. Oculomotor N.

IV. Trochlear N.

V. Trigeminal N.
VI. Abducent N.
VII. Facial N.
VIII. Auditory N.
IX. Glossopharyngeal N.
X. Vagus N.
XI. Accessory N.
XII. Hypoglossal N.

Figure 12:35 The inferior surface of the brain showing the cranial nerves.

7. Facial—mixed
8. Vestibulocochlear (auditory)—sensory
9. Glossopharyngeal—mixed
10. Vagus—mixed
11. Accessory—motor
12. Hypoglossal—motor

1. The olfactory nerves (sensory)

The olfactory nerves are the nerves of the *sense of smell*. Their sensory nerve endings and fibres arise in the upper part of the mucous membrane of the nose and pass upwards through the cribriform plate of the ethmoid bone (Fig. 12:36). These nerves pass to the *olfactory bulb*, a group of nerve cells where synapses occur and the impulse is passed to a second neurone. The nerves then proceed backwards as the olfactory tract, to the area for the perception of smell in the temporal lobe of the cerebrum.

2. The optic nerves (sensory) (Fig. 12:37)

The optic nerves are the nerves of the *sense of sight*. The fibres originate in the retinae of the eyes and they combine to form the optic nerves which are about 4 cm in length. They are directed backwards and medially through the posterior part of the orbital cavity. They then pass through the *optic foramina* of the sphenoid bone into the

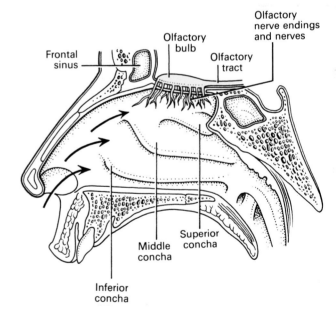

Frontal sinus

Olfactory bulb

Olfactory nerve endings and nerves

Olfactory tract

Middle concha

Superior concha

Inferior concha

Figure 12:36 The olfactory nerve.

cranial cavity and join at the *optic chiasma* just above the pituitary gland. The nerves proceed backwards as the *optic tracts* to the *lateral geniculate body*. Impulses pass from these to the centre for sight in the occipital lobes of the cerebrum and to the cerebellum. In the occipital lobe

5. The trigeminal nerves (mixed)

The trigeminal nerves contain motor and sensory fibres and are among the largest of the cranial nerves. They are the chief sensory nerves for the face and head, receiving impulses of *pain, temperature and touch*. The motor fibres stimulate the *muscles of mastication*.

There are three main branches of the trigeminal nerves and the distribution of the sensory fibres is shown in Figure 12:38.

(a) *The ophthalmic nerves* are sensory only and supply the *lacrimal glands*, the *conjunctiva of the eyes*, the *forehead*, the *eyelids*, the *anterior aspect* of the *scalp* and the *mucous membrane of the nose*.

(b) *The maxillary nerves* are sensory only and supply *the cheeks*, the *upper gums*, the *upper teeth* and the *lower eyelids*.

(c) *The mandibular nerves* contain both sensory and motor fibres. These are the largest of the three divisions and they supply the *teeth and gums of the lower jaw*, the *pinna of the ears*, the *lower lip* and the *tongue*. The motor fibres supply the *muscles of mastication*.

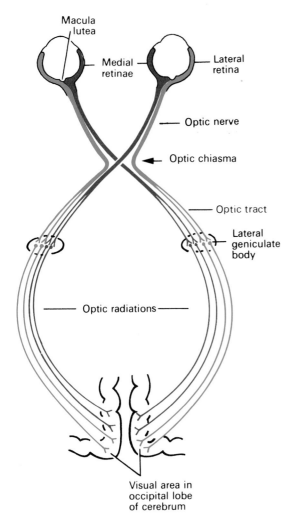

Figure 12:37   The visual pathway.

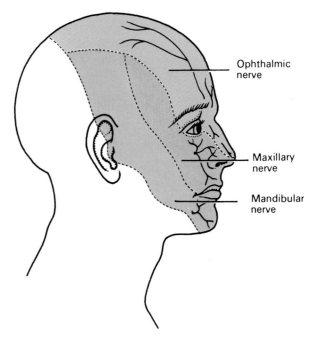

Figure 12:38   The cutaneous distribution of the main branches of the right trigeminal nerve.

sight is perceived, and in the cerebellum the impulses from the eyes contribute to the maintenance of balance (see p. 179).

3. The oculomotor nerves (motor)

The oculomotor nerves arise from nerve cells near the aqueduct of the midbrain. They supply four of the muscles which move the eyes, namely the *superior, inferior and medial recti* and the *inferior oblique muscles*. They also supply the *ciliary muscles* which control the focusing power of the eyes by altering the shape of the lens, the *circular muscles of the iris* producing constriction of the pupils and the muscles which control the movements of the upper eyelids (see p. 210).

4. The trochlear nerves (motor)

The trochlear nerves arise from nerve cells near the aqueduct of the midbrain. These nerves supply the *superior oblique muscles* of the eyes.

6. The abducent nerves (motor)

The abducent nerves arise from a group of nerve cells lying under the floor of the fourth ventricle. They supply the *lateral rectus muscles* of the eyeballs.

7. The facial nerves (mixed)

The facial nerves are composed of both motor and sensory nerve fibres. The fibres arise from nerve cells in the lower part of the pons varolii. The motor fibres supply the *muscles of facial expression*. The sensory fibres convey impulses from the *taste buds* in the anterior two-thirds of the tongue to the taste perception area in the cerebral cortex.

8. The vestibulocochlear (auditory) nerves (sensory)

The vestibulocochlear nerves are composed of two distinct sets of fibres which form:

The vestibular nerves
The cochlear nerves

The vestibular nerves arise from the semicircular canals of the inner ear and convey impulses to the cerebellum. They are associated with the *maintenance of posture and balance.*

The cochlear nerves originate in the cochlea or internal ear and convey impulses to the hearing areas in the cerebral cortex where *sound is perceived.*

9. The glossopharyngeal nerves (mixed)

The glossopharyngeal nerves are composed of motor and sensory fibres. The fibres arise from nuclei in the medulla oblongata. The motor fibres stimulate the *muscles of the pharynx* and the *secretory cells of the parotid glands.*

The sensory fibres convey impulses from the *taste buds* in the posterior third of the tongue and from the *tonsils and pharynx* to the cerebral cortex.

10. The vagus nerves (mixed) (Fig. 12:39)

The vagus nerves are composed of both motor and sensory fibres. They have a more extensive distribution than any other cranial nerves. They arise from nerve cells in the medulla oblongata and other nuclei, and pass through a foramen in the base of the skull then downwards through the neck into the thorax and the abdomen.

The motor fibres supply the *smooth muscles* and *secretory glands* of the pharynx, larynx, trachea, heart, oesophagus, stomach, intestine, pancreas, gall bladder, bile ducts, spleen, kidneys, ureter and blood vessels in the thoracic and abdominal cavities.

The sensory fibres convey impulses from the *lining membranes* of the same structures to the brain.

11. The accessory nerves (motor)

The accessory nerves arise from nerve cells in the medulla oblongata and in the spinal cord. The fibres supply the *sternocleidomastoid and trapezius muscles.* Branches join the vagus nerves and supply the *pharyngeal and laryngeal muscles.*

12. The hypoglossal nerves (motor)

The hypoglossal nerves arise from nerve cells in the medulla oblongata. They supply the *muscles of the tongue and muscles surrounding the hyoid bone.*

A summary of the cranial nerves is given in Table 12:3.

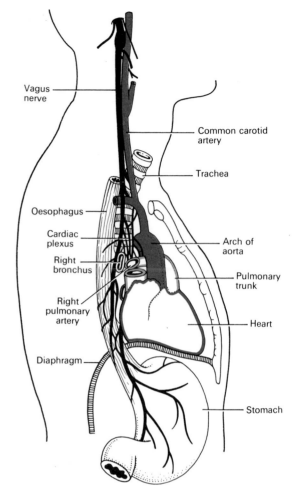

Figure 12:39 The vagus nerve in the thorax viewed from the side.

The Autonomic Nervous System

The autonomic or involuntary part of the nervous system controls the functions of the body which are carried out automatically. Autonomic activity is initiated at levels of the brain below the cerebrum. This means that stimulation does not occur voluntarily but the individual may be conscious of the effects of stimulation, such as an increase in the heart rate.

The following list provides *some examples* of physiological activities controlled by the autonomic nervous system:

The rate and force of the heart beat
The secretion of the glands of the alimentary tract
The contraction of involuntary muscle
The size of the pupils of the eyes

The *efferent nerves* of the autonomic nervous system

Table 12:3   Summary of the cranial nerves

Name and no.	Central connexion	Peripheral connexion	Function
1. Olfactory (sensory)	Smell area in temporal lobe of cerebrum through olfactory bulb	Mucous membrane in roof of nose	Sense of smell
2. Optic (sensory)	Sight area in occipital lobe of cerebrum Cerebellum	Retina of the eye	Sense of sight Balance
3. Oculomotor (motor)	Nerve cells near the floor of the aqueduct of the midbrain	Superior, inferior, and medial rectus muscles of the eye Ciliary muscles of the eye Circular muscle fibres of the iris	Moving the eyeball Focusing Regulating the size of the pupil
4. Trochlear (motor)	Nerve cells near floor of aqueduct of midbrain	Superior oblique muscles of the eyes	Movement of the eyeball
5. Trigeminal (mixed)	Motor fibres from the pons varolii Sensory fibres from the trigeminal ganglion	Muscles of mastication Sensory to gums, cheek, lower jaw, iris, cornea	Chewing Sensation from the face
6. Abducens (motor)	Floor of fourth ventricle	Lateral rectus muscle of the eye	Movement of the eye
7. Facial (mixed)	Pons varolii	Sensory fibres to the tongue Motor fibres to the muscles of the face	Sense of taste Movements of facial expression
8. Vestibulocochlear (sensory) (a) Vestibular (b) Cochlear	Cerebellum Hearing area of cerebrum	Semicircular canals in the inner ear Organ of Corti in cochlea	Maintenance of balance Sense of hearing
9. Glossopharyngeal (mixed)	Medulla oblongata	Parotid glands Back of tongue and pharynx	Secretion of saliva Sense of taste Movement of pharynx
10. Vagus (mixed)	Medulla oblongata	Pharynx, larynx; organs, glands, ducts, blood vessels in the thorax and abdomen	Movement and secretion
11. Accessory (motor)	Medulla oblongata	Sternocleidomastoid, trapezius, laryngeal and pharyngeal muscles	Movement of the head, shoulders, pharynx and larynx
12. Hypoglossal (motor)	Medulla oblongata	Tongue	Movement of tongue

arise from nerve cells in the brain. These emerge at various levels between the midbrain and the sacral region of the spinal cord. Many of them travel within the same nerve sheath as the peripheral nerves of the central nervous system to reach the organs which they innervate.

For descriptive convenience the autonomic nervous system is divided into two parts: *sympathetic and parasympathetic.*

THE SYMPATHETIC NERVOUS SYSTEM (THORACOLUMBAR OUTFLOW)

Three neurones are involved in conveying impulses from their origin in the *hypothalamus* and the *medulla oblongata* to the effector organs and tissues.

Neurone 1 has its cell in the *brain* and its fibre extends into the *spinal cord.*

Neurone 2 has its cell in the *lateral column of grey matter* in the spinal cord between the levels of the first thoracic and second or third lumbar vertebrae. The nerve fibre of this cell leaves the cord by the anterior root and terminates in one of the ganglia either in the *lateral chain of sympathetic ganglia* or passes through it to one of the *prevertebral ganglia.*

Neurone 3 has its cell in a ganglion and terminates in the organ or tissue being stimulated.

Sympathetic ganglia

The lateral chain of sympathetic ganglia. This is a chain of ganglia which extends from the upper cervical level to the sacrum. The ganglia are attached to each other by nerve fibres and one chain lies in close association with each side of the bodies of the vertebrae. *Preganglionic fibres* which emerge from the cord may synapse with a ganglion cell at the same level or it may pass up or down the chain through one or more ganglia before it synapses with the cell of the third neurone. For example, the nerve which dilates the pupil of the eye leaves the cord at the level of the first thoracic vertebra and passes through three ganglia before synapsing with the cell of neurone 3 in the superior cervical ganglion. The *postganglionic fibre* then passes to the eye.

The prevertebral ganglia. There are three prevertebral ganglia situated in the abdominal cavity close to the origins of arteries of the same names. They are:

The coeliac ganglion
The superior mesenteric ganglion
The inferior mesenteric ganglion

The ganglia consist of nerve cells rather diffusely distributed among a network of nerve fibres which form plexuses.

Second neurone sympathetic fibres pass through the lateral chain to reach these ganglia which consist of the cells and dendrites of the third neurone.

THE PARASYMPATHETIC NERVOUS SYSTEM (CRANIOSACRAL OUTFLOW)

Two neurones are involved in the transmission of impulses from their source to the effector organ.

Neurone 1 has its cell either in the brain or in the spinal cord. Those originating in the brain are the *cranial nerves* which have their cells in the midbrain and brain stem and their nerve fibres terminate outside the brain (see Fig. 12:41). The cells of the *sacral outflow* are in the lateral columns of grey matter at the distal end of the spinal cord and their fibres leave the cord in sacral segments 2, 3 and 4 and synapse with second neurone cells in the walls of the pelvic organs.

Neurone 2 has its cell either in a ganglion or in the wall of the organ supplied (see Fig. 12:41).

Functions of the autonomic nervous system

The autonomic nervous system is involved in a complex of reflex activity which, like the reflexes described previously, depend upon sensory input to the brain and spinal cord and motor output. In this case the reflex action is contraction of involuntary (smooth and cardiac)

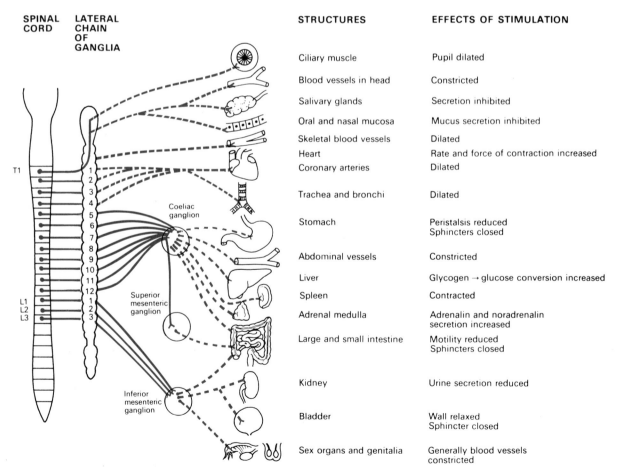

STRUCTURES	EFFECTS OF STIMULATION
Ciliary muscle	Pupil dilated
Blood vessels in head	Constricted
Salivary glands	Secretion inhibited
Oral and nasal mucosa	Mucus secretion inhibited
Skeletal blood vessels	Dilated
Heart	Rate and force of contraction increased
Coronary arteries	Dilated
Trachea and bronchi	Dilated
Stomach	Peristalsis reduced Sphincters closed
Abdominal vessels	Constricted
Liver	Glycogen → glucose conversion increased
Spleen	Contracted
Adrenal medulla	Adrenalin and noradrenalin secretion increased
Large and small intestine	Motility reduced Sphincters closed
Kidney	Urine secretion reduced
Bladder	Wall relaxed Sphincter closed
Sex organs and genitalia	Generally blood vessels constricted

Figure 12:40   The sympathetic outflow, the main structures supplied and the effects of stimulation. Solid red lines—preganglionic fibres, broken lines—postganglionic fibres.

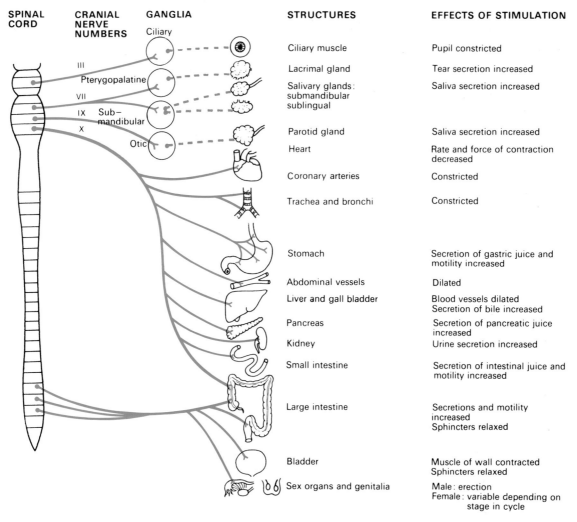

SPINAL CORD	CRANIAL NERVE NUMBERS	GANGLIA	STRUCTURES	EFFECTS OF STIMULATION
		Ciliary	Ciliary muscle	Pupil constricted
	III	Pterygopalatine	Lacrimal gland	Tear secretion increased
	VII		Salivary glands: submandibular sublingual	Saliva secretion increased
	IX	Sub-mandibular		
	X	Otic	Parotid gland	Saliva secretion increased
			Heart	Rate and force of contraction decreased
			Coronary arteries	Constricted
			Trachea and bronchi	Constricted
			Stomach	Secretion of gastric juice and motility increased
			Abdominal vessels	Dilated
			Liver and gall bladder	Blood vessels dilated Secretion of bile increased
			Pancreas	Secretion of pancreatic juice increased
			Kidney	Urine secretion increased
			Small intestine	Secretion of intestinal juice and motility increased
			Large intestine	Secretions and motility increased Sphincters relaxed
			Bladder	Muscle of wall contracted Sphincters relaxed
			Sex organs and genitalia	Male: erection Female: variable depending on stage in cycle

Figure 12:41   The parasympathetic outflow, the main structures supplied and the effects of stimulation. Solid blue lines—preganglionic fibres, broken lines—postganglionic fibres. Where there are no broken lines the 3rd neurone is in the wall of the structure.

muscle or the secretion of glands. These reflexes are co-ordinated in the brain below the level of consciousness, that is, below the level of the cerebrum. Some sensory input does reach consciousness and may result in temporary inhibition of the reflex action, for example, reflex micturition can be inhibited temporarily (see p. 158).

The majority of the organs of the body are supplied by *both* sympathetic and parasympathetic nerve fibres which have opposite effects. In health their effects are very delicately balanced which ensures the optimum functioning of the organ.

Sympathetic stimulation as a whole has similar effects on the body as those produced by the hormones *adrenalin* and *noradrenalin* secreted by the medulla of the adrenal glands. It prepares the body to deal with excitement and with stressful situations, for example strengthening its defences in danger and in extremes of environmental temperature. It is sometimes said that sympathetic stimulation mobilises the body for 'fight or flight'.

Parasympathetic stimulation has a tendency to slow down body processes except the digestion and absorption of food. Its effect is that of a 'peace maker' allowing restoration processes to occur quietly and peacefully.

Normally the two systems function simultaneously producing a regular heart beat, normal temperature and an internal environment compatible with the immediate external surroundings.

THE EFFECTS OF AUTONOMIC STIMULATION ON THE VARIOUS SYSTEMS OF THE BODY

ON THE CARDIOVASCULAR SYSTEM

Sympathetic stimulation

1. Exerts an accelerating effect upon the sinuatrial node in the heart, thus increasing the rate and force of the heart beat.

2. Causes dilatation of the coronary arteries, increasing the blood supply to the heart muscle.

3. Causes dilatation of the blood vessels supplying skeletal muscle, increasing the supply of oxygen and nutritional materials and the removal of metabolic waste products, thus increasing the capacity of the muscle to do work.

4. Causes sustained contraction of the spleen, thus increasing the volume of circulating blood.

5. Raises the blood pressure by constricting the small arteries and arterioles which supply the skin. In this way an increased blood supply is available for highly active tissue, such as skeletal muscle.

6. Constricts the blood vessels in the secretory glands of the digestive system thus reducing the flow of digestive juices.

Parasympathetic stimulation

1. Exerts an inhibitory action on the sinuatrial node in the heart, decreasing its rate and force of contraction.

2. Produces constriction of the coronary arteries thus reducing the flow of blood to the heart muscle.

ON THE RESPIRATORY SYSTEM

Sympathetic stimulation
Produces dilatation of the bronchi allowing a greater amount of air to enter the lungs at each inspiration, thus in conjunction with the increased heart rate, increasing the oxygen intake and carbon dioxide output of the body.

Parasympathetic stimulation
Produces constriction of the bronchi.

ON THE DIGESTIVE AND URINARY SYSTEMS

Sympathetic stimulation

1. *The liver* converts an increased amount of glycogen to glucose, thus making more carbohydrate immediately available to provide energy.

2. *The adrenal (suprarenal) glands* are stimulated to secrete adrenalin and noradrenalin which, by chemical means, potentiate the effects of sympathetic stimulation.

3. *The stomach and small intestine.* Muscular movement and the secretion of digestive juices are inhibited which delays the digestion, onward movement and absorption of food.

4. *Urethral and anal sphincters.* The muscle tone of the sphincters is increased which inhibits micturition and defaecation.

Parasympathetic stimulation

1. *The stomach and small intestine.* The activity of these two organs is increased, thus speeding up digestion and absorption of food.

2. *The pancreas.* There is an increase in the secretion of pancreatic juice and the hormone insulin.

3. *Urethral and anal sphincters.* Relaxation of the internal urethral sphincter is accompanied by contraction of the muscle of the bladder wall and micturition occurs.

Similarly relaxation of the internal anal sphincter is accompanied by contraction of the muscle of the rectum and defaecation occurs.

ON THE EYE

Sympathetic stimulation
This causes contraction of the radiating muscle fibres of the iris thus *dilatation* of the pupil occurs. Retraction of the eyelids takes place producing a look of alertness and excitement.

Parasympathetic stimulation
This causes contraction of the circular muscle fibres of the iris thus the pupil *constricts*. The eyelids tend to close producing the appearance of sleepiness.

Under normal healthy conditions there is a very fine balance between these two effects ensuring the optimum size of pupil.

ON THE SKIN

Sympathetic stimulation

1. Causes greater activity of the sweat glands, therefore more sweat is produced and heat loss from the body is increased.

2. Produces contraction of the arrectores pilorum (the muscles in the skin) causing heat production and giving the appearance of 'goose flesh'.

3. Causes constriction of the blood vessels preventing heat loss.

There is no parasympathetic nerve supply to the skin and, therefore, nerve supply which is anatomically sympathetic has the dual function of facilitating heat loss and increasing heat production.

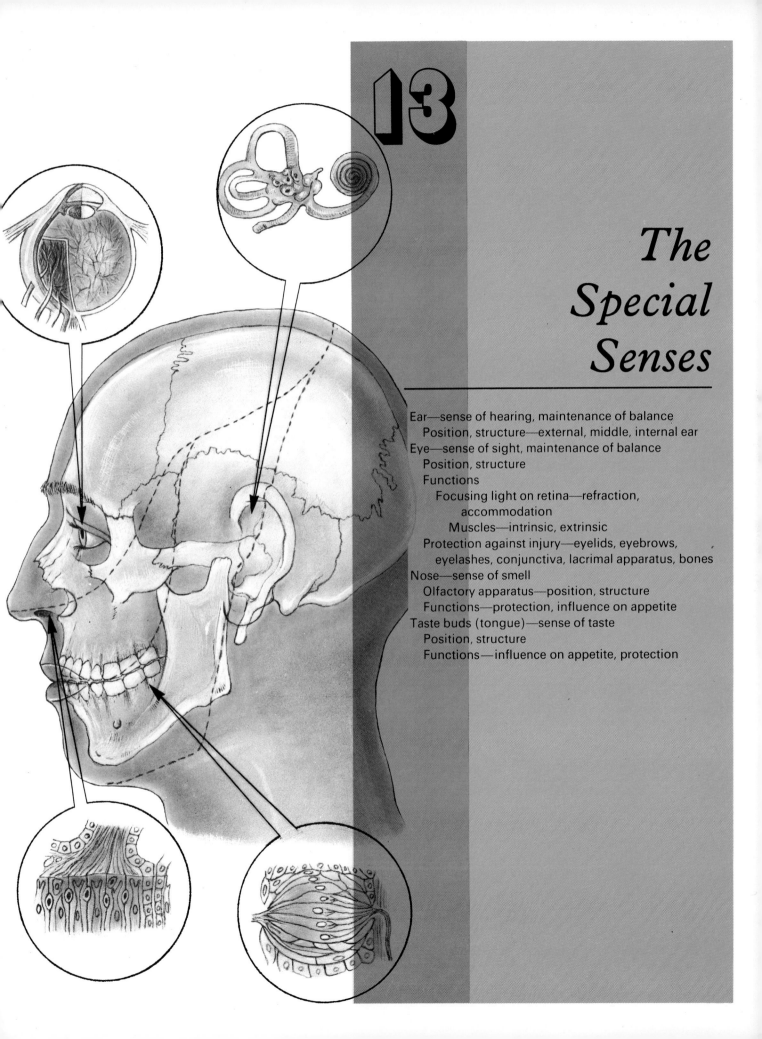

13

The Special Senses

Ear—sense of hearing, maintenance of balance
  Position, structure—external, middle, internal ear
Eye—sense of sight, maintenance of balance
  Position, structure
  Functions
    Focusing light on retina—refraction,
      accommodation
      Muscles—intrinsic, extrinsic
  Protection against injury—eyelids, eyebrows,
    eyelashes, conjunctiva, lacrimal apparatus, bones
Nose—sense of smell
  Olfactory apparatus—position, structure
  Functions—protection, influence on appetite
Taste buds (tongue)—sense of taste
  Position, structure
  Functions—influence on appetite, protection

13. *The Special Senses*

Hearing and the Ear

The ear is the organ of hearing. It is supplied by the *8th cranial nerve*, that is, the *cochlear part* of the *vestibulocochlear* nerve which is stimulated by vibrations caused by sound waves.

With the exception of the auricle or pinna the structures which form the ear are encased within the petrous portion of the temporal bone.

STRUCTURE
The ear is divided into three distinct parts (Figs. 13:1 and 13:6A):

    The external ear
    The tympanic cavity or middle ear
    The internal ear

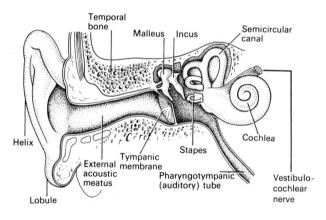

Figure 13:1   The parts of the ear.

THE EXTERNAL EAR
The external ear is divided, for descriptive purposes, into two parts:

    The auricle or pinna
    The external acoustic meatus

The auricle (Fig. 13:2)
The auricle is the expanded portion which projects from the side of the head. It is composed of *fibroelastic cartilage* covered with skin. The skin is covered with fine hairs

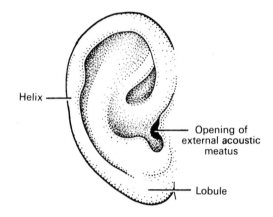

Figure 13:2   The auricle of the ear.

and there are many sebaceous glands opening into the hair follicles. It is deeply grooved and ridged and the most prominent outer ridge is called the *helix*.

The lobule is the soft pliable part at the lower extremity and is composed of fibrous and adipose tissue richly supplied with blood capillaries and covered with skin.

The external acoustic meatus
This is a tube from the auricle to the *tympanic membrane* or ear drum. It consists of a *cartilaginous portion* and an *osseous portion*.

The cartilaginous portion forms the lateral one-third of the canal and the osseous portion the medial two-thirds. The meatus is approximately 2·5 cm (1 inch) in length and is slightly 'S' shaped. When examining the eardrum or syringing the ear of an adult the canal can be straightened by gently pulling the auricle backwards and upwards.

The external acoustic meatus is lined with a continuation of the skin of the auricle. In the cartilaginous part there are modified sweat glands called *ceruminous glands* which secrete *cerumen* or wax. Wax has a protective function in that it traps foreign material, such as dust or insects, preventing them from reaching the tympanic membrane. The skin lining the meatus extends over the outer surface of the tympanic membrane.

The tympanic membrane (Fig. 13:3) completely separates the external acoustic meatus from the middle ear. It is oval in shape with the slightly broader edge upwards. It is formed by three separate types of tissue:

1. The outer covering of *stratified epithelium* is hairless skin and is a continuation of the lining membrane of the external acoustic meatus.
2. The middle coat consists of *fibrous tissue*.
3. The inner lining consists of *cuboidal epithelium* which is continuous with that of the middle ear.

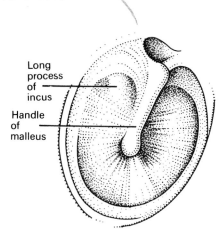

Figure 13:3   The tympanic membrane viewed from the outside showing the shadows cast by the malleus and the incus.

THE TYMPANIC CAVITY OR MIDDLE EAR

The tympanic cavity is an irregular shaped cavity within the petrous portion of the temporal bone. The cavity, its contents and the air sacs which open out of it are lined with mucous membrane. Air fills the cavity and reaches it through the *pharyngotympanic* or *auditory tube* which extends from the nasal part of the pharynx to the middle ear. It is about 4 cm long and is lined with ciliated epithelium. The presence of air at atmospheric pressure on both sides of the tympanic membrane enables it to vibrate when sound waves strike it.

The lateral wall of the middle ear is formed by the tympanic membrane.

The roof and floor are formed by the temporal bone.

The posterior wall is formed by the temporal bone and has an opening which leads to the *mastoid antrim* through which air passes to the mastoid air cells.

The medial wall of the middle ear is composed of a thin wall of temporal bone in which there are two openings:

The fenestra vestibuli or oval window
The fenestra cochleae or round window

The fenestra vestibuli is occluded by part of a small bone called *the stapes*. The fenestra cochlea is filled with a fine sheet of *fibrous tissue*.

The auditory ossicles (Fig. 13:4)
Within the middle ear there are three minute bones known as the *auditory ossicles*. The three bones extend across the cavity from the tympanic membrane to the fenestra vestibuli. They form a series of movable joints with the tympanic membrane, with each other and with the medial wall of the cavity at the fenestra vestibuli.

The auditory ossicles are: *the malleus, the incus* and *the stapes*.

The malleus lies in the lateral aspect of the middle ear. It is shaped rather like a hammer and presents *a head, neck* and *handle*.

The handle of the malleus is in contact with the medial wall of the tympanic membrane and the head forms a movable joint with the incus.

The incus is the intermediate bone of the three. It has been given its name because it is shaped like an anvil. It possesses a *body* and *two processes*. The body articulates with the malleus, the long process with the stapes, and the bone is stabilised by the short process which is fixed by fibrous tissue to the posterior wall of the cavity.

The stapes is the most medial of the ossicles. It is shaped like a stirrup presenting *a head, neck, two limbs and a base*.

The head articulates with the incus and its base fits into the fenestra vestibuli.

The three ossicles are maintained in position by fine ligaments.

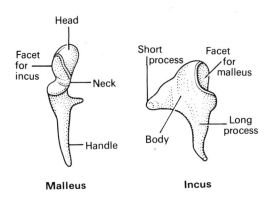

Malleus   **Incus**

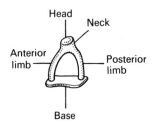

Stapes

Figure 13:4   The auditory ossicles.

THE INTERNAL EAR (Fig. 13:5)

The internal ear contains the organ of hearing and is generally described in two parts:

    The bony labyrinth
    The membranous labyrinth

The membranous labyrinth

The membranous labyrinth is the same shape as the bony labyrinth and is divided into the same parts: a vestibule which contains a *utricle* and a *saccule*, a cochlea and three semicircular canals. It is considerably smaller than

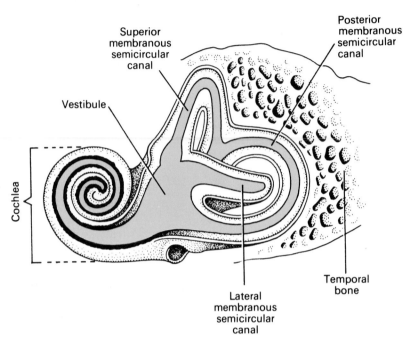

Figure 13:5   The internal ear. Membranous labyrinth coloured.

The bony labyrinth

This is a cavity within the temporal bone which is lined with periosteum. It is larger than the membranous labyrinth of the same shape which fits into it, like a tube within a tube. The space between the bony walls and the membranous tube is filled with fluid known as *perilymph*. The membranous labyrinth also contains fluid, called *endolymph*.

The bony labyrinth is described in three parts:

    One vestibule
    One cochlea
    Three semicircular canals

The vestibule is the expanded part nearest the middle ear. It contains the fenestra vestibuli and the fenestra cochleae.

The cochlea resembles a snails' shell. It has a broad *base* where it is continuous with the vestibule and a narrow *apex*. The cochlea spirals round a central piece of bone called the *modiolus*.

The semicircular canals are arranged so that one is situated in each of the three planes of space. They are continuous with the vestibule.

its bony counterpart and is separated from it by perilymph. It contains endolymph.

The membranous cochlea is called the *duct of the cochlea* and when it is cut in cross section it is triangular in shape (Fig. 13:7). The inferior aspect is formed by a membrane, known as the *basilar membrane*, upon which special neuroepithelial cells and nerve fibres lie.

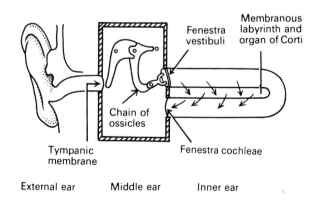

Figure 13:6 A   Diagram of the ear showing the passage of sound waves from the external to the inner ear.

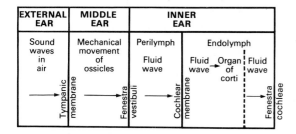

EXTERNAL EAR	MIDDLE EAR	INNER EAR		
Sound waves in air	Mechanical movement of ossicles	Perilymph Fluid wave	Endolymph Fluid wave → Organ of corti	Fluid wave

Figure 13:6B Summary of the transmission of sound through the ear.

Many of the cells are long and narrow and are arranged side by side. Some are surmounted by minute hair-like processes and are known as the *hair cells*. These cells and their nerve fibres form the true organ of hearing known as the *organ of Corti*. The nerve fibres combine to form the *auditory part of the vestibulocochlear nerve (8th cranial nerve)*, which passes through a foramen in the petrous portion of the temporal bone to reach the hearing area in the temporal lobe of the cerebrum.

THE PHYSIOLOGY OF HEARING

Every sound that is produced causes vibrations or disturbances in the atmospheric air. These vibrations travel as a succession of waves known as *sound waves*.

Sound waves travel at approximately 335 metres (1088 feet) per second. It is the function of the ear to pick up these vibrations and direct them towards the cochlea, where the organ of Corti is adapted to transmit them, as nerve impulses, by the 8th cranial nerve to the *hearing area* of the cerebral cortex where they are perceived as sound.

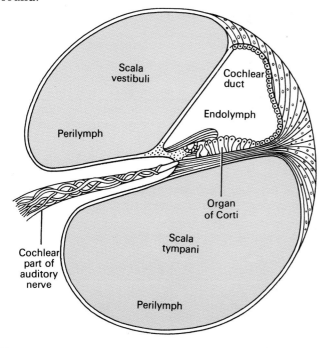

Figure 13:7 A section of the membranous cochlea showing the organ of Corti.

Function of the external ear

The auricle is formed in such a way that the vibrations produced by any sound can be concentrated and directed through the external acoustic meatus towards the tympanic membrane. The tympanic membrane vibrates in harmony with the sound waves.

The cerumen secreted by the lining of the external acoustic meatus protects the canal.

Function of the tympanic cavity or middle ear

The vibrations set up in the tympanic membrane are conveyed through the middle ear by the three auditory ossicles. The handle of the malleus is connected to the inner wall of the tympanic membrane and, as it vibrates, a corresponding movement of the malleus takes place. The movements of the malleus are transmitted via the incus to the stapes. The foot plate of the stapes fits accurately into the fenestra vestibuli. The vibrations transferred from the incus to the stapes cause the foot plate of the stapes to rock to and fro in the fenestra vestibuli. This carries the vibrations to the inner ear. The auditory ossicles can only move and transmit the vibrations of the tympanic membrane to the inner ear when air at atmospheric pressure is present in the middle ear. This air enters the middle ear from the nasal part of the pharynx through the pharyngotympanic tube.

Functions of the inner ear

On the inner aspect of the fenestra vestibuli there is perilymph which surrounds the membranous labyrinth. The rocking movement of the stapes sets up wave motion in the perilymph which in turn causes vibration of the membranous labyrinth. The vibrations of the membranous labyrinth stimulate movement of the endolymph, and subsequently the special neuroepithelial cells forming the *organ of Corti* are stimulated. High-pitched sounds stimulate the cells at the base of the cochlea, and low pitched sounds the cells at the apex.

These nerve impulses are transmitted via the nerve fibres of the *cochlear nerve* which subsequently becomes the *8th cranial nerve*. The *vestibulocochlear nerve* transmits the impulses to various nuclei in the pons varolii and midbrain. Some of the nerve fibres pass to the hearing area of the cerebral cortex on the same side and some cross over to the corresponding area in the opposite hemisphere. It is in the hearing area of the cerebral cortex that the nerve impulses are perceived as sound.

THE SEMICIRCULAR CANALS

It must be appreciated that the semicircular canals have no auditory function. Anatomically they are closely associated with the cochlea but their function is entirely different. Their function is to provide information about the position of the head in space and this is associated with the maintenance of equilibrium and balance.

There are three semicircular canals—the *superior*,

posterior and *lateral* which lie in the three planes of space. They are situated above and behind the vestibule of the inner ear and open into it.

Structure of the semicircular canals (Fig. 13:8)
The semicircular canals like the cochlea are composed of an outer bony wall and inner membranous tubes or *ducts*.

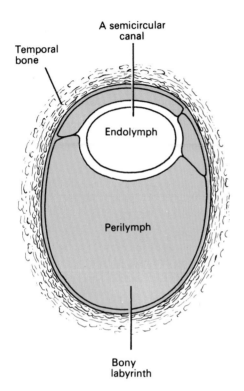

Figure 13:8 Transverse section of a semicircular canal.

The membranous ducts contain endolymph and they are separated from the bony wall by perilymph.

The utricle is a membranous sac which is part of the vestibule. The three membranous ducts open into it at enlargements called the *ampullae*.

The saccule is a part of the vestibule and communicates with the utricle and the cochlea.

Within the walls of the utricle, the saccule and the ampullae there are fine epithelial cells with minute hairlike projections known as the *hair cells*. Amongst these hair cells there are the minute nerve endings of the *vestibular part* of the vestibulocochlear nerve.

Functions of the semicircular canals
The semicircular canals, the utricle and the saccule are associated with the position of the head in space. Any change of position of the head causes movement in the perilymph and endolymph. This movement stimulates the nerve endings and the hair cells in the utricle, the saccule and the ampullae. The resultant nerve impulses are then transmitted by the vestibular nerve, which joins the nerve from the cochlea of the inner ear, to

form the *vestibulocochlear nerve*. The vestibular branch of the 8th cranial nerve passes through an important relay station, the *vestibular nucleus*, then to the *cerebellum*.

The cerebellum also receives nerve impulses from the eyes and the muscles and joints. Impulses from these three sources are coordinated and efferent impulses pass to the cerebrum where position in space is perceived, and to muscles to maintain posture and balance.

Sight and the Eye

The eye is a special organ of the sense of sight. It is supplied by the *optic nerve* (2nd cranial nerve).

The eye is situated in the orbital cavity. It is almost spherical in shape and is approximately 2·5 cm (1 inch) in diameter. It is smaller in diameter than the orbital cavity and the intervening space is occupied by fatty tissue. The bony walls of the orbit and the fat help to protect the eye from injury.

Structurally the two eyes are the same but, unlike the ear, some of their activities are coordinated so that they function as a pair. It is possible to see with only one eye but three dimensional vision is impaired when only one eye is used, especially in relation to the judgement of the distance of an object.

THE STRUCTURE OF THE EYE (Fig. 13:9)
There are three layers of tissue which constitute the walls of the eye. They are:
1. The outer fibrous layer—sclera, cornea
2. The middle vascular layer—choroid, ciliary body, iris
3. The inner nervous tissue layer—retina

Other structures include:
The lens
The aqueous fluid or humour
The vitreous body

THE SCLERA AND CORNEA
The sclera, or the white of the eye, forms the outermost layer of tissue of the posterior and lateral aspects of the eyeball and is continuous anteriorly with the transparent *cornea*.

The sclera is composed of *fibrous tissue*. It is a firm membrane and maintains the form and shape of the eye. It gives attachment to the *extraoccular or extrinsic muscles* of the eye (see p. 209).

Anteriorly the sclera continues as a clear transparent epithelial membrane called *the cornea*. Light rays pass through the cornea to reach the retina. It is convex anteriorly and is involved in refraction or bending of light rays to focus them on the retina.

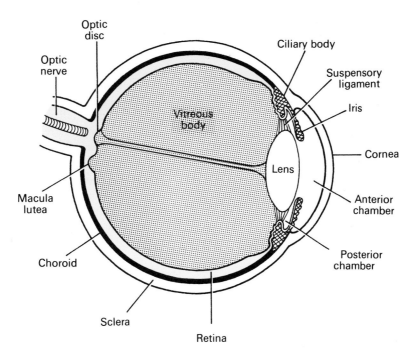

Figure 13:9 Section of the eye.

THE CHOROID

The choroid lines the posterior five-sixths of the inner surface of the sclera. It is very rich in blood vessels and is a deep chocolate brown in colour. Light enters the eye through the pupil, stimulates the nerve endings in the retina then is absorbed by the choroid.

The ciliary body

The ciliary body is a continuation of the choroid anteriorly, consisting of non-striated muscle fibres (*ciliary muscle*) and epithelial cells. It gives attachment to a fine ligament called the *suspensory ligament* which, at its other end, is attached to the capsule enclosing the lens. Contraction and relaxation of the ciliary muscle changes the thickness of the lens which bends the light rays entering the eye to focus them on the retina. The epithelial cells secrete *aqueous fluid* (humour) into the anterior segment of the eye, that is, the space in front of the lens. It may also be associated with the secretion of the *vitreous body* (humour) which occupies the space behind the lens.

The ciliary body is supplied by parasympathetic nerves which are branches of the occulomotor nerve (3rd cranial nerve). Stimulation causes contraction of the muscle and accommodation of the eye.

The iris

The iris extends anteriorly from the ciliary body covering the anterior one-sixth of the eye. It lies behind the cornea in front of the lens. It divides the anterior segment of

the eye into *anterior and posterior chambers* which contain *aqueous fluid* secreted by the ciliary body. It is a circular body composed of pigment cells and two layers of muscle fibres, one circular and the other radiating (Fig. 13:10). In the centre there is an aperture called the *pupil*.

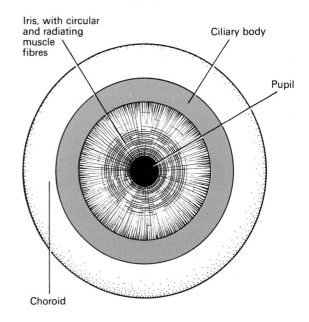

Figure 13:10 Drawing of the choroid, ciliary body and iris. Viewed from the front.

The pupil varies in size depending upon the intensity of light present. In bright light the circular muscle fibres contract and *constrict the pupil*. In dim light the radiating muscle fibres contract *dilating the pupil*.

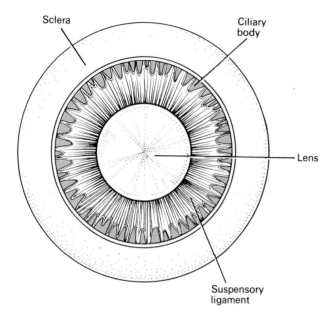

Figure 13:11   Drawing of the lens and suspensory ligament viewed from the front. The iris has been removed.

The iris is supplied by parasympathetic and sympathetic nerves. Parasympathetic stimulation, supplied by the occulomotor nerve, constricts the pupil and sympathetic stimulation from the superior cervical ganglion, dilates the pupil.

The iris is the coloured part of the eye and its colour depends on the number of pigment cells present. Albinos have no pigment cells and people with blue eyes have fewer than those with brown eyes.

The lens (Fig. 13:11)

The lens is described as a circular bi-convex transparent body. It is suspended from the ciliary body by the *suspensory ligament* and enclosed within a capsule.

The capsule and the lens are *highly elastic*. The lens lies immediately behind the pupil and its thickness is controlled by the ciliary muscle through the suspensory ligament.

THE RETINA (Figs. 13:12 and 13:13)

The retina is the innermost layer of the wall of the eye. It is an extremely delicate membrane and is especially adapted to be stimulated by light rays. It is composed of several layers of nerve cells and nerve fibres lying on a pigmented layer of epithelial cells which attach it to the choroid. The layer highly sensitive to light is known as the *layer of rods and cones*.

The retina lines about three-quarters of the eyeball and is thickest at the back and thins out anteriorly to end just behind the ciliary body. Near the centre of the posterior part of the retina there is a small depression. Here the cells appear yellow in colour, hence it is called the *macula lutea*. In the centre of the area there is a little depression called the *fovea centralis*. This is the most sensitive part of the retina and consists of only *cone-shaped cells*. Towards the anterior part of the retina there are fewer cone- than rod-shaped cells.

The retina has a purple tint due to the presence of *rhodopsin or visual purple* in the rods. This substance is bleached by bright light, and vitamin A is essential for its resynthesis.

About 0·5 cm to the nasal side of the macula lutea all the fibres of the retina converge to form the *optic nerve* which passes through the sphenoid bone, eventually to reach the cerebral cortex in the occipital lobe of the cerebrum.

The small area of retina where the optic nerve leaves the eye has no light-sensitive cells. It is called the *optic disc or the blind spot*.

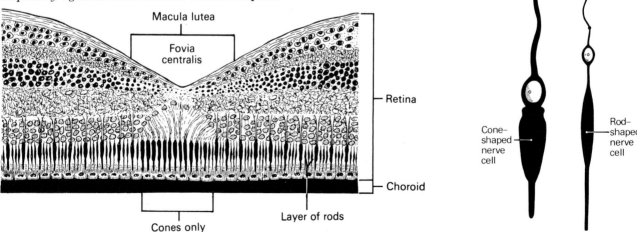

Figure 13:12A   Magnified section of the retina.

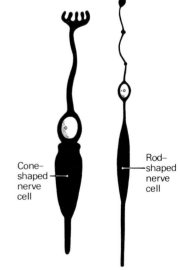

Figure 13:12B   Cone- and rod-shaped cells in the retina.

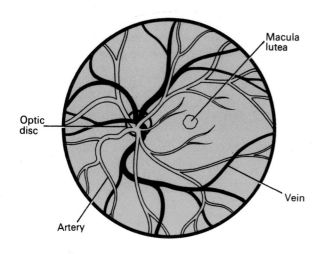

Figure 13:13   The retina as seen through the pupil.

Blood supply to the eye

The eye is supplied with arterial blood by the *ciliary arteries* and the *central arteries of the retina*. These are branches of the ophthalmic artery which is one of the branches of the internal carotid artery.

Venous drainage is by a number of veins which eventually empty into the cavernous sinus.

INTERIOR OF THE EYEBALL

The anterior segment of the eye, that is, the space between the cornea and the lens, is incompletely divided into *anterior* and *posterior chambers* by the iris. Both chambers contain a clear, watery fluid known as the *aqueous fluid (humour)*. This fluid filters into the posterior chamber from capillaries in the ciliary body and is drained back into the circulation through the tiny ducts at the angle of the sclera and ciliary body.

Behind the lens and filling the cavity of the eyeball is the *vitreous body (humour)*. This is a soft, colourless, transparent, jelly-like substance composed of 99 per cent water, some salts and mucoprotein. It maintains sufficient intraocular pressure to support the retina against the choroid and prevent the walls of the eyeball from collapsing.

The eye keeps its shape because of the intraocular pressure exerted by the vitreous body and the aqueous fluid. The pressure, between 10 and 25 mmHg, remains fairly constant throughout life.

THE OPTIC NERVES

The fibres of the optic nerve originate in the retina of the eye. All the fibres converge at a point about 0·5 cm to the nasal side of the macula lutea to form the optic nerve. The nerve pierces the choroid and sclera to pass backwards and medially through the orbital cavity. It then passes through the optic foramen of the sphenoid bone, backwards and medially to meet the nerve from the other eye at the *optic chiasma*.

The optic chiasma

This is situated immediately in front of and above the pituitary gland resting in the hypophyseal fossa. In the optic chiasma the nerve fibres of the optic nerve from the *nasal side* of each retina *cross over* to the *opposite side*. The fibres from the *temporal side* of the retina *do not cross over* but continue backwards on *the same side* (see Fig. 13:14).

The optic tract

This is the term used to describe the optic pathway posterior to the optic chiasma. Each tract consists of the nasal fibres from the retina of one eye and the temporal fibres of the retina from the other eye. The optic tracts pass backwards through the cerebrum to groups of nerve cells known as the *lateral geniculate bodies*. They lie just below and behind the thalamus and act as relay stations for the optic nerves. From here the nerve fibres proceed backwards and medially as the *optic radiations* to terminate

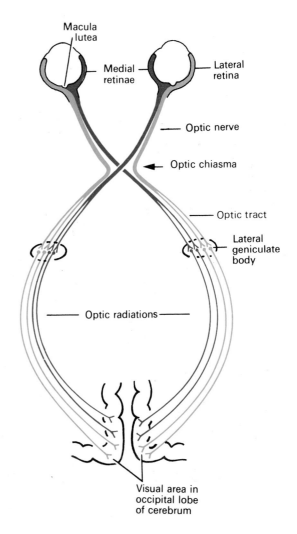

Figure 13:14   The optic nerves and their pathways.

in the *visual area* of the cerebral cortex in the occipital lobes of the cerebrum. Other neurones originating in the lateral geniculate body convey impulses from the eyes to *the cerebellum* where they contribute to the maintenance of balance.

Fig. 13:15). This range of colour is called the *spectrum of visible light*. A rainbow is an example of white light from the sun being broken up by raindrops which act as prisms and reflectors.

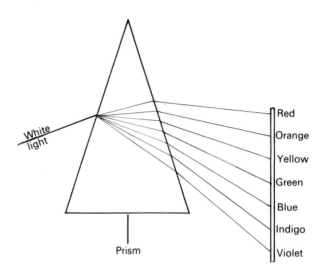

Figure 13:15   White light broken into the colours of the visible spectrum when passing through a prism.

THE PHYSIOLOGY OF SIGHT

Light waves travel at a speed of 186 000 miles (297 600 kilometres) per second. Light is reflected into the eyes by the various objects within the field of vision. White light is actually a combination of all the colours of the visual spectrum or rainbow, that is, red, orange, yellow, green, blue, indigo and violet. This can be demonstrated by passing white light through a glass prism which refracts or bends the rays of the different colours to a greater or lesser extent. The different colours have different wavelengths, therefore when white light is passed through a prism the different colours take different paths. Red light has the longest wavelength and violet the shortest (see

The spectrum of light

The spectrum of light is extremely broad but only a small part of it is visible to the human eye (Fig. 13:16). Beyond the short end there are the cosmic rays, X-rays and ultraviolet rays. Beyond the long end there are the infra-red heat waves, radar, and radio waves.

A specific colour of the visible spectrum is perceived when only one wavelength of light *is reflected* by an object and all other wavelengths are *absorbed*. For example, when an object appears *white* all wavelengths are being *reflected*, when it appears *black* all wavelengths are being *absorbed* and it appears *red* when only the *red wavelength is reflected* and *all others are absorbed*. Black and white are not regarded as colours.

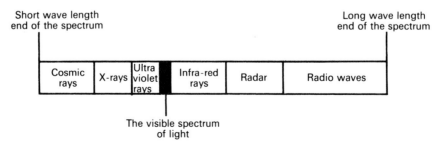

Figure 13:16   The spectrum of light.

In order to achieve clear vision light reflected from objects within the visual field is focused on to the retinae of both eyes.

Two processes are involved in producing a clear image:

Refraction of the light rays

Accommodation of the eyes

Although these may be considered as two separate processes effective vision is dependent upon their co-ordination.

REFRACTION OF THE LIGHT RAYS

When light rays pass from a medium of one density to a medium of a different density they are refracted or bent (Fig. 13:17). This principle is used in the eye to focus light on the retina.

Several structures which are more dense than air are involved in the refraction of light *in the eye*.

Before reaching the retina light rays pass successively through the cornea, the aqueous fluid, the lens and the vitreous body.

The lens is the only structure within the eye which can change the amount by which it bends light rays. Light rays from objects 20 feet or more from the eyes travel in almost parallel lines and require the least amount of refraction by the lens. As objects move nearer to the eyes the rays of light reflected by the object become more divergent and therefore require to be refracted more acutely.

The lens is a bi-convex, elastic, transparent body suspended by the suspensory ligament from the ciliary muscle. When light rays enter the eye from near objects the function of the lens is to refract these rays on to the macula lutea. To do this the *ciliary muscle contracts forwards* releasing its pull on the suspensory ligament. Due to the presence of elastic fibres the lens becomes thicker, increasing its convexity and therefore its refracting power.

It will be appreciated that looking at near objects will 'tire' the eyes more quickly due to the continuous use of the ciliary muscle.

ACCOMMODATION OF THE EYES TO LIGHT
(Figs. 13:18 and 13:19)

Three factors are involved in accommodation:

The pupils

The movement of the eyeballs, called convergence

The lens

The size of the pupils

Pupil size influences accommodation by controlling the amount of light entering the eye. *In a bright light the pupils are constricted. In a dim light they are dilated.*

If the pupils were dilated in a bright light too much light may enter the eye and damage the retina. In a dim light if the pupils were constricted, insufficient light may enter the eye to stimulate the nerve endings in the retina and nothing would be seen.

The iris consists of a layer of circular and a layer of radiating smooth muscle fibres. Contraction of the circular fibres constricts the pupil, and contraction of the

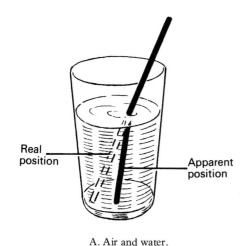

A. Air and water.

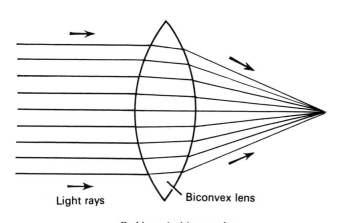

B. Air and a biconvex lens.

Figure 13:17 Bending or refraction of light rays passing from a less dense to a more dense medium.

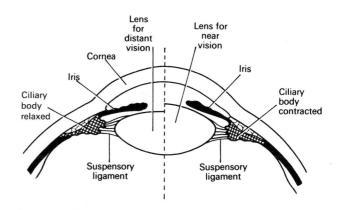

Figure 13:18 Diagram of the difference in the shape of the lens for near and distant vision.

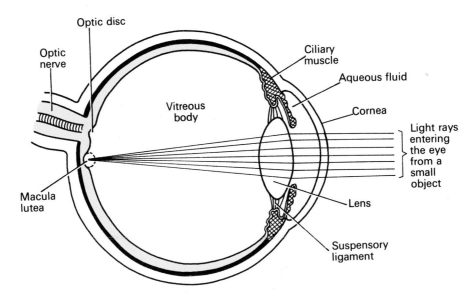

Figure 13:19 Section of the eye showing the focusing of light rays on the retina.

radiating fibres dilates it. The size of the pupil is controlled by nerves of the autonomic nervous system. *Sympathetic stimulation dilates the pupils. Parasympathetic stimulation constricts the pupils.*

The movements of the eyeballs—convergence

Light rays from objects enter the two eyes at different angles. It is important that the rays entering should stimulate the corresponding areas of the two retinae. If this is not achieved the individual complains of double vision. Movement of the eyeballs is necessary to ensure this. For near objects both eyeballs move inwards so that their visual axes *converge* on the object to be seen. The nearer the object the greater the convergence. For a distant object less convergence is necessary. The movements of the eyeballs occur due to the contraction of the *extrinsic muscles* of the eyes which will be discussed shortly. Convergence of the eyes is not under voluntary control. When voluntary movement of the eyes occurs both eyes move and convergence is maintained.

FUNCTION OF THE RETINA

The retina is described as the *photosensitive* part of the eye. The light-sensitive cells in the retina are the *rods and cones.* Light rays cause chemical changes in these cells and they emit nerve impulses which pass to the optic nerves.

The rods are more sensitive than the cones. They are stimulated by *low intensity or dim light,* for example, by the dim light in the interior of a darkened room.

The cones are sensitive to *bright light and colour.* The effect of different wavelengths of light on the cones results in the perception of different colours. In a bright light the light rays are focused on the macula lutea.

The rods are more numerous towards the periphery of the retina. *Visual purple or rhodopsin* is a pigment present only in the rods. It is bleached by bright light and when this occurs the rods can not be stimulated. Rhodopsin is quickly reconstituted when the intensity of light is decreased and an adequate supply of vitamin A is available. When the individual moves from an area of bright light to one of dim light there is a variable period of time when it is difficult to see. The rate at which *dark adaptation* takes place is dependent upon the reconstitution of rhodopsin.

The reader may have noticed that it is easier to see a star in the sky at night if the head is turned slightly away from it. When this happens the dim light rays from the star stimulate the area of the retina where there is the greatest concentration of rods. If looked at directly the light rays are not of sufficient intensity to stimulate the less sensitive cones in the area of the macula lutea. It may also have been noticed that in dim evening light different colours cannot be distinguished. This is because the low intensity light rays cannot stimulate the cones which are the only cells in the retina sensitive to colour.

BINOCULAR VISION (Fig. 13:20)

Binocular or stereoscopic vision has certain advantages. Each eye 'sees' a scene slightly differently. There is an overlap in the middle but the left eye sees more on the left than can be seen by the right eye and vice versa. The images from the two eyes are fused in the cerebrum so that one image is perceived.

Binocular vision provides much more accurate assessment of the distance, depth, height and width of an object. Some people with monocular vision may find it

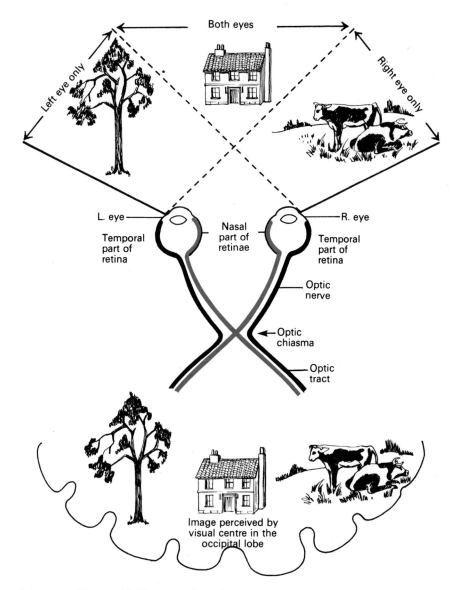

Figure 13:20   Diagram of the parts of the visual field—monocular and binocular.

difficult to judge the speed with which a vehicle is approaching.

THE EXTRINSIC MUSCLES OF THE EYE

There are six extrinsic muscles situated round the eyeball. Four of these are *straight* and two *oblique* (Fig. 13:21). They are named as follows:

Medial rectus
Lateral rectus
Superior rectus
Inferior rectus
Superior oblique
Inferior oblique

These muscles have their origin in the bony walls of the orbital cavity and are inserted into the sclera of the eyeball. They are responsible for moving the eye. They consist of striated muscle fibres. The movement of the eyes to look in a particular direction is under voluntary control but the co-ordination of movement of the eyes during convergence and accommodation to near or distant vision is controlled by reflex action.

The medial rectus rotates the eyeball *inwards.*
The lateral rectus rotates the eyeball *outwards.*
The superior rectus rotates the eyeball *upwards.*
The inferior rectus rotates the eyeball *downwards.*
The superior oblique rotates the eyeball so that the cornea turns in a *downwards and outwards* direction.
The inferior oblique rotates the eyeball so that the cornea turns *upwards and outwards.*

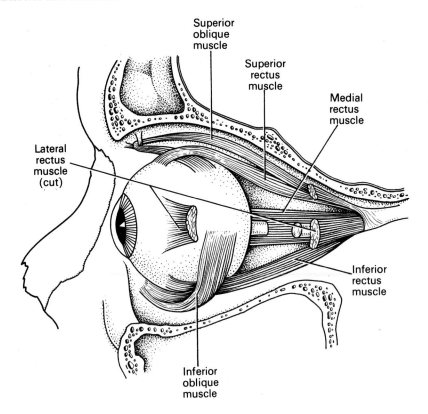

Superior
oblique
muscle

Superior
rectus
muscle

Medial
rectus
muscle

Lateral
rectus
muscle
(cut)

Inferior
rectus
muscle

Inferior
oblique
muscle

Figure 13:21   The extrinsic muscles of the eye.

It is to be appreciated that rarely do any of the extrinsic muscles of the eyeball contract singly. The free, quick movements of the eyeball occur due to the contraction of more than one muscle at a time. To achieve the co-ordinated movement of the two eyes an orderly contraction and relaxation of these muscles is necessary. For example, when the eyes turn to the right the lateral rectus of the right eye and the medial rectus of the left eye contract and their opposing muscles relax.

NERVE SUPPLY TO THE MUSCLES OF THE EYE
The oculomotor (3rd cranial nerve) supplies:

Superior rectus ⎫
Inferior rectus ⎪
Medial rectus ⎬ extrinsic muscles
Inferior oblique ⎭
Iris ⎫
Ciliary muscle ⎬ intrinsic muscles

The trochlear (4th cranial nerve) supplies the superior oblique.

The abducens (6th cranial nerve) supplies the lateral rectus.

THE ACCESSORY ORGANS OF THE EYE
The eye is a delicate organ which is protected by several structures (Figs. 13:22 and 13:23):

The eyebrows
The eyelids and eyelashes
The lacrimal apparatus

THE EYEBROWS
These are two arched eminences of skin surmounting the supra-orbital margins of the frontal bone. Numerous hairs project obliquely from the surface of the skin.

The function of the eyebrows is to protect the anterior aspect of the eyeball from sweat, dust and other foreign bodies.

THE EYELIDS AND EYELASHES
The eyelids are two movable folds situated above and below the front of the eye and on their free edges there are outgrowths of hair—the eyelashes. The layers of tissue which form the eyelids are:

A thin covering of skin
A thin sheet of areolar tissue
Two muscles—the *orbicularis oculi and levator palpebrae superioris*
A thin sheet of dense connective tissue called the *tarsal plate*—in the upper eyelids only—supports the other structures and maintains their shape and form
A lining of *conjunctiva*.

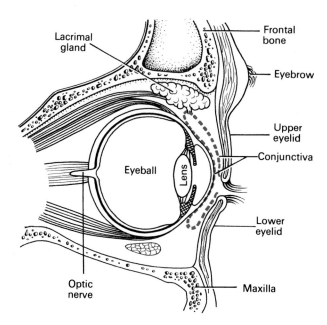

Figure 13:22   Section of the eye and its accessory structures.

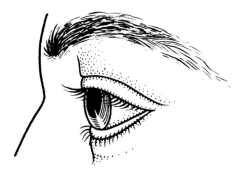

Figure 13:23   Side view of some structures which protect the eye.

The conjunctiva

This is a fine transparent membrane which *lines the eyelids* and is *reflected over the front of the eyeball*. When the eyelids are closed this becomes a closed sac. It protects the delicate cornea and the front of the eye. When drops of a drug are put into the eye they are placed in the lower conjunctival sac. The medial and lateral angles of the eye where the upper and lower lids come together are called respectively the *medial canthus* and the *lateral canthus*.

Function

The eyelids and eyelashes protect the eye from injury. If injury is feared or the conjunctiva touched very lightly the eyelids close. This is termed the conjunctival or *corneal reflex*.

When the *orbicularis oculi* contracts the eyes close. When the *levator palpabrae* contract the eyelids open.

THE LACRIMAL APPARATUS (Fig. 13:24)

For each eye this consists of:

The lacrimal gland and its ducts
The lacrimal canaliculi
The lacrimal sac
The nasolacrimal duct

The lacrimal glands are situated in recesses in the frontal bones on the lateral aspect of each eye just behind the supraorbital margin. Each gland is approximately the size and shape of an almond, and is composed of *secretory epithelial cells*. The glands secrete *tears* which are composed of water, salts and a bactericidal protein called *lysozyme*.

The tears leave the gland by several small ducts and pass over the front of the eye under the lids towards the medial canthus where they drain into the *two lacrimal canaliculi*. One canaliculus lies above the other, separated by the *caruncle* which is a small red body situated at the medial canthus. The tears then drain into the *lacrimal sac*.

The lacrimal sac is situated in a fossa in the lacrimal bone and is the upper expanded end of the *nasolacrimal duct*.

The nasolacrimal duct is a membranous canal approximately 2 cm long and extends from the lower part of the lacrimal sac to open into the nasal cavity, at the level of the inferior concha.

The tears pass over the anterior aspect of the eyeball, flow into the lacrimal canaliculi, then to the lacrimal sac and through the nasolacrimal duct into the nasal cavity.

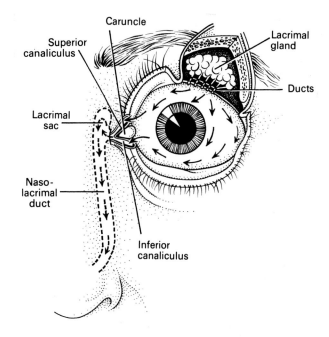

Figure 13:24   The lacrimal apparatus. Arrows show the direction of flow of tears.

Tears are secreted by the lacrimal glands continuously and their function is to bathe the front of the eyeball, washing away any dust, grit and micro-organisms. Due to the presence of the lysozyme, micro-organisms present on the front of the eyeball are destroyed. In emotional states the secretion of tears may be increased and, if the nasolacrimal duct cannot convey them all into the nasal cavity, they overflow.

Smell and the Nose

The anatomy of the nose was discussed in Chapter 7 dealing with the respiratory system (Fig. 13:25). The nose, therefore, is an organ with a dual function:

An organ of the respiratory system

An organ of special sense

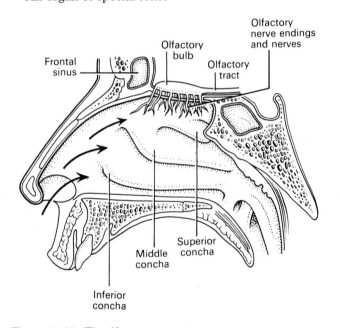

Figure 13:25   The olfactory structures.

THE OLFACTORY NERVES (1st CRANIAL NERVES)

These are the sensory nerves of smell. They have their origins in special cells in the mucous membrane of the roof of the nose and the superior nasal concha. Nerve fibres from the cells are gathered into bundles then pass through the *cribriform plate of the ethmoid bone* to the *olfactory bulb*. In the *olfactory bulb* inter-connections and synapses occur. From the bulb, bundles of nerve fibres form the *olfactory tracts* which pass backwards to the olfactory area in the *temporal lobes* of the cerebral cortex in each hemisphere where the impulses are interpreted (Fig. 13:26).

PHYSIOLOGY OF SMELL

The sense of smell in human beings is generally less acute than in other animals.

All odorous materials give off particles of their substance and these chemical particles are carried into the nose with the inhaled air. They dissolve in the secretions of the mucous membrane and will only stimulate the nerve cells of the olfactory region when in solution.

When an individual is continuously exposed to an odour, perception of the odour decreases and eventually ceases. This loss of perception only affects that specific odour and adaptation probably occurs both in the cerebrum and in the nerve endings in the nose.

This modification of perception may be dangerous when it is associated with the escape of unpleasant smelling poisonous gases.

The air entering the nose is heated and convection currents carry the inspired air to the roof of the nose. 'Sniffing' concentrates more particles more quickly in the roof of the nose. This increases the number of special cells stimulated and the perception of the smell. The sense of smell may affect the appetite. If the odours are pleasant the appetite may improve and vice versa.

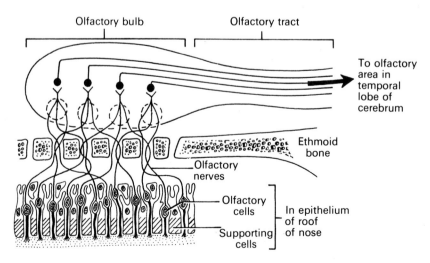

Figure 13:26   An enlarged section of the olfactory apparatus in the nose and the inferior surface of the cerebrum.

Taste and the Tongue

The anatomy of the tongue was described in Chapter 9 on the digestive system. *The tongue is the organ of taste* and is supplied by the *facial nerve* (7th cranial nerve) and the *glossopharyngeal* nerve (9th cranial nerve).

THE TASTE BUDS

The taste buds are the end organs of the sense of taste. They are most numerous in the papillae of the tongue but they are also found in the epithelium of the soft palate, the pharynx and the epiglottis (Fig. 13:27). A taste bud consists of a small bundle of cells which have hair-like processes protruding through a tiny pore in the mucous membrane. At their other end the cells are continuous with the nerve fibres which eventually form the sensory portions of the 7th and 9th cranial nerves.

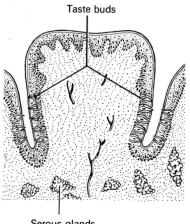

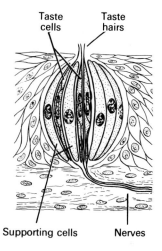

Figure 13:28   *Top:* A section of the vallate papilla. *Bottom:* A section of a taste bud.

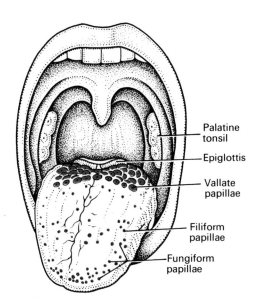

Figure 13:27   The sites of the nerve endings of the sense of taste in the tongue.

The three types of papillae in the tongue are (Fig. 13:28):

    Vallate—which form an inverted 'V'-shape at the base of the tongue

    Fungiform—situated on the sides and the tip of the tongue

    Filiform—found mainly on the anterior two-thirds of the tongue

THE PHYSIOLOGY OF TASTE

There are four fundamental sensations of taste, *sweet, sour, bitter and salt.* The large variety of other 'tastes' which are experienced are either a combination of two or more of the fundamental tastes, or are associated with the sense of smell.

Taste is a chemical sensation, therefore substances must be in solution before they can be appreciated. The substances in solution enter the pores of the taste buds and stimulate the hair-like endings of the cells. The impulses are transmitted to the thalamus and then to the *taste area* in the cerebral cortex, one in each hemisphere, where taste is perceived.

The taste buds are not evenly distributed over the whole area of the tongue, and it is thought that stimulation in relation to the four tastes take place at different parts of the tongue:

    Sweet and salty, mainly at the *tip of the tongue*

    Sour, at the *sides of the tongue*

    Bitter, at the *back of the tongue*

The tongue is also sensitive to *temperature, pain, touch and pressure.* Highly spiced substances are appreciated due to temperature and touch rather than taste.

The tip of the tongue is highly sensitive to *pain, temperature, touch and pressure.* These are not special senses but they are very important functions of the tongue.

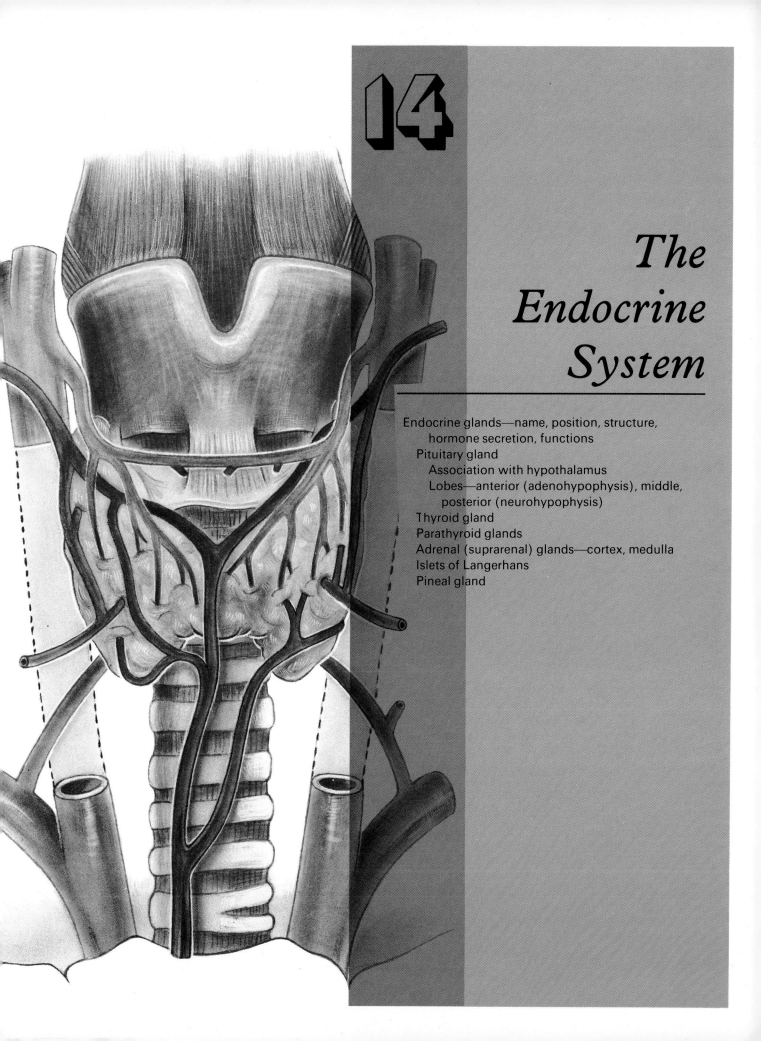

14

The Endocrine System

Endocrine glands—name, position, structure,
    hormone secretion, functions
Pituitary gland
    Association with hypothalamus
    Lobes—anterior (adenohypophysis), middle,
        posterior (neurohypophysis)
Thyroid gland
Parathyroid glands
Adrenal (suprarenal) glands—cortex, medulla
Islets of Langerhans
Pineal gland

14. *The Endocrine System*

The endocrine system consists of glands widely separated from one another and possessing no direct anatomical relationship. The glands are commonly referred to as the *ductless glands* because the secretions which they produce *do not leave* the gland through a duct or a canal but pass *directly* from the cells into the blood stream.

The secretions produced by the endocrine glands are called *hormones*.

A hormone can be described as a chemical substance which, having been formed in one particular organ or gland is carried in the blood-stream to another organ (target organ), probably quite far distant, where it has its effect, influencing its activity, its growth, its nutrition.

The internal environment of the body is controlled partly by the autonomic nervous system and partly by the endocrine glands. The hormones secreted by the ductless glands are mainly excitatory in nature, stimulating activity in other organs or glands.

The endocrine system consists of the following glands (Fig. 14:2):

1 pituitary gland
1 thyroid gland
4 parathyroid glands
2 adrenal or suprarenal glands
The islets of Langerhans in the pancreas
1 pineal gland or body
2 ovaries in the female
2 testes in the male

The ovaries and the testes secrete hormones associated with the reproductive system, therefore their functions will be understood more readily if they are studied in Chapter 15.

The Pituitary Gland (Fig. 14:1)

The pituitary gland is reddish-grey in colour and roughly oval in shape. It is situated in the *hypophyseal fossa* (sella turcica) of the sphenoid bone at the base of the brain. It is approximately 15 mm long and 5 mm across.

The gland is attached to the brain by a stalk which is continuous with the *hypothalamus* above. There is communication between the hypothalamus and the pituitary gland by means of nerve fibres and a complex arrangement of blood vessels which pass between them through the *pituitary stalk*. The anterior lobe of the gland is in association with the hypothalamus through the network of blood vessels, and nerve fibres connect it with the posterior lobe (Fig. 14:3).

The blood supply to the gland is by the *superior and inferior hypophyseal arteries* which are branches of the internal carotid arteries. Venous drainage is into the *venous sinuses* of the brain.

The pituitary gland has three distinct parts:

The anterior lobe or pars anterior or adenohypophysis
The middle lobe or pars intermedia
The posterior lobe or pars nervosa or neurohypophysis

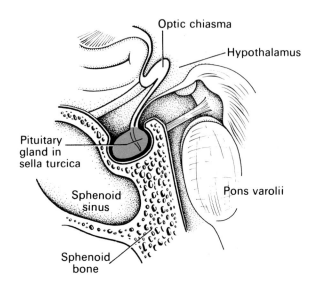

Figure 14:1 The position of the pituitary gland and its associated structures.

THE ANTERIOR LOBE (ADENOHYPOPHYSIS)

This lobe is composed of several different types of epithelial cells which can be distinguished microscopically by their size and shape.

The hormones produced by the anterior pituitary stimulate the production of hormones by other endocrine glands or growth of the body as a whole. The release of

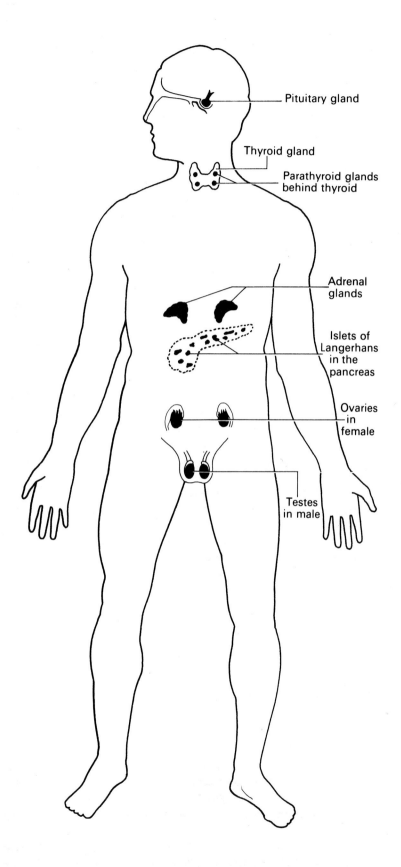

Pituitary gland

Thyroid gland

Parathyroid glands
behind thyroid

Adrenal
glands

Islets of
Langerhans
in the
pancreas

Ovaries
in
female

Testes
in male

Figure 14:2   Diagram of the positions in the body of the endocrine glands.

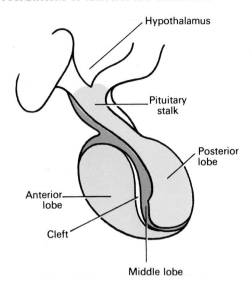

Figure 14:3   The parts of the pituitary gland and its relation to the hypothalamus.

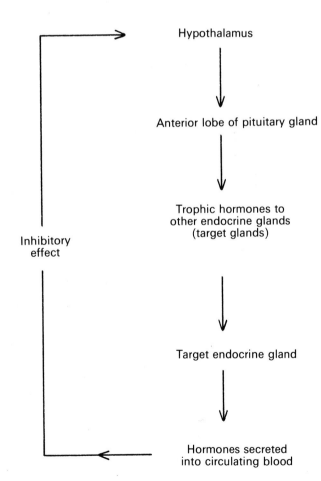

Figure 14:4   Diagram of the negative feedback control of the secretions of hormones by the anterior lobe of the pituitary gland.

anterior pituitary hormones follows stimulation of the gland by '*releasing factors*' produced by the hypothalamus and conveyed to the gland through the special network of blood vessels mentioned above. The whole system is controlled by a *negative feedback mechanism*. That is, when there is a low level of a hormone in the blood supplying the hypothalamus it produces the appropriate releasing factor which stimulates release of hormone by the anterior pituitary which in turn stimulates the target gland to produce and release its hormone. As a result the blood level of that hormone rises and inhibits the secretion of releasing factor by the hypothalamus (see Fig. 14:4).

FUNCTIONS OF THE ANTERIOR LOBE

The anterior lobe of the pituitary gland secretes several hormones:

Growth hormone or somatotrophic hormone
Thyrotrophic hormone or thyroid stimulating
  hormone (TSH)
Adrenocorticotrophic hormone (ACTH)
Lactogenic hormone (prolactin)

Gonadotrophic hormones
  —follicle stimulating hormone (FSH)
  —luteinising hormone (LH)

Functions of the growth hormone

This hormone stimulates growth directly and in conjunction with other hormones. It affects the growth in length of the long bones by promoting the growth of the epiphyseal cartilage. Growth of long bones stops when ossification overtakes the growth of this cartilage.

Growth hormone promotes protein anabolism, the absorption of calcium from the bowel and the conversion of glycogen to glucose. It is in highest concentration in the blood until the individual has reached his full stature, but a continual supply is necessary to stimulate the repair and replacement of body tissue throughout life.

Secretion of the hormone is controlled by the hypothalamus.

Functions of the thyrotrophic hormone (TSH)

This hormone controls the growth and activity of the *thyroid gland*. It influences the uptake of iodine, the synthesis of the hormones *thyroxine* and *triiodothyronine* by the thyroid gland and the release of stored hormones into the bloodstream.

The amount of thyrotrophic hormone secreted varies from time to time and its secretion is influenced by the amount of thyroid hormones circulating in the blood at any given time. If the concentration of these hormones in the blood falls the hypothalamus stimulates the anterior lobe of the pituitary to secrete more of the thyrotrophic hormone and this stimulates the thyroid gland into activity. On the other hand if the blood level of thyroxine and triiodothyronine is high the pituitary gland decreases its output of TSH. The hypothalamus has a stimulating effect on the production of thyrotrophic hormone especially by increasing the secretion when the body is cold and decreasing secretion when it is hot.

Functions of the adrenocorticotrophic hormone (ACTH)

This hormone stimulates the *cortex* of the *adrenal gland* to produce its hormones. The amount of ACTH secreted depends upon the concentration in the blood of the hormones from the adrenal cortex and on stimulation of the pituitary gland by the hypothalamus. This latter influence on the production of ACTH is particularly important during emotional and physical stress.

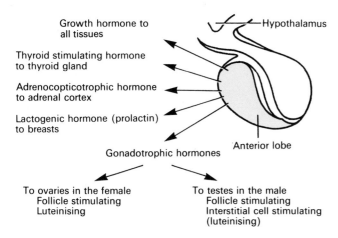

Figure 14:5 The hormones secreted by the anterior lobe of the pituitary gland and their target structures.

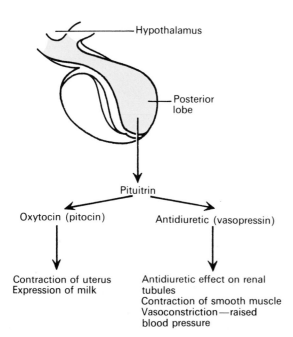

Figure 14:6 The hormones secreted by the posterior lobe of the pituitary gland and their main functions.

Functions of the lactogenic hormone (prolactin)

The lactogenic hormone has a direct effect upon the *breasts or mammary glands* immediately after the delivery of a baby and the expulsion of the placenta (after-birth). In conjunction with other hormones it stimulates the breasts to secrete milk.

During the months of pregnancy the breasts develop in preparation for lactation. This development is *not due to* prolactin, but to the activity of the *ovarian hormones*. It is only after the delivery of the baby that the lactogenic hormone is secreted. During pregnancy the ovarian hormones inhibit the secretion of prolactin. Also involved in the function of the breast in the postpartum period is a hormone called *oxytocin*, secreted from the *posterior lobe* of the pituitary gland. *Oxytocin*, amongst other functions, stimulates the *flow* of milk through the ducts of the breast.

Three main processes are associated with lactation.

1. *The growth and development of the breasts* during pregnancy is influenced mainly by *ovarian hormones*.
2. *The secretion of milk* involves a group of hormones believed to include prolactin, the growth hormone, the adrenocorticotrophic hormone and the thyrotrophic hormone.
3. *The flow of milk* from the breast is only possible when *oxytocin*, secreted by the posterior lobe of the pituitary is present. Sucking plays an essential part in reflex stimulation of the pituitary gland. When sucking stops lactation will continue only for a short time.

Functions of the gonadotrophic hormones

The anterior lobe secretes *two gonadotrophic or sex hormones* in both the female and the male:

The follicle stimulating hormone (FSH)

The luteinising hormone (LH), sometimes called the *interstitial cell stimulating hormone* in the male (ICSH)

Female gonadotrophic hormones

The follicle stimulating hormone. The target organs are the ovaries where FSH stimulates the development and ripening of the ovarian follicle (see Ch. 15). During its development the ovarian follicle secretes its own hormone, *oestrogen*. As the level of oestrogen increases in the blood so the follicle stimulating hormone secretion is reduced.

The luteinising hormone. This hormone promotes the final maturation of the ovarian follicle, ovulation (discharge of the mature ovum) and the formation of the *corpus luteum* which secretes the second ovarian hormone known as *progesterone*. As the level of progesterone in the blood increases there is a gradual reduction in the production of the luteinising hormone.

Male gonadotrophic hormones

The follicle stimulating hormone. This has an effect on the epithelial tissue of the *seminiferous tubules* in the testes. Under its influence they produce *spermatozoa* (male germ cells).

The interstitial cell stimulating hormone. This stimulates the *interstitial cells* in the testes to secrete the hormone *testosterone.*

THE POSTERIOR LOBE (NEUROHYPOPHYSIS)

The posterior lobe of the pituitary gland is composed of secretory cells called *pituicytes* and nerve fibres which arise from cells in two areas of the hypothalamus called the *supraoptic nucleus* and the *paraventricular nucleus.*

FUNCTIONS OF THE POSTERIOR LOBE

The posterior lobe and the hypothalamus form a single functional unit which controls several activities in the body.

The secretion of the posterior lobe is known as *pituitrin* and it contains two hormones:

Oxytocin or pitocin
Antidiuretic hormone (ADH) or vasopressin

The hormones are secreted by the cells of the supraoptic and paraventricular nuclei and they migrate along the nerve fibres which extend from these nuclei to the posterior pituitary where they are stored in the nerve endings. Each hormone is released in response to a different stimulus.

Oxytocin

Oxytocin promotes contraction of the uterine muscle and contraction of the myoepithelial cells of the lactating breast, squeezing milk into the large ducts behind the nipple. In late pregnancy the uterus becomes very sensitive to oxytocin. The amount secreted is increased just before and during labour and by the sucking of the baby.

Antidiuretic hormone (ADH) or vasopressin

This hormone has two main functions.

1. *Antidiuretic effect.* It increases the permeability to water of the distal and collecting tubules of the nephrons of the kidneys. As a result the reabsorption of water from the glomerular filtrate is increased. The amount of ADH secreted is influenced by the osmotic pressure of the blood circulating to the osmoreceptors in the hypothalamus. As the osmotic pressure rises the secretion of ADH increases and more water is reabsorbed. Conversely, when the osmotic pressure of the blood is low, the secretion of ADH is reduced, less water is reabsorbed and more urine is produced.

2. *Pressor effect.* Involuntary muscle in the walls of the intestine, gall bladder, urinary bladder and blood vessels is stimulated to contract by ADH. Contraction of the walls of the blood vessels raises the blood pressure and this may be its most important pressor effect.

THE MIDDLE LOBE OF THE PITUITARY GLAND

This is the smallest part of the pituitary gland and consists of epithelial cells. It is believed that it secretes a hormone which is associated with the growth and development of melanocytes which give the skin its colour.

The Thyroid Gland (Fig. 14:7)

The thyroid gland is situated in the neck in association with the larynx and trachea at the level of the fifth, sixth and seventh cervical and the first thoracic vertebrae. It is a highly vascular gland, brownish red in colour and surrounded by a fibrous capsule. It consists of *two lobes* which lie one on either side of the lower half of the thyroid cartilage and the upper four or five cartilaginous rings of the trachea. The lobes are joined anteriorly by a narrow portion, *the isthmus,* which lies in front of the second and third cartilaginous rings of the trachea.

The lobes are roughly conical in shape. *The apex* is the upper narrow part, and the lower broad part, the *base.* The lobes are approximately 5 cm (2 inches) long and 3 cm (1¼ inches) wide. The isthmus is about 1·25 cm (½ inch) long and 1·25 cm wide.

The arterial blood supply to the gland is through the superior and inferior thyroid arteries. The superior

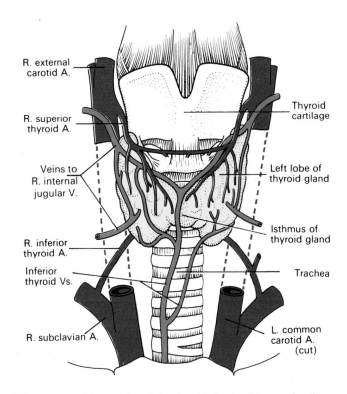

Figure 14:7 The position of the thyroid gland and its associated structures.

thyroid artery is a branch of the external carotid artery and the inferior thyroid artery is a branch of the subclavian artery.

The venous return is by the thyroid veins which drain into the internal jugular veins.

The gland is composed of *epithelial cells* which vary in shape depending upon the activity of the gland. They are *cubical in quiescent periods* to *columnar in active conditions*. The cells are arranged in a single layer in a roughly spherical or oval fashion round a central space, *the alveolus*. This space contains a thick sticky fluid known as *colloid*. The cells are in direct contact with supporting connective tissue and blood capillaries (Fig. 14:8).

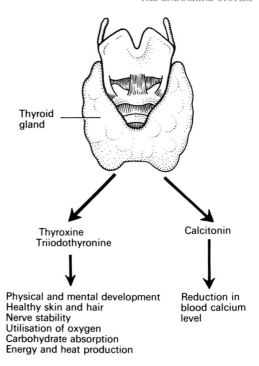

Figure 14:9   The thyroid gland, its hormones and their main functions.

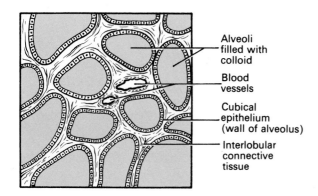

Figure 14:8   The minute structure of the thyroid gland.

FUNCTION (Fig. 14:9)

Iodine present in the blood is taken up by the gland and the hormones *thyroxine* and *triiodothyronine* are formed. These hormones are stored in the gland in the form of thyroglobin and are released into the bloodstream as required. Most of the *iodine* ingested as a constituent of food is either utilised in the formation of thyroxine and triiodothyronine or stored in the gland for future use. The uptake of iodine from the blood, the formation of the hormones and their release into the blood is stimulated by the *thyrotrophic hormone* from the anterior lobe of the pituitary gland.

The effects of thyroxine and triiodothyronine on other body functions are widespread.

1. They are essential for normal mental and physical development.
2. They are responsible for the maintenance of healthy skin and hair.
3. They are associated with nerve stability, probably influencing the excitability of the nerve fibres.
4. They control the utilisation of oxygen in the body. In this way they have a controlling influence on the basal metabolic rate.
5. They stimulate the absorption of carbohydrate from the small intestine.
6. They influence heat production during the catabolism of nutrient materials in the cells.

Calcitonin is a hormone produced by the thyroid gland. Its function is to reduce the blood level of calcium by inhibiting the reabsorption of calcium from bones. It has the opposite effect from *parathormone* from the parathyroid glands. Its secretion is increased by a rise in blood calcium and by gastrointestinal hormones following a meal.

The Parathyroid Glands (Fig. 14:10)

There are four parathyroid glands, two of which are embedded in the posterior surface of each lobe of the thyroid gland.

The parathyroids are quite small, approximately 6 mm ($\frac{1}{4}$ inch) in length and slightly less in breadth. They are roughly oval in shape and yellowish-brown in colour.

The arterial supply is through the inferior thyroid arteries and the venous return is through the middle thyroid veins.

The glands are surrounded by fine connective tissue capsules. The cells forming the glands are spherical in shape and are arranged in columns with channels containing blood between the columns.

FUNCTIONS

The function of the parathyroid glands is to secrete the hormone *parathormone*. The amount secreted is strongly influenced by the level of calcium in the blood. When hypocalcaemia occurs more hormone is secreted while in

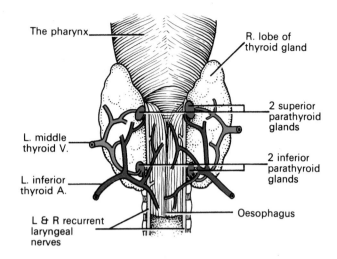

Figure 14:10   The positions of the parathyroid glands and their related structures viewed from behind.

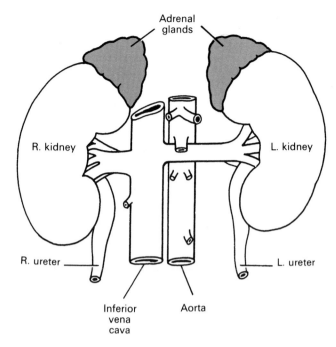

Figure 14:11   The positions of the adrenal glands and some of their associated structures.

hypercalcaemia secretion is inhibited. The normal blood calcium level is between 2·1 and 2·6 mmol/l (8·5 to 10·5 mg/100 ml).

The main functions of parathormone are to maintain the blood concentration of calcium within normal limits. This is achieved by influencing the amount of calcium absorbed from the small intestine and reabsorbed from the renal tubules. If these sources provide inadequate supplies parathormone stimulates the mobilisation of calcium from bones.

Parathormone and calcitonin from the thyroid gland act together to maintain a normal blood calcium level under varying conditions.

The Adrenal or Suprarenal Glands (Fig. 14:11)

There are two adrenal glands, one situated on the upper pole of each kidney. The adrenal glands are surrounded by a capsule of areolar tissue containing fat and are enclosed within the renal fascia.

The right adrenal gland is roughly pyramidal in shape, the left is semilunar, and both are yellowish-brown in colour. Each gland is approximately 4 cm (1½ inches) long and 3 cm (1¼ inches) thick.

The arterial blood supply to the glands is by branches from the abdominal aorta and renal arteries.

The venous return is through the suprarenal veins. The right gland drains into the inferior vena cava and the left into the left renal vein.

The glands are composed of two distinct parts which differ both anatomically and physiologically. The outer part is called the *cortex* and the inner part the *medulla*.

THE ADRENAL CORTEX
The cortex is yellowish in colour and completely sur-

rounds the medulla. Microscopically three layers of cells can be distinguished but it is not yet quite clear whether their functions are distinct or interlinked.

FUNCTIONS (Fig. 14:12)
The adrenal cortex produces a considerable number of different substances which have been classified into three groups:
1. Glucocorticoids
2. Mineralocorticoids
3. Sex hormones

The glucocorticoids
Cortisol (hydrocortisone) and *corticosterone* are the names given to the glucocorticoids. Secretion is stimulated by ACTH from the anterior lobe of the pituitary gland and their function is to *regulate carbohydrate metabolism*.

1. They are antagonistic to insulin. Under their influence the blood glucose level is maintained, and even raised in times of stress. They are responsible for *changing glycogen to glucose*.

2. The blood glucose level may be raised during stress by the process of *gluconeogenesis*. In this case the nitrogenous portion is removed from amino acids and the residue converted to glucose. It also influences the mobilisation of fat.

3. These hormones tend to decrease the number of eosinophils and lymphocytes in the blood, and to increase the neutrophil count.

4. They reduce the body's response to antigens.

5. They decrease the absorption of calcium from the small intestine by negating the effect of vitamin D.

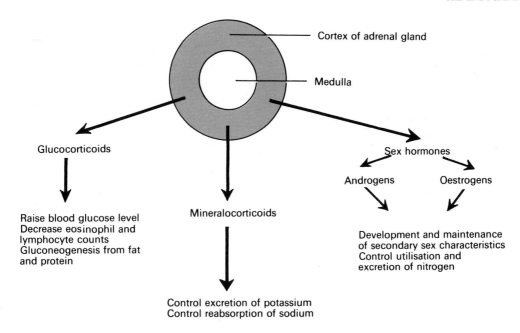

Figure 14:12 The hormones secreted by the cortex of the adrenal gland and their main functions.

The mineralocorticoids

Aldosterone is the name given to the main mineralo-corticoid. Its functions are associated with the maintenance of the electrolyte balance in the body.

Aldosterone stimulates the reabsorption of sodium by the renal tubules and when the amount of sodium *reabsorbed* is increased the amount of potassium *excreted* is increased. Indirectly this affects water excretion as the amount of water excreted is related to the amount of electrolyte excreted.

The amount of aldosterone produced is influenced by the sodium level in the blood. If there is a fall in the sodium blood level more aldosterone is secreted and more sodium reabsorbed.

Angiotensin produced by the kidneys stimulates the secretion of aldosterone. The amount of angiotensin produced is influenced by the amount of blood flow through the kidney and the sodium level in the blood. A fall in the blood sodium (normal 138 to 148 mmol/l) stimulates the production of angiotensin which stimulates an increase in the production of aldosterone. The result is an increase in the amount of sodium reabsorbed from the renal tubules and a corresponding decrease in the re-absorption of potassium.

The sex hormones

The sex hormones produced by the adrenal cortex of both males and females are testosterone, oestrogens and progesterone. The adrenal gland is the source of sex hormones until the testes and the ovaries mature at puberty.

The secretion of these hormones by the adrenal cortex is controlled by ACTH and not by gonadotropins which stimulate the testes and the ovaries.

Their functions are:

1. To influence the development and maintenance of the secondary sex characteristics in both male and female
2. To increase the deposition of protein in muscles and reduce the excretion of nitrogen especially in the male

THE ADRENAL MEDULLA

The medulla is completely surrounded by the cortex. It is an outgrowth of tissue from the same source as the nervous system and its functions are closely allied to those of the sympathetic part of the autonomic nervous system (Fig. 14:13).

When stimulated by sympathetic neurones the adrenal medulla produces *catecholamines* called *adrenalin* and *noradrenalin*. Noradrenalin is the chemical transmitter of the sympathetic nervous system; therefore when it is produced by the adrenal medulla and circulates in the blood it potentiates the effects of sympathetic stimulation without continued neurological activity.

The following are the main effects of adrenalin and noradrenalin:

1. Dilatation of the coronary arteries, thus increasing the blood supply to the heart muscle.
2. Dilatation of the bronchi allowing a greater amount of air to enter the lungs at each inspiration.
3. Dilatation of the blood vessels to the skeletal muscles increasing the supply of oxygen and nutritional

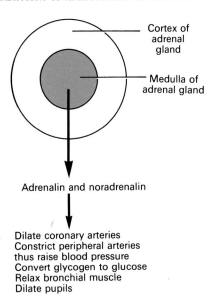

Adrenalin and noradrenalin

Dilate coronary arteries
Constrict peripheral arteries
thus raise blood pressure
Convert glycogen to glucose
Relax bronchial muscle
Dilate pupils

Figure 14:13   The hormones secreted by the medulla of the adrenal gland and their main functions.

There are two main types of cell in the islets of Langerhans: the α cells which produce a hormone called *glucagon* and the β cells which produce *insulin*. Both hormones influence the level of glucose in the blood, each balancing the effects of the other. Glucagon tends to raise the blood glucose level and insulin reduces it (Fig. 14:14).

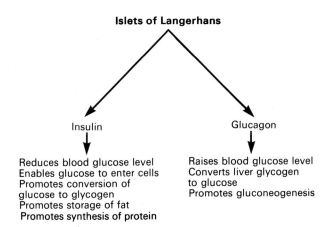

Figure 14:14   The hormones secreted by the islets of Langerhans and their main functions.

material to the muscles. This enables muscle activity to be sustained.

4. Constriction of the blood vessels to the skin, thus raising the blood pressure.
5. Increasing the rate of change of glycogen to glucose, thus ensuring sufficient glucose for sustained muscle contraction.
6. Dilatation of the pupil of the eye due to stimulation of the radiating muscle fibres of the iris.
7. Slowing down of peristalsis in the alimentary tract.
8. Reducing the secretion of saliva and other digestive juices.
9. Increasing the tone of the anal and urethral sphincter muscles, thus inhibiting micturition and defaecation.
10. Increasing the activity of the sweat glands and contraction of the arrectores pilorum causing 'goose flesh'.

Adrenalin and noradrenalin prepare the body to deal with abnormal conditions so that it responds effectively to fear, excitement and danger.

The Islets of Langerhans

The cells which make up the islets of Langerhans are found in clusters irregularly distributed throughout the substance of the pancreas. Unlike the pancreatic tissue which produces a digestive juice there are no ducts leading from the clusters of islet cells. Their secretion passes directly into the pancreatic veins and so circulates throughout the body in the blood.

The normal blood glucose level is between 3·5 and 5·5 mmol/litre (60 to 100 mg/100 ml). When there has been excessive exercise or an insufficient intake of carbohydrate foods it may fall to the low end of the normal range or even below. When this happens glucagon has the effect of *raising the blood glucose level* by mobilising the glycogen stores *in the liver*.

Insulin has the opposite effect, *it reduces the blood glucose level*.

1. It facilitates the movement of glucose across cell walls. After entering the cells glucose is catabolised to release energy and heat.
2. It promotes the conversion of glucose to glycogen inside muscle and liver cells. This occurs when glucose is in excess of immediate requirements.
3. It promotes the storage of fat and may prevent its mobilisation.
4. It promotes the synthesis of protein.

An insufficiency of insulin in the body leads to the development of *diabetes mellitus* which is characterised by disturbances in both glucose and fat metabolism. Patients suffering from this condition have a blood glucose level which exceeds the normal renal threshold level and so glucose is found in the urine. Because of the high concentration of glucose in the urine an excessive amount of water is excreted leading to dehydration and polydypsia. Excessive amounts of fat are partially metabolised to the stage of keto acids. These accumulate in excess upsetting the acid-base balance in the body.

The Pineal Gland or Body

The pineal gland is a small body situated in the brain below the corpus collosum and posterior to the third ventricle. It is reddish-grey in colour and is approximately 10 mm in length. It is surrounded by a fine capsule and is composed of epithelial cells arranged to form lobules which are surrounded by fine connective tissue.

Its functions are as yet unclear. It may be associated with the development of the gonads by influencing the release of gonadotrophic hormones from the anterior pituitary.

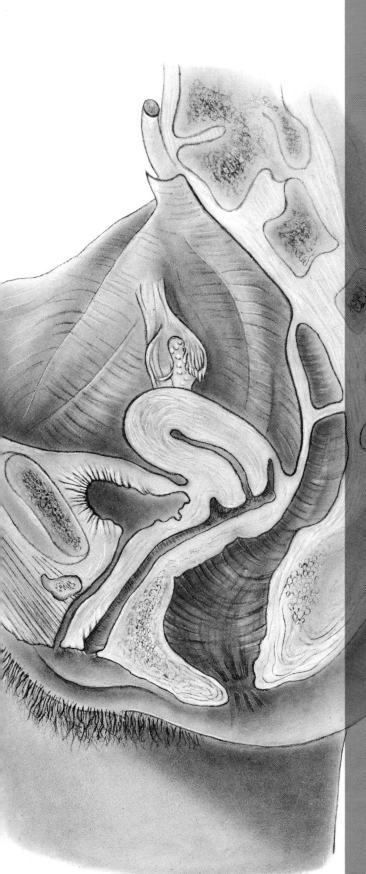

The Reproductive Systems

15. The Reproductive Systems

The ability to reproduce is one of the properties which distinguishes living from non-living matter. The more primitive the animal the simpler is the process of reproduction. In human beings the process is one of sexual reproduction.

The reproductive organs of the male and the female differ anatomically and physiologically. The female produces an egg cell or *ovum* which is fertilised by the germ cell or *spermatozoon* produced by the male. The resultant *zygote* embeds itself in the wall of the *uterus* in the female, where it grows and develops until the mature baby is born after a *gestation period* of 40 weeks. The function of the female reproductive system is, therefore, to form the ovum and if it is fertilised, to nurture it until it is born. The function of the male reproductive system is to form and transmit the spermatozoa to the female.

The Female Reproductive System

The female reproductive organs are divided into two groups (Fig. 15:1):

The external organs or genitalia
The internal organs or genitalia

THE EXTERNAL GENITALIA (Fig. 15:2)
The external genitalia are known collectively as the *vulva* and consists of several structures:

The labia majora (labia majus)
The labia minora (labia minus)
The clitoris
The vestibule
The hymen
The greater vestibular glands

THE LABIA MAJORA
These are the two large folds of skin which form the boundary of the vulva. They are composed of skin, fibrous tissue and fat and they contain large numbers of sebaceous glands in the medial layer of the folds. Anteriorly the two folds join in front of the symphysis pubis, and posteriorly they merge with the skin of the perineum. At puberty hair grows on the mons pubis and on the lateral aspect of the labia majora.

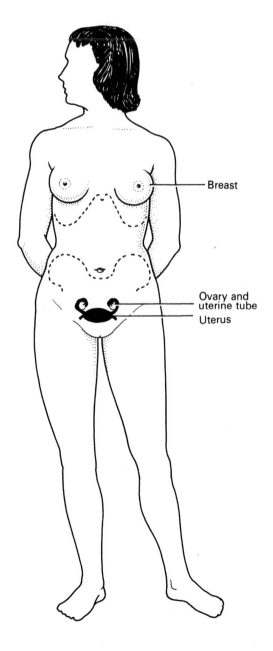

Breast

Ovary and
uterine tube

Uterus

Figure 15:1 The female reproductive organs. Dotted lines denote the positions of the lower ribs and the pelvis.

228

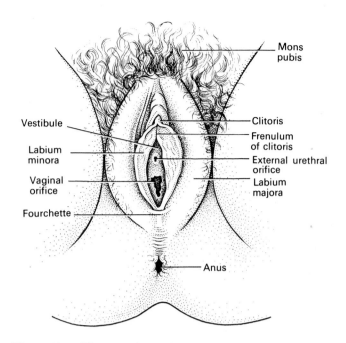

Figure 15:2 The external genitalia in the female.

THE LABIA MINORA

The labia minora are two smaller folds of skin containing numerous sebaceous glands which lie between the labia majora. Anteriorly they are divided into two parts, one stretching in front of the clitoris to form the *prepuce*, the other passing behind it to form the *frenulum*.

Posteriorly the labia minora are fused together forming the *fourchette*.

The area between the labia minora is called the *vestibule*. The vagina, the urethra and the ducts of the greater vestibular glands open into the vestibule.

THE CLITORIS

The clitoris corresponds to the penis in the male and contains erectile tissue. It is attached to the symphysis pubis by a suspensory ligament and lies between the prepuce and the frenulum.

THE HYMEN

The hymen is a thin layer of mucous membrane which partially occludes the opening of the vagina.

THE GREATER VESTIBULAR GLANDS

The greater vestibular glands lie in the labia majora, one on each side near the vaginal opening. They are about the size of a small pea and have ducts about 2 cm long which open into the vestibule. The glands secrete mucus which lubricates the vulva.

BLOOD SUPPLY, LYMPH DRAINAGE AND NERVE SUPPLY

Arterial supply is provided by branches from the *internal pudendal arteries* which arise from the internal iliac arteries, and by the *external pudendal arteries*, branches from the femoral arteries.

The venous drainage forms a large venous plexus which eventually drains into the internal iliac vein.

The lymph drainage is through the superficial inguinal glands in the groin.

Nerve supply. The pudendal nerve gives off various branches which supply the genitalia.

THE PERINEUM

The perineum is the area extending from the fourchette to the anal canal. It is roughly triangular in shape and is composed of connective tissue, muscle and fat. It gives attachment to the muscles of the pelvic floor.

THE INTERNAL ORGANS (Figs. 15:3 and 15:4)

The internal organs of the female reproductive system lie in the pelvic cavity and consist of:

The vagina
The uterus
2 uterine tubes
2 ovaries

THE VAGINA

The vagina is a fibromuscular tube connecting the internal and external organs of generation. It runs obliquely upwards and backwards at an angle of approximately 45°. The anterior and posterior walls are in apposition except at the vault where it is divided into *four fornices* by the protruding *cervix of the uterus*.

The anterior fornix lies in front of the cervix and is in contact with the base of the bladder.

The posterior fornix lies behind the cervix.

The lateral fornices are the areas of the vagina on either side of the cervix.

In the adult the anterior wall of the vagina measures about 7·5 cm (3 inches) and the posterior wall about 9 cm (3½ inches).

Structure

The vagina has three layers of tissue:

1. An outer covering of *areolar and elastic tissue* containing bundles of nerves and many blood vessels
2. A middle layer of *smooth muscle tissue* which consists of longitudinal and circular fibres
3. An inner lining of *stratified squamous epithelium* arranged in transverse folds, or *rugae*

Blood supply, lymph drainage and nerve supply

The arterial blood supply is in the form of a plexus round the vagina derived from the uterine and vaginal arteries which are branches of the internal iliac arteries.

The venous drainage is by a venous plexus situated in the muscular wall of the vagina which drains into the internal iliac veins.

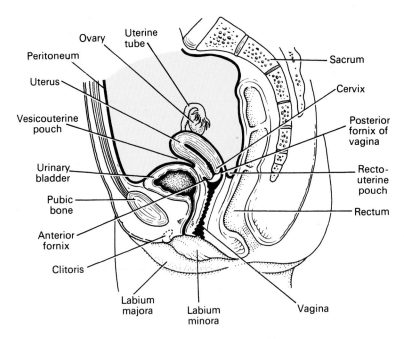

Figure 15:3 The female reproductive organs in the pelvis and their associated structures. Lateral view.

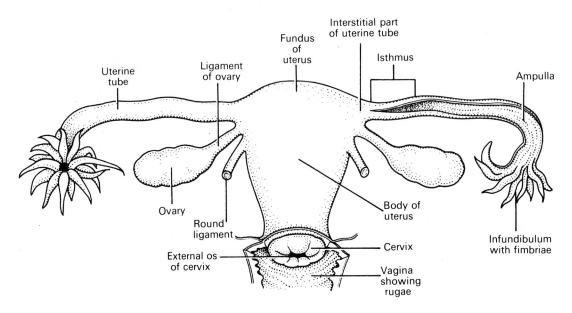

Figure 15:4 The female reproductive organs in the pelvis: posterior wall of the vagina and right uterine tube removed.

The lymph drainage is through the deep and superficial iliac glands.

The nerve supply. There are two sources of nerve supply: the autonomic nervous system provides parasympathetic and sympathetic fibres and the pudendal nerves provide sensory fibres to the lower part.

THE UTERUS
The uterus is a hollow muscular organ shaped like a pear which is flattened anteroposteriorly. It lies in the pelvic cavity between the urinary bladder and the rectum and its position is one of *anteversion and anteflexion*.

Anteversion means that the uterus *leans forward*.

Anteflexion means that it is bent forward almost at right angles to the vagina with its anterior surface resting on the urinary bladder. As the bladder fills the degree of anteflexion is reduced slightly.

When the body is in the anatomical position the uterus

lies in an almost horizontal position. The uterus is about 7·5 cm (3 inches) long, 5 cm (2 inches) wide and its walls are about 2·5 cm (1 inch) thick. It weighs from 30 to 40 grams (1 to 1½ oz). The uterus is described in three parts (see Fig. 15:5):

The fundus
The body
The cervix

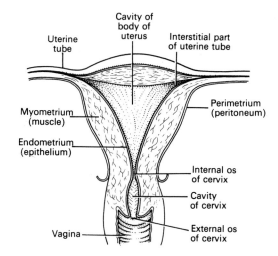

Figure 15:5 Diagram of a section of the uterus.

The fundus is the dome-shaped part of the uterus above the openings of the uterine tubes.

The body is the main part which is narrowest inferiorly at the *internal os* where it is continuous with the cervix.

The cervix is the part which protrudes through the anterior wall of the vagina and opens into it at the *external os*.

The structure

The walls of the uterus are composed of three layers of tissue:

The perimetrium—an outer covering of peritoneum
The myometrium—a middle layer of smooth muscle fibres
The endometrium—a mucous membrane lining

The perimetrium consists of peritoneum which is distributed differently on the various surfaces of the uterus.

Anteriorly it extends over the fundus and the body to the level of the internal os where it is reflected on to the upper surface of the urinary bladder. This fold of peritoneum is called the *vesicouterine pouch*.

Posteriorly the peritoneum extends over the fundus, the body and the cervix except for the part of the cervix which is within the wall of the vagina. It is then reflected on to the rectum to form the *rectouterine pouch*.

Laterally only the fundus is covered because the peritoneum forms a double fold with the uterine tubes in the upper free border. This double fold of peritoneum is called the *broad ligament* which, at its lateral ends, attaches the uterus to the sides of the pelvis.

The myometrium is the thickest layer of tissue in the uterine wall. It consists of a mass of smooth muscle fibres interlaced with areolar tissue, blood vessels and nerves. It is arranged in three layers:

The inner layer of circular and longitudinal fibres
The middle layer of longitudinal, oblique and transverse fibres
The outer layer of longitudinal fibres

The endometrium consists of columnar epithelium. It contains large numbers of mucus-secreting tubular glands which open into the lumen of the uterus. Near the external os the lining of the cervix changes to stratified squamous epithelium of the same type as the lining of the vagina.

Blood supply, lymph drainage and nerve supply

The arterial supply is by the *uterine arteries* which are branches of the internal iliac arteries. They pass up the lateral aspects of the uterus between the two layers of peritoneum which form the broad ligaments. They supply the uterus and the uterine tubes and join with the ovarian arteries to supply the ovaries. Branches pass downwards to anastomose with the vaginal arteries to supply the vagina.

Venous drainage. The veins form the same pattern as the arteries and eventually drain into the internal iliac veins.

Lymph drainage. There are deep and superficial lymph vessels which drain lymph from the uterus and the uterine tubes to the lateral aortic lymph nodes and groups of nodes associated with the iliac blood vessels.

Nerve supply. The nerves supplying the uterus and the uterine tubes consist of parasympathetic fibres from the sacral outflow and sympathetic fibres from the lumbar outflow.

SUPPORTS OF THE UTERUS

The uterus is supported in the pelvic cavity by a number of structures:

The surrounding organs
The muscles of the pelvic floor
Ligaments derived from folds of peritoneum and connective tissue which suspend it from the walls of the cavity

The supporting ligaments (Fig. 15:6)

1. *Two broad ligaments.* These are formed by a double fold of peritoneum, one on each side of the uterus. They hang down from the uterine tubes as though draped over them and at their lateral ends they are attached to the sides of the pelvis.

2. *The round ligaments.* These are bands of fibrous

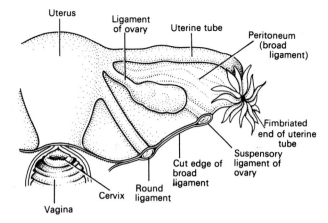

Figure 15:6   The main ligaments supporting the uterus. Only one side shown.

THE UTERINE TUBES (Fig. 15:7)

The uterine tubes lie one on each side of the uterus in the upper free border of the broad ligament. They are about 10 cm (4 inches) long and extend laterally from the wall of the uterus to penetrate the posterior wall of the broad ligament opening into the peritoneal cavity near the ovaries. They are described in three parts.

The uterine (intramural) part lies within the wall of the uterus between the fundus and the body.

The isthmus is the straight narrow part just lateral to the wall of the uterus.

The ampulla is the widest part of the tube (diameter about 3 mm) and is long and tortuous.

The infundibulum is a dilated trumpet-like portion opening into the peritoneal cavity. The end of the tube has finger-like projections called *fimbriae*, one of which is longer than the others and is called the *ovarian fimbria*

tissue which lie between the two layers of broad ligament, one on each side of the uterus. They are attached to the body of the uterus just below the uterine tubes and pass to the sides of the pelvis then through the *inguinal canal* to end by fusing with the labia majora.

3. *Two uterosacral ligaments*. These ligaments originate from the posterior wall of the cervix and the posterior fornix of the vagina. They extend backwards, one on each side of the rectum, to be inserted into the sacrum.

4. *Two transverse cervical ligaments* (cardinal ligaments). These extend from the sides of the cervix and the lateral fornices of the vagina to the side walls of the pelvis.

5. *The pubocervical fascia*. This extends forward from the transverse cervical ligaments on each side of the bladder and is attached to the posterior surface of the pubis.

Functions

After puberty the uterus goes through a regular cycle of changes which prepares it to receive, nourish and protect a fertilised ovum. It provides the environment for the growing fetus during the 40 week gestation period, at the end of which the baby is born. The cycle is usually regular, lasting between 26 and 30 days, and is called the *menstrual cycle*. If the ovum is not fertilised the cycle ends with a short period of bleeding (see p. 234).

During pregnancy the walls of the uterus relax to accommodate the growing fetus. At the end of the gestation period *labour* begins and is concluded when the baby is born and the placenta extruded. During labour the muscle of the fundus and body of the uterus contracts intermittently and the cervix relaxes and dilates. As labour progresses the uterine contractions become stronger and more frequent. When the cervix is fully dilated the mother assists the birth of the baby by holding her breath and bearing down during the contractions.

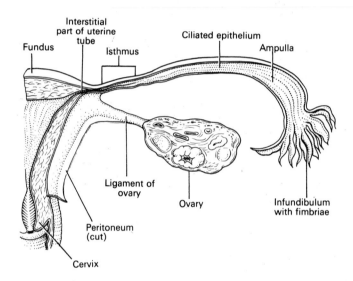

Figure 15:7   A section of a uterine tube and its related structures.

Structure

The uterine tubes are thin muscular tubes composed of three layers of tissue.

An inner lining of ciliated epithelium is arranged in longitudinal folds.

A middle coat of muscular tissue is a continuation of the uterine muscle. It is thinner than that of the uterus, and has only a longitudinal and a circular layer of fibres.

An outer layer of peritoneum is formed by the broad ligament.

Blood supply, lymph drainage and nerve supply

These are the same as for the uterus (see p. 231).

Function

The function of the uterine tubes is to convey the ovum from the ovary to the uterus. This is done by peristalsis assisted by the movement of the cilia of the lining epithelium.

Fertilisation of the ovum usually takes place in the uterine tube.

THE OVARIES

The ovaries are the female gonads or sex glands. They lie in a shallow fossa on the lateral walls of the pelvis, and are attached to the posterior layer of the broad ligament by a band of peritoneum called the *mesovarium*. The size of the ovaries varies in different individuals. Their length is between 2·5 cm and 3·5 cm, their breadth is about 2 cm and they are about 1 cm thick.

Structure and functions

The ovaries are formed by two distinct layers of tissue: the medulla and the cortex.

The medulla is composed of fibrous tissue and contains blood vessels and nerves. It is in the centre of the gland and is surrounded by the cortex.

The cortex is composed of a framework of connective tissue called the *stroma* which is covered by cubical epithelium called *germinal epithelium*. This layer contains the *ovarian follicles* and each follicle contains an *ovum*. Before puberty the ovaries are inactive but the stroma already contains immature follicles called *primordial follicles*, each of which contains an ovum. During the child-bearing years one ovarian follicle matures during each menstrual cycle. It reaches the surface of the ovary, ruptures and releases its ovum into the peritoneal cavity (Fig. 15:8).

This stage in the cycle is stimulated by the *follicle stimulating hormone* (FSH) from the anterior pituitary. While maturing the follicle produces the hormone *oestrogen*. After the follicle has ruptured its lining cells develop tissue which fills the cavity, called the *corpus luteum* (yellow body), under the influence of the *luteinising hormone* (LH) from the anterior pituitary. The corpus luteum produces a hormone called *progesterone*. If the ovum is fertilised it embeds in the wall of the uterus where it grows and develops, and for the first three months of the pregnancy, the corpus luteum continues to produce progesterone (Fig. 15:9). If the ovum is not fertilised the corpus luteum degenerates, menstruation occurs and the next cycle begins. Sometimes more than one follicle matures at a time which means that two or more ova are released in the same cycle. When this happens and the ova are fertilised the result is a multiple pregnancy.

Blood supply, lymph drainage and nerve supply

Arterial supply is by the *ovarian arteries* which branch from the abdominal aorta just below the renal arteries.

Venous drainage is into a plexus of veins behind the uterus from which the ovarian veins arise. The right ovarian vein opens into the inferior vena cava and the left into the left renal vein.

Lymph drainage is to the lateral aortic and pre-aortic lymph nodes. The lymph vessels follow the same route as the arteries.

Nerve supply. The ovaries are supplied by para-

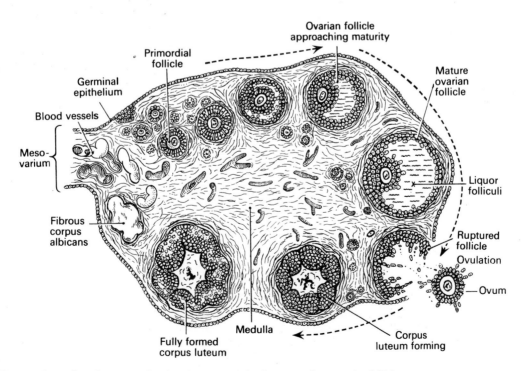

Figure 15:8 Diagram of a section of an ovary showing the stages of development of one ovarian follicle.

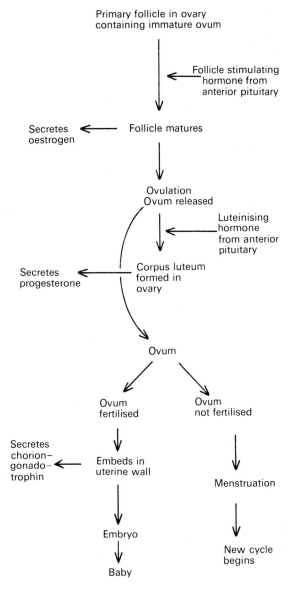

Figure 15:9 A summary of the stages of development of the ovum and the associated hormones.

sympathetic nerves from the sacral outflow and sympathetic nerves from the lumbar outflow. Their precise functions are not yet fully understood.

PUBERTY IN THE FEMALE

Puberty is the age at which the internal reproductive organs reach maturity. The ovaries are stimulated by the gonadotrophins from the anterior pituitary, that is, the *follicle stimulating hormone* and the *luteinising hormone*.

The age of puberty varies between 10 and 14 years and a number of physical and psychological changes take place at this time. The physical changes include:

1. The uterus, uterine tubes and the ovaries reach maturity.
2. The menstrual cycle and ovulation begin.

3. The breasts develop and enlarge.
4. Pubic and axillary hair begins to grow.
5. There is an increase in the rate of growth in height and widening of the pelvis.
6. There is an increase in the amount of fat deposited in the subcutaneous tissue.

THE MENSTRUAL CYCLE

The menstrual cycle includes a series of events which occur about every 28 days throughout the entire childbearing period of about 35 years. The cycle consists of a series of changes which take place concurrently in the ovaries and in the walls of the uterus. The uterus responds to changes in the amounts of hormones produced by the ovaries which, in turn, respond to stimulation by gonadotrophins produced by the anterior lobe of the pituitary gland.

The menstrual cycle (Fig. 15:10) is described as having three phases, named according to the changes in the state of the uterine wall. They are:

The proliferative phase	14 days
The secretory phase	10 days
The menstrual phase	4 days

The proliferative phase

At this stage in the cycle an ovarian follicle is growing towards maturity under the influence of the *follicle stimulating hormone* from the anterior pituitary and the follicle is producing *oestrogen*. Oestrogen stimulates the proliferation of the endometrium in preparation for the reception of the fertilised ovum. The endometrium becomes thicker by rapid cell multiplication and this is accompanied by an increase in the numbers of mucus-secreting tubular glands and blood capillaries. This phase ends when *ovulation* occurs and at the same time the production of oestrogen stops, i.e., when the ovarian follicle ruptures.

The secretory phase

Immediately after ovulation occurs the lining cells of the ovarian follicle are stimulated by the *luteinising hormone* from the anterior pituitary to develop the *corpus luteum*, which produces *progesterone*. Under the influence of progesterone the endometrium becomes oedematous and the secretory glands produce increased amounts of watery mucus, which may assist the passage of the spermatozoa through the uterus to the uterine tubes where the ovum is usually fertilised.

During the secretory phase there is a similar increase in the secretion of watery mucus by the vaginal glands and by the glands of the lining of the uterine tubes.

The ovum may survive in a fertilisable form for as few as 8 hours after ovulation. The spermatozoa, deposited in the female genital tract during coitus, may be able to fertilise the ovum for only about 24 hours although they may survive for several days. This means that the period in

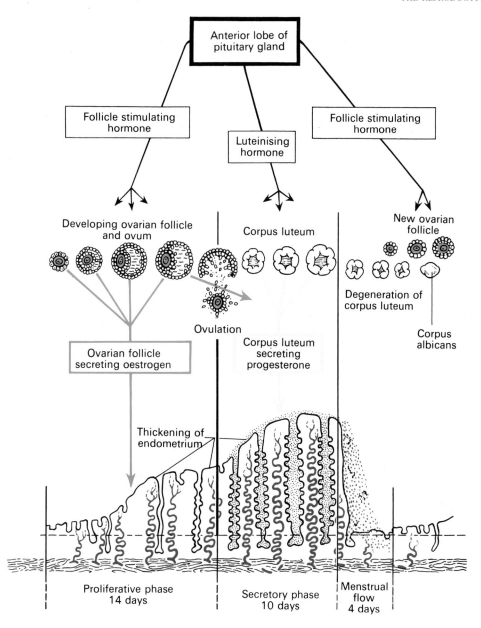

Figure 15:10 Diagram showing the endometrium at the various stages of the menstrual cycle and the associated hormones.

each cycle during which fertilisation can occur is relatively short. However, the date of ovulation cannot be predicted with certainty, even when cycles are regular.

If fertilisation of the ovum does not occur the cycle goes into the third phase.

The menstrual phase

If the ovum is not fertilised, the high level of progesterone in the blood inhibits the activity of the pituitary gland and the production of luteinising hormone is considerably reduced. The withdrawal of this hormone causes degeneration of the corpus luteum and thus progesterone production is decreased. After about 10 days, when the

progesterone in the blood has been reduced to a critical level, the lining of the uterus degenerates and breaks down and menstruation begins.

The extra secretions, lining cells, blood from the broken down capillaries and the unfertilised ovum constitute the menstrual flow.

When the amount of progesterone in the blood falls to a critical level another ovarian follicle is stimulated by the FSH and the next cycle begins.

If the ovum is fertilised there is no breakdown of the endometrium and no menstrual flow. The fertilised ovum travels through the uterine tube to the uterus where it becomes embedded in the uterine wall and produces a hormone called *chorion gonadotrophin*. This hormone keeps the corpus luteum intact enabling it to continue to secrete progesterone for the first 3 to 4 months of the pregnancy. During that time the *placenta* develops and produces oestrogen, progesterone and gonadotrophins. The placenta provides an indirect link between the circulation of the mother and that of the fetus. Through the placenta the fetus obtains nutritional materials and oxygen and gets rid of carbon dioxide and other waste products.

MENOPAUSE (CLIMACTERIC)

The child bearing period usually lasts for about 35 years. The menopause is the name given to the time when the processes which occur at puberty are reversed. These changes usually take place over a period of years, sometimes as long as 10 years and they are caused by the changes in the concentration of the sex hormones.

The ovaries gradually become less responsive to the FSH and LH from the anterior pituitary. During this period ovulation and the menstrual cycle become irregular and eventually cease. Several other phenomena may occur at the same time. These include:

1. Short-term unpredictable vasodilatation with flushing, sweating and palpitations which may disturb the normal sleep pattern are the most common temporary side-effects.
2. The breasts shrink.
3. Axillary and pubic hair become sparse.
4. The sex organs atrophy.
5. Episodes of unpredictable behaviour sometimes occur.

THE BREASTS OR MAMMARY GLANDS

The breasts or mammary glands are accessory glands of the female reproductive system. They exist in the male also but only in a rudimentary form.

In the female the breasts are quite small until puberty. Thereafter they grow and develop under the influence of oestrogen and progesterone. During pregnancy these hormones stimulate still further growth. After the baby is born the hormone *prolactin* from the anterior pituitary stimulates the production of milk and *oxytocin* from the posterior pituitary stimulates the release of milk in response to the stimulation of the nipple by the sucking of the baby.

Structure (Fig. 15:11)

The mammary glands consist of the following tissues:
  Glandular tissue
  Fibrous tissue
  Fatty or adipose tissue

The glandular tissue consists of about 20 lobes in each breast. Each lobe is made up of a number of lobules. The lobules consist of a cluster of alveoli which open into small ducts which unite to form large excretory ducts

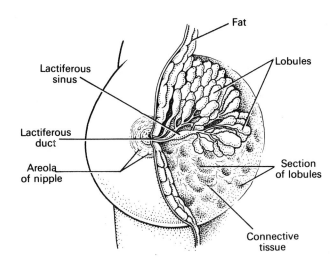

Figure 15:11   The breast.

called the *lactiferous ducts*. The lactiferous ducts converge towards the centre of the breast where they form dilatations or reservoirs for milk during lactation. Leading from these dilatations there are narrow ducts which open on to the surface at the nipple.

The fibrous tissue supports the glandular tissue and the ducts.

The fatty or adipose tissue covers the surface of the gland and is found between the lobes. The amount of fatty tissue determines the size of the breasts.

At the centre there is a small conical eminence, *the nipple*. The base of the nipple is surrounded by a pigmented area, *the areola*, which varies in colour from a deep pink to a light brown colour. On the surface of the areola there are numerous sebaceous glands called areolar glands (Montgomery's tubules) which in pregnancy lubricate the nipple.

Blood supply, lymph drainage and nerve supply

Arterial blood supply. The breasts are supplied with blood from the thoracic branches of the axillary arteries and from the internal mammary and intercostal arteries.

Venous drainage. This describes an anastomotic circle round the base of the nipple called the *circulus venosus*. Branches from this carry the venous blood to the circumference and end in the axillary and mammary veins.

Lymph drainage (Fig. 15:12). This is mainly into the axillary lymph vessels and glands.

Nerve supply. The breasts are supplied by branches from the fourth, fifth and sixth thoracic nerves which contain sympathetic fibres. There are numerous *sensory nerve endings* in the breast especially around the nipple. When these *touch receptors* are stimulated by sucking the impulses pass to the paraventricular nucleus of the hypothalamus and the flow of the hormone oxytocin is increased, promoting the release of milk.

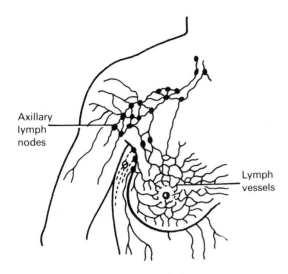

Figure 15:12   Lymph drainage from the breast.

Function

The mammary glands are active only during pregnancy and after the birth of a baby when they produce milk.

The Male Reproductive System

The male reproductive system consists of the following organs (Fig. 15:14):

- 2 testes
    (singular: testis)
- 2 epididymides
    (singular: epididymis)   } within the scrotum
- 2 deferent ducts and spermatic cords
- 2 seminal vesicles
- 2 ejaculatory ducts
- 1 prostate gland
- 1 penis

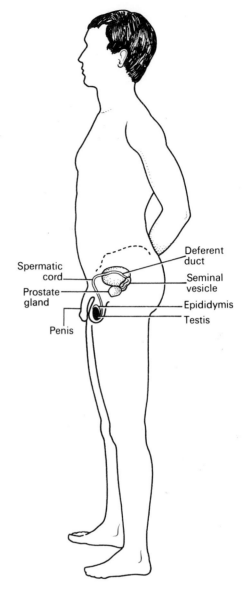

Figure 15:13   The male reproductive organs.

THE SCROTUM

The scrotum is a pouch of deeply pigmented skin divided into two compartments each of which contains one testis, one epididymis and the testicular end of a spermatic cord. It lies below the symphysis pubis, and in front of the upper parts of the thighs behind the penis.

THE TESTES

The testes are the reproductive glands of the male and are the equivalent of the ovaries in the female. They are suspended in the *scrotum* by the spermatic cords. The testes are approximately 4·5 cm long, 2·5 cm wide and 3 cm thick.

The testes are surrounded by three layers of tissue.

The tunica vaginalis is the outer covering of the testes and is a downgrowth of the abdominal and pelvic peritoneum. During early fetal life the testes develop in the lumbar region just below the kidneys then they descend into the scrotum taking coverings of peritoneum with them. This peritoneum eventually surrounds the testes in the scrotum becoming detached from the abdominal peritoneum.

The tunica albuginea is a fibrous covering surrounding the testes situated under the tunica vaginalis. Ingrowths of the tunica albuginea form septa dividing the glandular structure of the testes into *lobules*.

The tunica vasculosa consists of a network of capillaries supported by delicate connective tissue which lines the tunica albuginea. Therefore each lobule is surrounded by a fine network of blood capillaries.

Structure of the testes

Each testis consists of from 200 to 300 *lobules* composed of

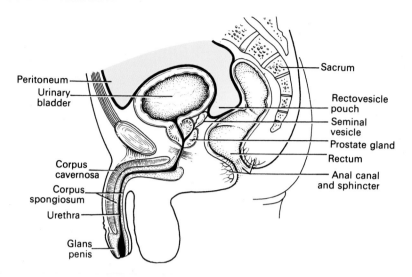

Figure 15:14 The male reproductive organs and their associated structures.

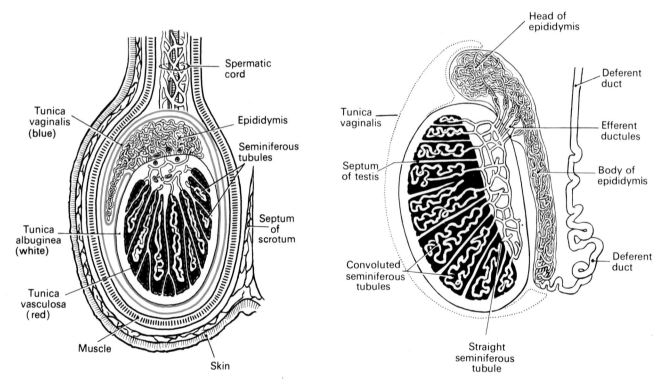

Figure 15:15 A longitudinal section of a testis and its coverings.

Figure 15:16 A longitudinal section of a testis and deferent duct.

germinal epithelial cells which are formed into the *convoluted seminiferous tubules*. Between the tubules there are groups of secretory cells known as the *interstitial cells*. The tubules eventually straighten out to become the straight seminiferous tubules. The straight tubules ascend to the upper pole of the testis and join together to become the *efferent ducts* which again join to become a complicated tortuous tubule, *the epididymis* (see Fig. 15:15).

The epididymis is a convoluted tubule which is folded upon itself and descends to the lower pole of the testis

lying posterior to the seminiferous tubules. At the lower border of the testis the epididymis continues as the *deferent duct* which passes upwards on the posterior wall of the testis. At first it is a coiled tube then it straightens out and leaves the testis and the scrotum enclosed within the spermatic cord (Fig. 15:16).

THE SPERMATIC CORDS

There are *two spermatic cords*, one leading from each

testis. Each spermatic cord is composed of the following structures:

1 testicular artery
1 testicular venous plexus
Lymph vessels
1 deferent duct
Nerves

The spermatic cord suspends the testis in the scrotum. It is composed of a thin sheet of fibrous tissue covering an inner layer of muscle tissue. Fine connective tissue surrounds the blood and lymph vessels, nerves and the deferent duct.

The spermatic cord passes through the inguinal canal. At the deep inguinal ring the structures within the cord diverge.

The testicular artery is a branch of the abdominal aorta and arises from it just below the renal arteries.

The testicular vein passes upwards through the pelvic cavity into the abdominal cavity. The left vein opens into the left renal vein and the right into the inferior vena cava.

The lymph drainage from the testis is into the lymph nodes around the aorta.

The deferent duct passes upwards from the testis through the inguinal canal and ascends medially towards the posterior wall of the bladder where it is joined by the duct from the *seminal vesicle* to form the *ejaculatory duct*.

The nerve supply is provided by branches from the 10th and 11th thoracic nerves.

THE SEMINAL VESICLES

The seminal vesicles are two pouches which lie on the posterior aspect of the bladder. They are approximately 5 cm long and roughly pyramidal in shape.

The seminal vesicles are composed of:

An outer coat of fibrous tissue
A middle coat of muscular tissue
An inner lining of columnar epithelium

At its lower end each seminal vesicle opens into a short duct which joins with the corresponding deferent duct to form an ejaculatory duct.

THE EJACULATORY DUCTS

The ejaculatory ducts are two short tubes approximately 2 cm long, each formed by the union of the duct from a seminal vesicle and a deferent duct. They pass through the prostate gland and join the prostatic part of the urethra.

The ejaculatory ducts are composed of the same layers of tissue as the seminal vesicles, i.e., fibrous, muscular and columnar epithelial tissue.

THE PROSTATE GLAND (Fig. 15:17 and 15:18)

The prostate gland lies in the pelvic cavity in front of the rectum and behind the symphysis pubis. It surrounds the

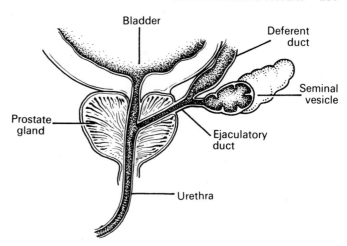

Figure 15:17   Section of the prostate gland and associated reproductive structures on one side.

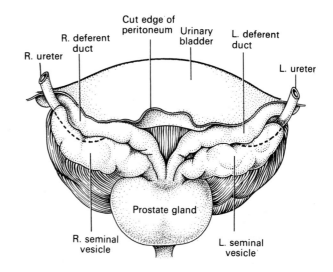

Figure 15:18   Structures associated with the prostate gland. Posterior view.

first part of the urethra and is about the size of a chestnut. It consists of an outer fibrous covering, a layer of smooth muscle and glandular substance composed of columnar epithelial cells.

The secretion of the prostate gland consists of a thin lubricating fluid which passes into the urethra through numerous ducts.

THE URETHRA AND PENIS

URETHRA

The male urethra provides a common pathway for the flow of urine and the secretions of the male reproductive organs called *semen*. It is much longer than in the female, measuring about 19 to 20 cm (7·5 to 8 inches). The first part is surrounded by the prostate gland and is called the *prostatic urethra*. The second part is the *membranous urethra* which is situated in the perineum behind the lower part of the symphysis pubis. The third part is the *spongy urethra* which is situated in the penis.

There are two urethral sphincters. The *internal sphincter* consists of smooth muscle fibres situated at the neck of the bladder above the prostate gland. The *external sphincter* consists of striated muscle fibres surrounding the membranous part of the urethra.

PENIS (Fig. 15:19)
The penis is composed of a *root* and a *body*. The root lies in the perineum and the body surrounds the urethra. It is formed by three elongated masses of *erectile tissue* and involuntary muscle very rich in blood vessels. The erectile tissue is supported by fibrous tissue and covered with skin.

the deferent duct, the seminal vesicle, the ejaculatory duct and the urethra to be implanted in the female vagina during coitus.

In the epididymis and the deferent duct the spermatozoa become more mature and mobile. They are now capable of independent movement through a liquid medium. If they are not ejaculated they are reabsorbed by these tubules (Fig. 15:21).

In man successful spermatogenesis takes place at a temperature about 3C° lower than normal body temperature. This lower temperature is achieved because the testes in the scrotum are covered by only a thin layer of tissue containing very little fat.

Semen is the name given to the fluid ejaculated from

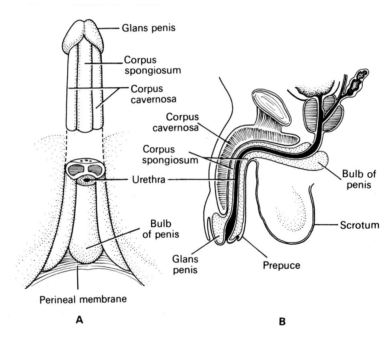

Figure 15:19　The penis. A. Viewed from below. B. Viewed from the side.

The two lateral columns of tissue are called the *corpora cavernosa* and the column between them the *corpus spongiosum* and at its tip it is expanded into a triangular structure known as the *glans penis*. Just above the glans the skin is folded upon itself and forms a movable double fold known as the *foreskin* or *prepuce*.

FUNCTIONS OF THE MALE REPRODUCTIVE SYSTEM
As in the female, the male reproductive organs are stimulated by the *gonadotrophic hormones* from the anterior lobe of the pituitary gland.

The follicle stimulating hormone stimulates the *seminiferous tubules* of the testes to produce the male germ cell—the spermatozoa (Fig. 15:20).

The spermatozoa then pass through the epididymis,

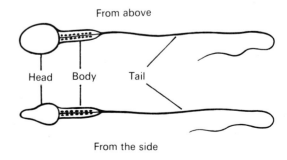

Figure 15:20　A spermatozoon.

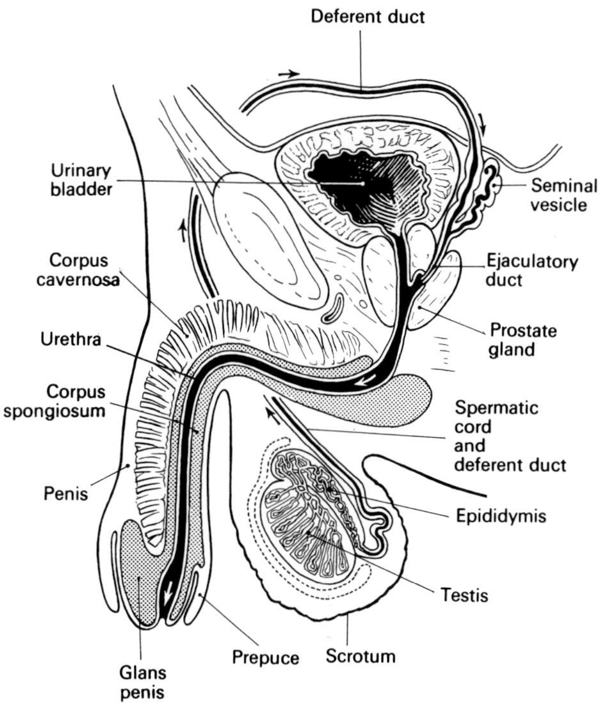

the urethra during coitus and consists of:

1. Spermatozoa
2. A viscid fluid which helps to nourish the spermatozoa, secreted by the seminal vesicles
3. A thin lubricating fluid produced by the prostate gland
4. Mucus secreted by glands in the lining membrane of the urethra

PUBERTY IN THE MALE

This occurs between the ages of 13 and 16 years. Luteinising hormone or, as it is called in the male, the *interstitial cell stimulating hormone* (ICSH) from the anterior lobe of the pituitary gland stimulates the *interstitial cells* of the testes to produce the hormone *testosterone*. This hormone influences the development of the body to sexual maturity. The changes which occur at puberty are:

1. Growth of muscle and bone and a marked increase in height
2. The voice 'breaks' due to enlargement of the larynx
3. Growth of hair on the face, on the axillae, the chest, the abdomen and the pubis
4. Enlargement of the penis, the scrotum and the prostate gland
5. Maturation of the seminiferous tubules and the production of spermatozoa

In the male fertility and sexual ability tend to decline gradually with ageing. There is no period comparable to the menopause in the female.

Figure 15:21 Section of the male reproductive organs. Arrows show the structures through which the spermatozoa pass.

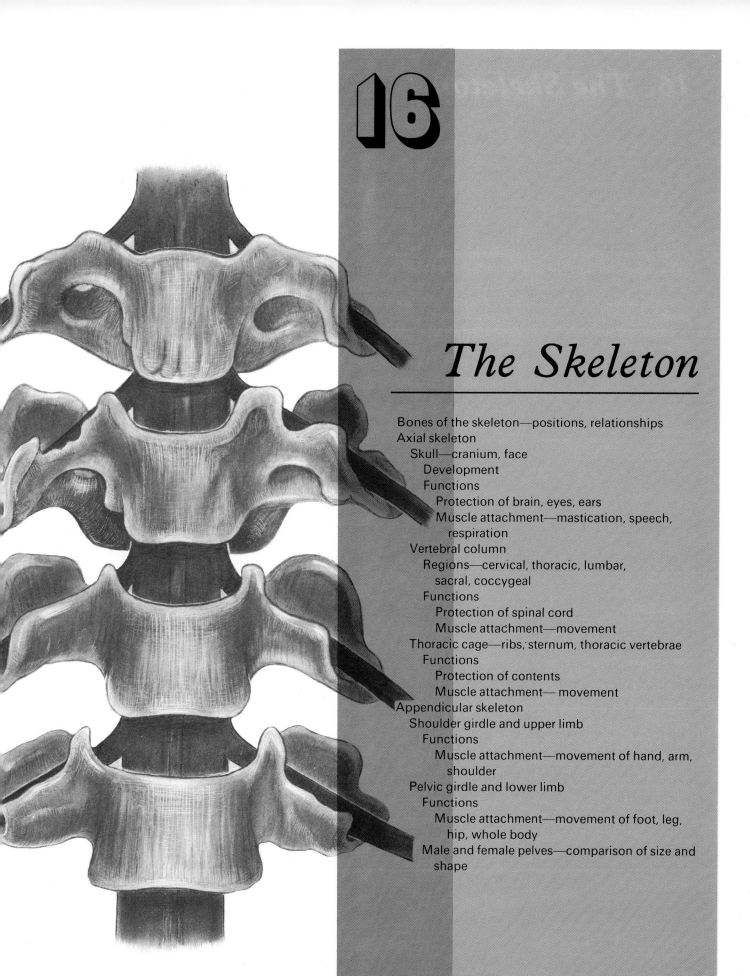

The Skeleton

16. The Skeleton

There are two different types of bone tissue, *cancellous* and *compact*, and they form bones which are classified as *long bones*, *short bones*, *flat bones*, *irregular bones* and *sesamoid bones*. All these types of bones form what is known as the *skeleton* or *bony framework* of the body.

The bones of the skeleton are divided into two groups (Figs. 16:1 and 16:2):

The axial skeleton consisting of the bones of the upright parts or axis of the body, i.e., the skull, vertebral column, ribs and sternum.

The appendicular skeleton consisting of the shoulder girdles, the upper limbs, the pelvic girdles and the lower limbs—the appendages.

The Axial Skeleton

This part consists of *the skull, the vertebral column, the ribs* and *the sternum.* Together the bones forming these structures constitute the central bony core of the body.

THE SKULL (Fig. 16:3)
The skull rests upon the upper end of the vertebral column and its bony structure is divided into two parts: the cranium and the face.

THE CRANIUM (Fig. 16:4)
The cranium is formed by a number of bones and it provides a bony protection for the brain. It is described in two parts: the *base* and the *vault*. The base is the part on which the brain rests and the vault the part which surrounds and covers it. The base of the cranium is divided into the *anterior*, *middle* and *posterior cranial fossae* which are deeply indented by the parts of the brain which rest on them and are perforated by many foramina for the passage of blood vessels and nerves.

The inner surfaces of all the bones of the cranium are markedly grooved by blood vessels with which they are in contact.

The bones which form the cranium are:

1 frontal bone
2 parietal bones
2 temporal bones
1 occipital bone
1 sphenoid bone
1 ethmoid bone

The frontal bone
This is the bone of the forehead. It forms part of the orbital cavities and the prominent ridges above the eyes are called the *supraorbital margins*. These roughened ridges give attachment to the muscles which raise the eyebrows. Just above the supraorbital margins, within the bone, there are two hollow spaces or sinuses which contain air, are lined with ciliated mucous membrane and communicate with the nasal cavities. The inner surface is concave and is grooved by the brain and blood vessels with which it is in close contact.

The joint formed between the frontal bone and the parietal bones is called the *coronal suture*. The frontal bone forms other immovable joints with the sphenoid, zygomatic, lacrimal, nasal and ethmoid bones. At birth the bone consists of two parts separated by the *frontal suture*, but union is usually complete by the eighth year of life (see Fig. 16:11).

Parietal bones
These two bones form the sides and roof of the skull. They articulate with each other at the *sagittal suture*, with the frontal bone at the coronal suture, with the occipital bone at the *lambdoidal suture* and with the temporal bones at the *squamous sutures*. The inner surface is concave and is grooved by the brain and blood vessels.

The temporal bones (Fig. 16:5)
These bones lie one on each side of the head and form immovable joints with the parietal, occipital, sphenoid and zygomatic bones. Each temporal bone is divided into four parts.

The squamous part is the fan-shaped portion which articulates with the parietal bone at the squamous suture.

The mastoid process is a thickened part of bone and can be felt just behind the ear. It contains a large number of very small air sinuses which communicate with the middle ear and are lined with simple squamous cells. A styloid process projects downwards from the mastoid process and gives attachment to muscles.

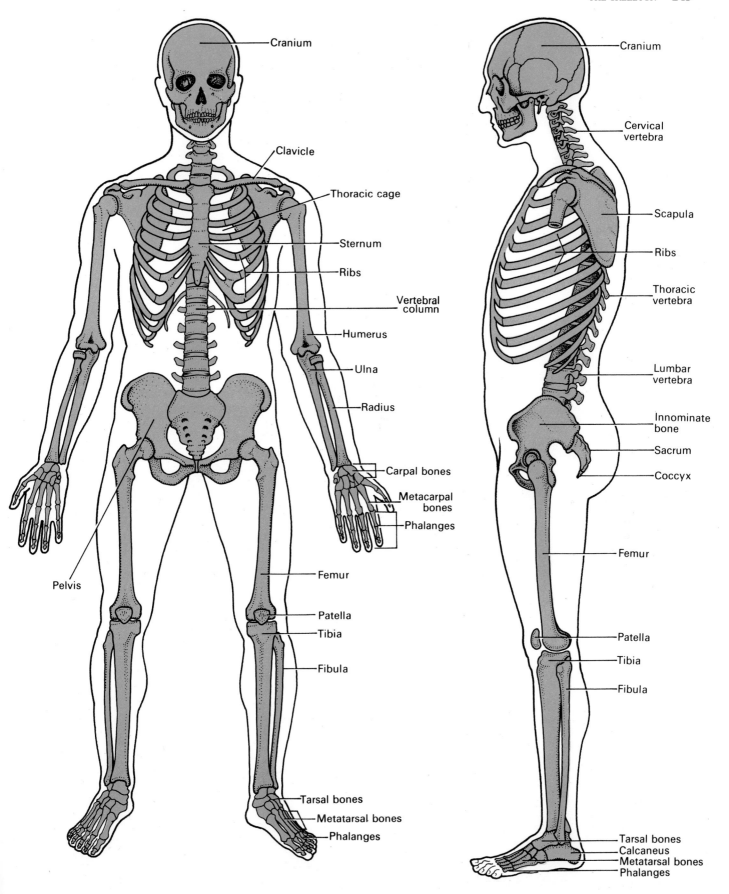

Figure 16:1 The skeleton. Anterior view.

Figure 16:2 The skeleton. Lateral view.

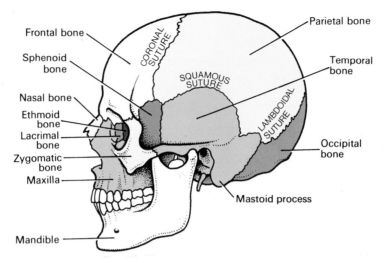

Figure 16:3   The bones of the skull and their joints or sutures.

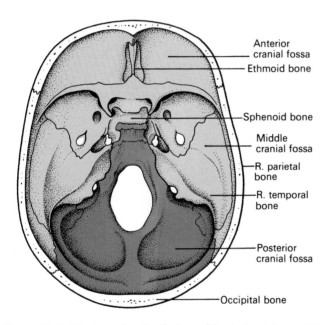

Figure 16:4   The bones forming the base of the skull and the cranial fossae. Viewed from above. Anterior cranial fossa—green. Middle cranial fossa—orange. Posterior cranial fossa—purple.

The petrous portion forms part of the base or floor of the skull and contains the organ of hearing, the organ of Corti (p. 201).

The zygomatic process is directed forward and articulates with the zygomatic bone to form the zygomatic arch.

The temporal bone has an articulating surface for the only movable bone of the skull, the mandible, at the *temporomandibular joint.* Immediately behind this articulating surface is the *auditory meatus* which passes inwards from the exterior towards the petrous portion of the bone.

The inner concave surface of the bone is deeply ridged by the brain and large blood vessels.

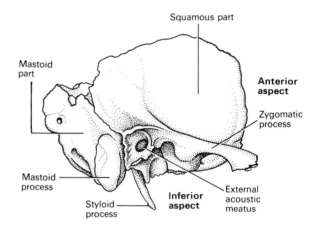

Figure 16:5   The right temporal bone. Lateral view.

The occipital bone (Fig. 16:6)

This is the bone which forms the back of the head and part of the base of the skull. It forms immovable joints with the parietal, temporal, and sphenoid bones. Its inner surface is deeply concave and the concavity is occupied by the occipital lobes of the cerebrum, the cerebellum, the posterior part of the superior sagittal sinus and the transverse sinuses. On the outer surface there is a roughened area called the *occipital protuberance* which gives attachment to muscles. The occiput has two articular condyles where it forms a hinge joint with the first bone of the vertebral column, the atlas. Between these condyles there is a large foramen called the *foramen magnum* through which the spinal cord passes.

The sphenoid bone (Fig. 16:7)

This bone is in the shape of a bat with its wings outstretched, and it occupies the middle portion of the base of

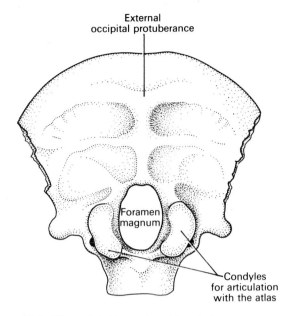

Figure 16:6 The occipital bone viewed from below.

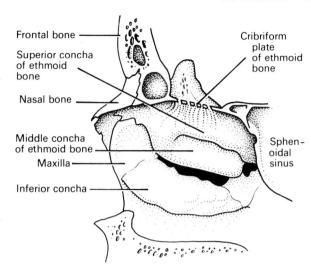

Figure 16:8 The ethmoid bone and its related structures.

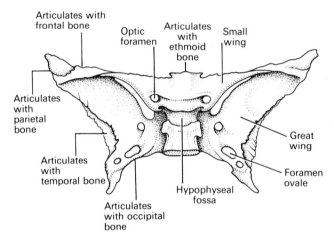

Figure 16:7 The sphenoid bone viewed from above.

the skull. The 'wings' extend outwards to the sides of the cranium articulating with the temporal, parietal and frontal bones. On the superior surface of the middle of the bone there is a little saddle-shaped depression called the *hypophyseal fossa (sella turcica)* in which the *pituitary gland* rests. The body of the bone contains foramina for the passage of blood vessels and nerves and some fairly large sinuses, which are lined by ciliated mucous membrane and communicate with the nasal cavities.

The ethmoid bone (Fig. 16:8)

The ethmoid bone occupies the anterior part of the base of the skull and helps to form the orbital cavity, the nasal septum and the lateral walls of the nasal cavity. On each

side it presents two projections into the nasal cavities called the *upper* and *middle conchae* or *turbinated processes*. It is a very delicate bone containing many air sinuses which have the same characteristics as those of the sphenoid bone. It has a horizontal flattened part called the *cribriform plate* which forms the roof of the nasal cavities and has numerous small foramina through which nerve fibres of the *olfactory nerve* (the nerve of the sense of smell) pass upwards from the nasal cavities to the brain. There is also a very fine *perpendicular plate* of bone which acts as the upper part of the *nasal septum*.

THE FACE

There are 13 bones which form the skeleton of the face but for completeness, the frontal bone which has already been described should be added (Fig. 16:9).

2 zygomatic or cheek bones
1 maxilla (originated as 2)
2 nasal bones
2 lacrimal bones
1 vomer
2 palatine bones
2 inferior conchae or turbinated bones
1 mandible (originated as 2)

Zygomatic or cheek bone

Each zygomatic bone forms the prominence of the cheek and part of the floor and lateral walls of the orbital cavity. It articulates with the zygomatic process of the temporal bone to form the zygomatic arch.

Maxilla or upper jaw bone

This originates as two bones but fusion takes place before birth. The maxilla forms the upper jaw, the anterior part of the roof of the mouth, the lateral walls of the

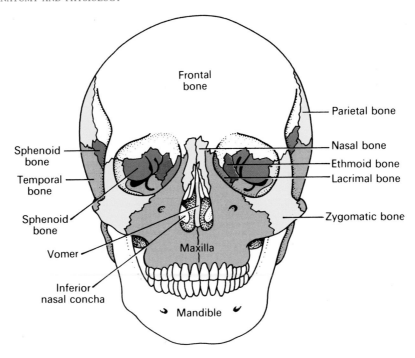

Figure 16:9 The bones of the face. Anterior view.

nasal cavities and part of the floor of the orbital cavities. It presents the *alveolar ridge* or *process* which projects downwards and carries the upper teeth. The teeth are present in the alveolar ridge before birth. On each side there is a large air sinus, the *maxillary sinus* or *antrum of Highmore,* which is lined with ciliated mucous membrane and communicates with the nasal cavity.

Nasal bones
These are two small flat bones which form the greater part of the lateral and superior surfaces of the bridge of the nose. They articulate with each other medially.

Lacrimal bones
These two small bones are in a position posterior and lateral to the nasal bones and form part of the medial walls of the orbital cavities. Each is pierced by a foramen for the passage of the *nasolacrimal duct* which carries the tears from the medial canthus of the eye to the nasal cavity.

Vomer
The vomer is a thin flat bone which extends upwards from the middle of the hard palate to separate the two nasal cavities. Superiorly it articulates with the perpendicular plate of the ethmoid bone.

Palatine bones
These are two L-shaped bones. The horizontal parts unite to form the posterior part of the hard palate and the

perpendicular parts project upwards to help to form the lateral walls of the nasal cavities. At their upper extremities they form part of the orbital cavities.

Inferior conchae or turbinated bones
Each concha is a long scroll-shaped bone which forms part of the lateral wall of the nasal cavity and projects into it below the middle concha. The superior and middle conchae are parts of the ethmoid bone.

Mandible (Fig. 16:10)
This is the strongest bone of the face and is the only movable bone of the skull. The bone originates as two

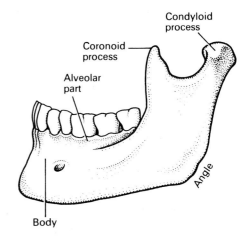

Figure 16:10 The mandible. Lateral view.

identical parts which unite at the midline. Each half consists of two main parts:

A *curved body* on the superior surface of which there is the *alveolar ridge* containing the lower teeth

A *ramus* which projects upwards almost at right angles to the body at its posterior end

At the upper end the ramus divides into two processes, the condyloid and coronoid processes. The *condyloid process* articulates with the temporal bone to form the *temporomandibular* joint and the *coronoid process* gives attachment to muscles and ligaments.

The point where the ramus joins the body is called the *angle* of the jaw.

Hyoid bone
This is an isolated horse-shoe-shaped bone lying in the soft tissues of the neck just above the *larynx* and below the *mandible*. It does not articulate with any other bone but is attached to the styloid process of the temporal bone by ligaments. It gives attachment to the base of the tongue.

THE SINUSES
Sinuses containing air are to be found in sphenoid, ethmoid and maxillary bones. They are all in communication with the upper air passages and are lined with ciliated mucous membrane. They serve two important purposes:

1. To give resonance to the voice
2. To lighten the bones of the face and cranium, making it easier for the head to balance on top of the vertebral column

FONTANELLES OF THE SKULL (Fig. 16:11)
It has already been stated that the immovable joints between the bones of the skull are called sutures. When ossification of the bones is complete these sutures are very closely knit; however, at birth ossification is not complete and there are several membranous areas called *fontanelles*. The two largest are the *anterior* and *posterior fontanelles*.

Anterior fontanelle
This is diamond-shaped and is situated at the junction of the frontal, coronal and sagittal sutures. It is the largest fontanelle and is not fully ossified until the child is 12 to 18 months old.

Posterior fontanelle
This is a smaller triangle-shaped membranous part at the junction of the sagittal and lambdoidal sutures and is usually ossified two to three months after birth.

THE VERTEBRAL COLUMN (Fig. 16:12)
The vertebral column consists of 24 separate movable, irregular bones plus the *sacrum* consisting of 5 fused bones, and the *coccyx* consisting of 4 fused bones. The 24 separate bones are divided into three groups:

7 cervical
12 thoracic
5 lumbar

All the movable vertebrae have certain characteristics in common but each group has its own special distinguishing features.

CHARACTERISTICS OF A TYPICAL VERTEBRA

A body
Each vertebra has a body which is situated anteriorly. The size of the body varies with the situation of the

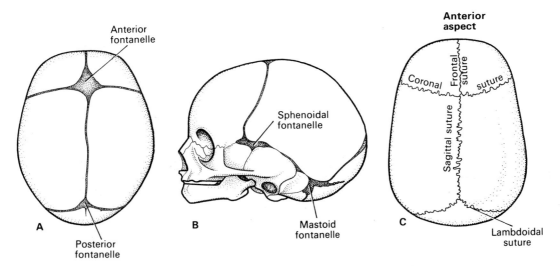

Figure 16:11 The skull showing the fontanelles and sutures. A. Fontanelles viewed from above. B. Fontanelles viewed from the side. C. Main sutures viewed from above when ossification is complete.

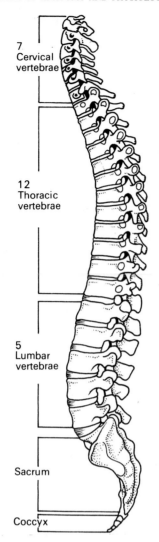

Figure 16:12 The vertebral column. Lateral view.

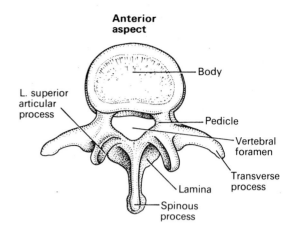

Figure 16:13 A lumbar vertebra showing the features of a typical vertebra—viewed from above.

SPECIAL FEATURES OF VERTEBRAE IN DIFFERENT PARTS OF VERTEBRAL COLUMN

Cervical vertebrae (Fig. 16:14)

Each transverse process exhibits a foramen through which a vertebral artery passes upwards to the brain. The spinous processes are forked or bifid at the ends giving attachment of muscles and ligaments. The bodies of this group are relatively small and the vertebral foramina relatively large. The first two cervical vertebrae are atypical and therefore must be described separately.

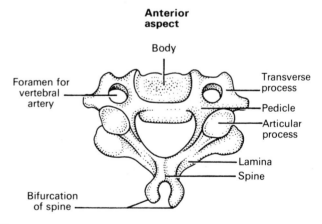

Figure 16:14 A cervical vertebra viewed from above.

vertebra, it is smaller in the cervical region and becomes larger towards the lumbar end of the column.

A neural arch

The neural arch encloses a large foramen, the *vertebral foramen*, which contains the spinal cord. The *pedicles* or roots of the neural arch project backwards from the body and the *laminae*, broad flattened plates of bone, project medially from the posterior ends of the pedicles to complete the neural arch. In the midline, where the laminae meet, there is a *spinous process* which projects backwards. At the points where the pedicles and laminae unite there are two lateral processes called *transverse processes*. On the superior and inferior surfaces of the neural arch, there are *two articular processes* for articulation with the vertebra above and the one below (Fig. 16:13).

The atlas (Fig. 16:15). The first cervical vertebra consists simply of a ring of bone with two short transverse processes. The ring is divided into two parts. The anterior part is occupied by the odontoid process of the axis, which is held in position by a *transverse ligament* (Fig. 16:16). Thus the odontoid process represents the body of the atlas. The posterior part of the ring is the true vertebral foramen and is occupied by the spinal cord. On its superior surface the bone has two articular facets

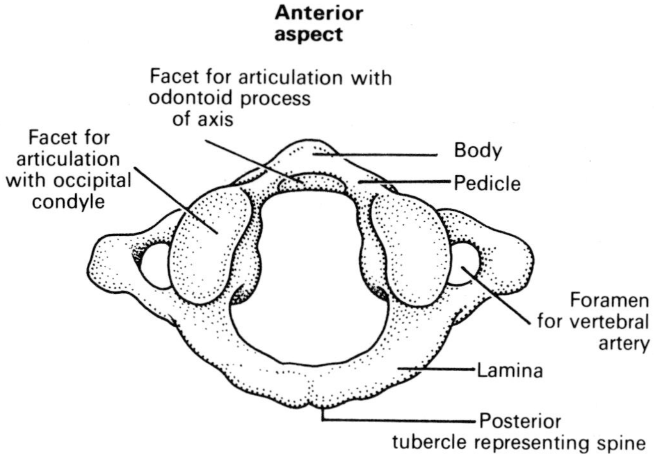

Anterior aspect

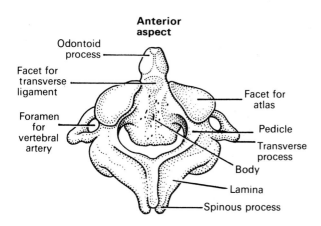

Figure 16:15A The atlas viewed from above.

which form joints with the condyles of the occipital bone of the skull. The nodding movement of the head takes place at these joints.

The axis. This is the second cervical vertebra. The body is small and has an upward projecting process called the *odontoid process* or the *dens*. This tooth-like process articulates with the first cervical vertebra and the movement at this joint is rotation, i.e., the turning of the head from side to side.

Thoracic vertebrae (Fig. 16:17)

The spinous processes are long and point downwards so that they partly overlap each other. The thoracic vertebrae articulate with the ribs and so present two half-facets on each lateral surface of the body and one articular surface on each transverse process.

Figure 16:15B The axis viewed from above.

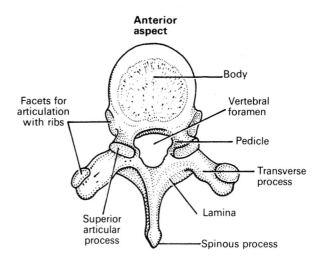

Figure 16:17A A thoracic vertebra. Viewed from above.

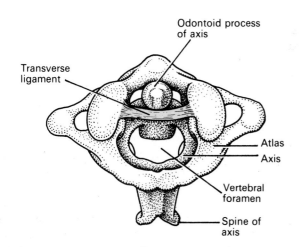

Figure 16:16 The atlas and axis in position showing the transverse ligament. Viewed from above.

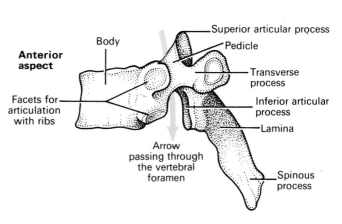

Figure 16:17B A thoracic vertebra. Viewed from the side.

Lumbar vertebrae

The bodies of the lumbar vertebrae are the largest and the vertebral foramina are the smallest. The spinous processes are short, flat-sided and project straight back (see Fig. 16:13).

The sacrum (Fig. 16:18)

This consists of five rudimentary vertebrae fused to form a wedge-shaped bone with a concave anterior surface. The upper part, or the base of the bone, articulates with the fifth lumbar vertebra. On each side it articulates with the ilium to form a *sacroiliac joint*, and at its inferior tip it articulates with the *coccyx*. The bodies of the individual vertebrae can still be distinguished although fused together. The anterior edge of the first of these bones protrudes into the pelvic cavity and is called the *promontory* of the sacrum. The vertebral foramina which form the neural canal are present, and on each side of the bone there is a series of foramina, one below the other, for the passage of nerves.

thinnest in the cervical region and become progressively thicker towards the lumbar region. The intervertebral discs are attached to the bodies of the adjacent vertebrae and the posterior longitudinal ligament in the vertebral canal helps to keep them in place. They have a shock-absorbing function and contribute to the flexibility of the vertebral column as a whole.

Intervertebral foramina

When two adjacent vertebrae are viewed from the side a foramen can be seen. Half of the wall of the foramen is formed by the vertebra above and half by the one below (Fig. 16:19). This is the intervertebral foramen and the spinal nerve passes through it. Throughout the length of the column there is an intervertebral foramen on each side between every pair of vertebrae, i.e., each vertebra contributes to the formation of four intervertebral foramina, two above and two below.

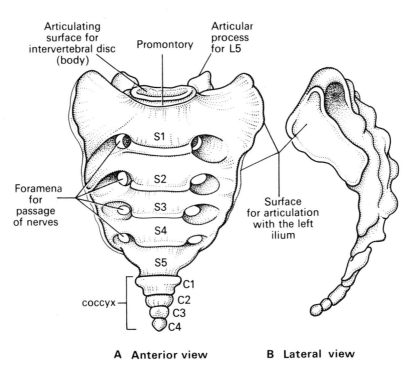

Figure 16:18   The sacrum and coccyx.

The coccyx (Fig. 16:18)

This consists of the four terminal vertebrae fused together to form a very small triangular bone the broad base of which articulates with the tip of the sacrum.

THE VERTEBRAL COLUMN AS A WHOLE

Intervertebral discs

The bodies of adjacent vertebrae are separated from each other by a pad of white fibrocartilage. These discs are

The curves of the vertebral column (Fig. 16:20)

When viewed from the side it will be seen that the vertebral column presents four curves, two of which are described as *primary* and two *secondary*.

When the fetus (the infant before birth) is in the uterus it lies curled up with the vertebral column bent so that the head and the knees are more or less touching. This position shows the *primary curvature* of the column. After birth this curvature is maintained until, at

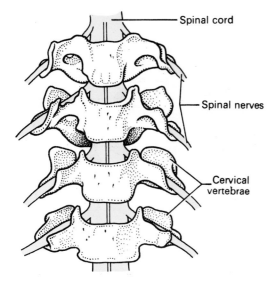

Figure 16:19 Cervical vertebrae separated to show the spinal cord and the spinal nerves emerging through the intervertebral foramina.

about three months old, the child can control the movements of his head. This control causes the development of the first secondary curve, *the cervical curve*. At the age of about 12 to 18 months the child begins to walk thus forming the other secondary curve, *the lumbar curve*. The primary curves which remain are the thoracic and sacral curves. When standing upright the vertebral column presents four curves:

> *Thoracic and sacral primary curves*, which are concave anteriorly
>
> *Cervical and lumbar secondary curves*, which are convex anteriorly

Ligaments of the vertebral column (Fig. 16:21)

The transverse ligament maintains the odontoid process of the axis in the correct position in relation to the atlas.

The anterior longitudinal ligament extends the whole length of the column and is situated on the anterior aspect of the bodies of the vertebrae. Its deeper layers are attached to each bone and therefore holds it firmly in position.

The posterior longitudinal ligament lies within the vertebral canal and extends the whole length of the vertebral column in close contact with the posterior surface of the bodies of the bones. It is attached to each bone and plays an important part in maintaining the intervertebral discs and the bones in their correct position.

The ligamenta flava connect the laminae of adjacent vertebrae.

The supraspinous ligament extends from the seventh cervical vertebra to the sacrum connecting the tips of the spinous processes.

The ligamentum nuchae extends from the occiput to the seventh cervical vertebra connecting the bifid spinous processes. It is a continuation of the supraspinous ligament.

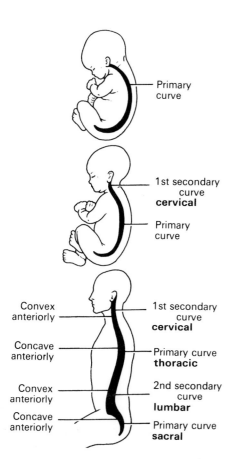

Figure 16:20 Diagram showing the order of development of the curves of the spine.

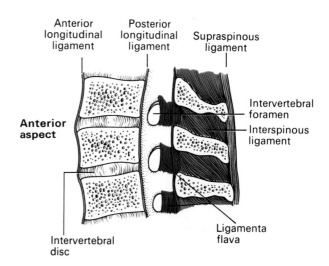

Figure 16:21 Section of the vertebral column showing the ligaments, intervertebral discs and intervertebral foramina.

Movements of the vertebral column

The movements between the individual bones of the vertebral column are very limited. However, the movements of the column as a whole are quite extensive and include *flexion* or bending forward, *extension* or bending backward, *lateral flexion* or bending to the side and *rotation* or turning round. There is more movement in the cervical and lumbar regions than elsewhere.

Functions of the vertebral column

1. Collectively the vertebral foramina form the vertebral canal which provides a strong bony protection for the delicate spinal cord lying within it.

2. The pedicles of adjacent vertebrae form intervertebral foramina on each side of the vertebral column. Spinal nerves, blood vessels and lymph vessels pass through these foramina.

3. Because of the numerous individual bones which make it up a certain amount of movement is possible.

4. It supports the skull which is protected from shock by the presence of the intervertebral discs.

5. It forms the axis of the trunk and gives attachment to the ribs, the shoulder girdle and the upper limbs, and the pelvic girdle and lower limbs.

THE THORACIC CAGE (Fig. 16:22)

The bones of the thorax or thoracic cage are:

    1 sternum
    12 pairs of ribs
    12 thoracic vertebrae

The sternum or breast bone (Fig. 16:23)

This is a *flat bone* which can be felt just under the skin in the middle of the front of the chest. It is about 15 cm (six inches) long, shaped like a dagger and is described in three parts.

The manubrium is the uppermost part and presents two articular facets on its lateral aspects for articulation with the clavicles at the *sternoclavicular joints*. The first two pairs of ribs also articulate with the manubrium just below the sternoclavicular joints.

The body or middle portion present facets on its lateral borders for the attachment of the ribs.

The xiphoid process is the tip of the bone. In some cases this part of the bone is never completely ossified. It gives attachment to the diaphragm and muscles of the anterior abdominal wall.

The ribs (Fig. 16:22)

There are 12 pairs of ribs which form the bony lateral walls of the thoracic cage. The first seven pairs are described as *true ribs* because their anterior ends are attached directly to the sternum by *costal cartilages*. The remaining five pairs are called *false ribs*. The eighth, ninth and tenth pairs are attached by their costal cartilages to the costal cartilage immediately above so that

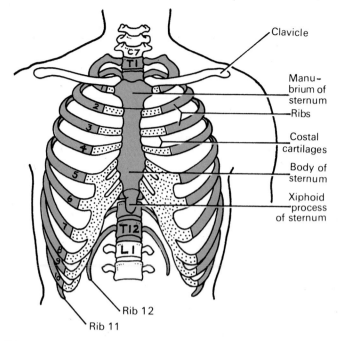

Figure 16:22A The thoracic cage. Anterior view.

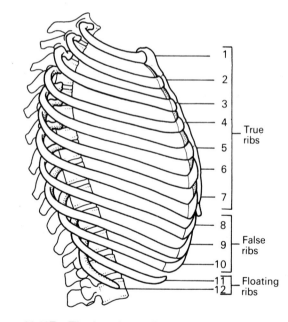

Figure 16:22B The thoracic cage. Lateral view.

they are only indirectly attached to the sternum (Fig. 16:23). The last two pairs have no anterior attachment and are therefore called *floating ribs*. All 12 pairs of ribs articulate posteriorly with the thoracic vertebrae.

Characteristics of a rib (Fig. 16:24). It is a flat curved bone which is described as having a head, neck, tubercle,

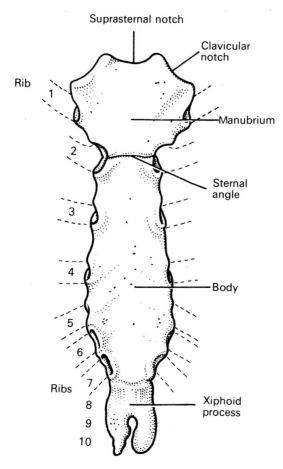

Figure 16:23   The sternum and its attachments.

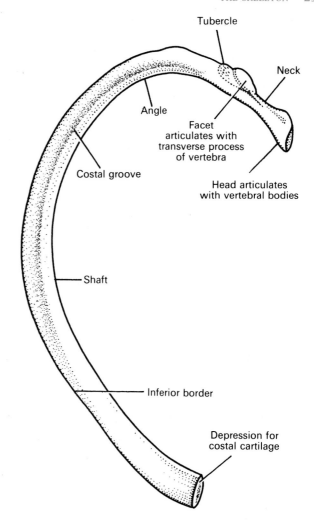

Figure 16:24   A typical rib.

angle, sternal end, anterior and posterior surface and a superior and inferior border.

The head articulates posteriorly with the bodies of two adjacent vertebrae. The neck is a constricted portion immediately anterior to the head and between the head and the tubercle. The tubercle articulates with the transverse process of a thoracic vertebra. The angle is the point at which the bone bends. The sternal end is attached to the sternum by a costal cartilage, i.e., a band of hyaline cartilage. The superior border is rounded and smooth while the inferior border exhibits a marked groove occupied by blood vessels and nerves.

The first rib is different from the others in that its broad surfaces are superior and inferior while its borders are anterior and posterior. It does not move during respiration.

The spaces between the ribs are occupied by the intercostal muscles. During inspiration, when these muscles contract, the ribs and sternum are lifted upwards and outwards increasing the capacity of the thoracic cavity.

Thoracic vertebrae
The 12 thoracic vertebrae have already been described.

The Appendicular Skeleton

The appendicular skeleton consists of the shoulder girdles with the upper limbs and the pelvic girdle with the lower limbs (Fig. 16:25).

THE SHOULDER GIRDLE AND UPPER LIMB
Each shoulder girdle consists of the following bones:
    1 clavicle
    1 scapula

Each upper extremity consists of the following bones:
    1 humerus
    1 radius
    1 ulna
    8 carpal bones
    5 metacarpal bones
    14 phalanges

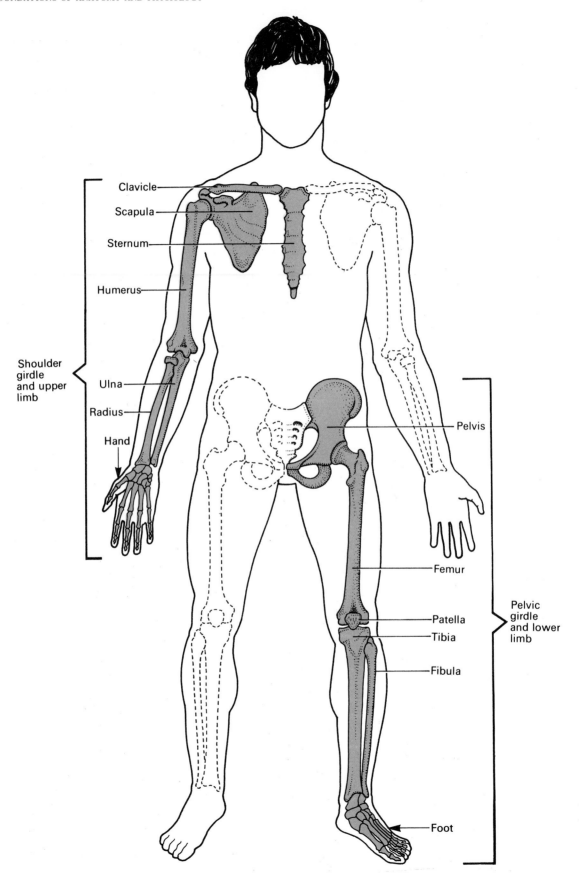

Figure 16:25   The bones of the appendicular skeleton.

The clavicle or collar bone (Fig. 16:26)
The clavicle is a long bone which has a double curve. At its rounded medial extremity it articulates with the manubrium of the sternum at the sternoclavicular joint and at its flattened lateral extremity it forms the *acromio-clavicular* joint with the *acromion process* of the scapula. The shaft of the bone is roughened for the attachment of muscles. The clavicle provides the only bony link between the upper extremity and the axial skeleton.

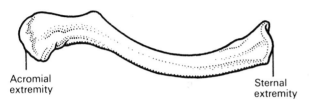

Figure 16:26 The right clavicle.

The scapula or shoulder blade (Fig. 16:27)
The scapula is a flat triangular-shaped bone which lies on the posterior chest wall superficial to the ribs and separated from them by muscles. It is described as having:

3 borders
3 angles
3 fossae

The three borders of the bone form its outer extremities and meet at its three angles. The borders are called medial, superior and lateral. The superior angle is where the medial and superior borders meet and the inferior angle is where the medial and lateral borders meet. At the point where the superior and lateral borders meet there is the lateral angle which presents a shallow articular surface called the *glenoid cavity* which, together with the *head of the humerus*, forms the *shoulder joint*.

Projecting forward from the superior border of the bone is the *coracoid process* which gives attachment to muscles and ligaments. The posterior surface of the bone is divided into two fossae by a *spine*, these are the *supraspinous fossa* and the *infraspinous fossa*. The spinous process projects beyond the lateral angle of the bone as the *acromion process* and overhangs the shoulder joint. It articulates with the lateral extremity of the clavicle at the *acromio-clavicular joint*. On the anterior surface of the bone there is the *subscapular fossa*. These fossae and borders are grooved and ridged for the attachment of muscles.

The humerus (Fig. 16:28)
This is a long bone and is the bone of the upper arm. It consists of a proximal end or head, neck, shaft and distal end.

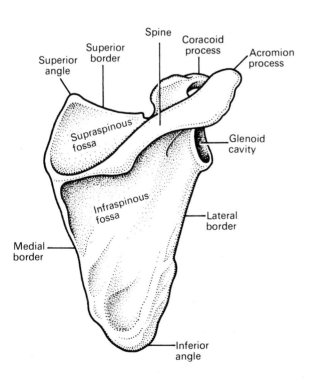

Figure 16:27 The right scapula. Posterior view.

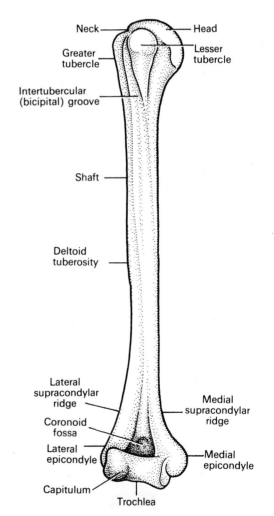

Figure 16:28 The right humerus. Anterior view.

The head is smooth and rounded and takes part in the formation of the shoulder joint where it articulates with the glenoid cavity of the scapula.

The neck is a slightly constricted part immediately distal to the head. Between the neck and the shaft there are two roughened projections of bone, the *greater and lesser tubercles*. Between them there is a deep groove, the *bicipital groove* or *intertubercular sulcus*, which is occupied by one of the tendons of the biceps muscle.

At its proximal end the shaft is cylindrical in shape but becomes flattened on its anterior and posterior surfaces towards the distal end.

The distal end of the bone presents two articular surfaces, the rounded *capitulum* lying laterally and the more rectangular *trochlea* medially. Immediately above the articular surfaces on the anterior aspect of the bone there is a deep fossa called the *coronoid fossa* and the corresponding fossa on the posterior surface of the bone is the *olecranon fossa*. In a position above the articular surfaces on each side, and projecting outwards from them, there are the *medial and lateral epicondyles*. The epicondyles and their *supracondylar ridges* give attachment to muscles.

The ulna and radius (Fig. 16:29)

These are the two bones of the forearm. The ulna is medial to the radius and when the arm is in the anatomical position, i.e., with the palm of the hand facing forward, the two bones are parallel. There is an *interosseous membrane* between the bones.

The ulna

This is a long bone consisting of a proximal end, a shaft and a distal end or head. The proximal end of the bone presents two articular fossae and two processes. The *trochlear notch* is a semilunar notch which articulates with the trochlea of the humerus. The proximal extremity of this notch is called the *olecranon process* which forms the point of the elbow and fits into the olecranon fossa of the humerus when the arm is straight. At the distal end of the trochlear notch there is another process, the *coronoid process* which in turn fits into the coronoid fossa of the humerus when the arm is bent. On the lateral surface of the proximal end of the bone there is the other articular surface, the *radial notch*. It is with this notch that the head of the radius articulates to form the *proximal radioulnar joint*.

The shaft of the bone is triangular in shape and is roughened for the attachment of muscles.

The distal end or *head of the ulna* is separated from the wrist joint by a pad of white fibrocartilage. On the lateral aspect there is a smooth surface for articulation with the radius at the *distal radioulnar joint*. The head presents a *styloid process* projecting from its posterior aspect which gives attachment to ligaments.

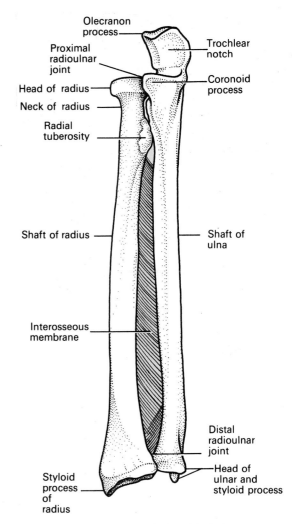

Figure 16:29 The right radius and ulna with the interosseous membrane. Anterior view.

The radius

The radius is the lateral bone of the forearm and presents a head, neck, tuberosity, shaft and distal extremity.

The head is disc-shaped and flat on top except for a very shallow depression which articulates with the capitulum of the humerus. The circumference of the head articulates with the *radial notch* of the ulna to form the proximal radioulnar joint.

The neck is a constricted portion of the bone which joins the head to the shaft.

Medially, at the upper end of the shaft, there is the radial tuberosity which gives attachment to muscles. The shaft is roughly triangular in shape being smooth on its lateral surface and more angular and roughened medially.

The distal end of the bone is expanded. It articulates with the carpal bones to form the wrist joint and with the ulna to form the distal radioulnar joint. On its lateral aspect it presents a styloid process which gives attachment of ligaments and muscles.

The carpal bones or the bones of the wrist (Fig. 16:30)

The carpal bones are eight in number and are arranged in two rows of four. Their names, from without inwards are:

Proximal row: scaphoid, lunate, triquetral, pisiform
Distal row: trapezium, trapezoid, capitate, hamate

These bones are all closely fitted together and held in position by ligaments which allow for a certain amount of movement between them. In addition to forming joints with each other, the proximal row is associated with the wrist joint and the distal row forms joints with the metacarpal bones. They give attachment to the short muscles of the hand which move the fingers and to the muscles of the forearm which move the wrist joint.

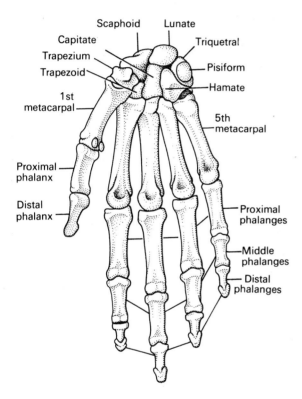

Figure 16:30   The bones of the wrist, hand and fingers. Anterior aspect.

The metacarpal bones or the bones of the hand

These are five in number and they form the structure of the palm of the hand. They are not given names but are numbered from the thumb side inwards. They are long slender bones, the proximal ends of which articulate with the carpal bones and the distal ends with the phalanges.

The phalanges or bones of the fingers

There are 14 phalanges arranged so that there are three in each finger and two in the thumb. The proximal phalanx of each finger is the largest and articulates with the cor-responding metacarpal bone at one end and with the middle phalanx at the other end. The distal phalanx is the smallest and forms the tip of the finger. The thumb which is the shortest of the digits has only two phalanges.

THE PELVIC GIRDLE AND LOWER LIMB

The bones which make up the pelvic girdle or pelvis are:

    2 innominate bones
    1 sacrum

The bones which make up the lower extremity are:

    1 femur
    1 tibia
    1 fibula
    1 patella
    7 tarsal bones
    5 metatarsal bones
    14 phalanges

The innominate or hip bones (Fig. 16:31)

Each hip bone consists of three separate bones, the *ilium*, the *ischium* and the *pubis*, fused together to form one large irregular bone. On its outer surface it has a deep depression called the *acetabulum* which, with the almost spherical head of the femur, forms the hip joint. Above and behind the acetabulum there is a large notch in the bone called the *great sciatic notch* and in front and below there is the *obturator foramen* occupied by a membrane.

The ilium is the upper flattened part of the bone and it presents the *iliac crest*, the anterior point of which is called the *anterior superior iliac spine* and below it the *anterior inferior iliac spine*.

The pubis is the anterior part of the bone and articulates with the pubis of the other hip bone at a cartilaginous joint called the *symphysis pubis*.

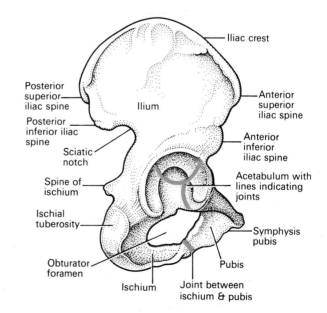

Figure 16:31   The right innominate bone. Lateral view.

The ischium is the inferior and posterior part of the bone, the most dependent part of which is the *ischial tuberosity*.

The union of the three parts takes place in the *acetabulum*.

The external surfaces of the innominate bones is markedly ridged for the attachment of muscles.

The pelvis (Fig. 16:32)

The pelvis is formed by the innominate bones which articulate anteriorly at the symphysis pubis and posteriorly with the sacrum to form the sacroiliac joints. It is divided into the *greater or false pelvis* above and the *lesser or true pelvis* below. Looking down on the pelvis from above there is a ridge of bone, called the *brim of the pelvis*, projecting into it all the way round. Posteriorly this ridge is formed by the promontory of the sacrum and laterally and anteriorly by the *iliopectineal lines* of the innominate bones (see Fig. 16:32). The part of the pelvis above this ridge is the false pelvis and the part below, the true pelvis.

DIFFERENCES BETWEEN THE MALE AND FEMALE PELVES

There are distinct differences between the pelves of the male and the female (Fig. 16:33). The female pelvis is shaped to allow for the passage of the baby during childbirth. The fundamental differences can be summarised in the following manner.

	Female	Male
Bones	Lighter and smaller	Heavier and longer
Cavity	Shallow and round	Deep and funnel-shaped
Sacrum	More concave anteriorly, making the true pelvis roomier	Less concave, making the true pelvis narrower at the outlet
Pubic arch	The angle made by the two pubic bones at the symphysis pubis is wider	The angle of the pubic arch is narrower

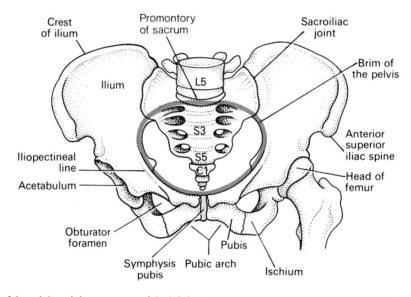

Figure 16:32 The bones of the pelvis and the upper part of the left femur.

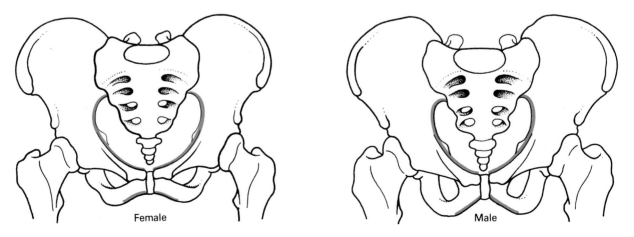

Figure 16:33 Diagram showing the differences in shape of the male and female pelves.

The femur or thigh bone (Fig. 16:34)

The femur is the longest and strongest bone of the body and can be described as having two extremities and a shaft.

The proximal extremity consists of a *head*, *neck* and *greater and lesser trochanters*. The head is almost spherical in shape and fits into the *acetabulum* of the hip bone to form the *hip joint*. In the centre of the head there is a small depression for the attachment of the *ligament of the head of the femur*. This extends from the acetabulum to the femur conveying a blood vessel which supplies blood to an area of the head of the bone. The neck of the bone extends outwards and slightly downwards from the head to the shaft. The greater and lesser trochanters are two large eminences situated at the junction of the neck and the shaft, with the *intertrochanteric line* between them on the posterior surface of the bone. These two trochanters and the intertrochanteric line give attachment to muscles which move the hip joint.

The shaft of the bone is smooth and rounded on its anterior surface but on its posterior surface it has a roughened ridge extending the length of the shaft, called the *linea aspera* which gives attachment to muscles. The shaft of the bone is slightly convex anteriorly and becomes broader towards its distal end. The posterior surface of the lower third forms a flat triangular area called the *popliteal surface*.

The distal extremity presents two articular *condyles* which take part in the formation of the knee joint. Between the condyles there is a deep depression called the *intercondylar fossa*. The anterior aspect of the lower end of the bone has an articular surface for the patella called the *patellar surface*.

The tibia or shin bone (Fig. 16:35)

The tibia is the medial of the two bones of the leg. It is a long bone and presents two extremities and a shaft.

The proximal extremity is broad and flat and presents two *condyles* for articulation with the condyles of the femur at the knee joint. Between the condyles there is a ridge called the *intercondylar eminence*. Distal to the articular surfaces on the anterior aspect of the bone there

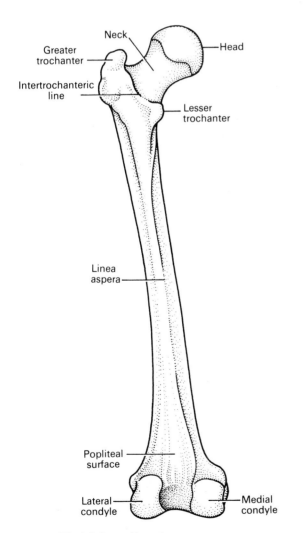

Figure 16:34 The left femur. Posterior aspect.

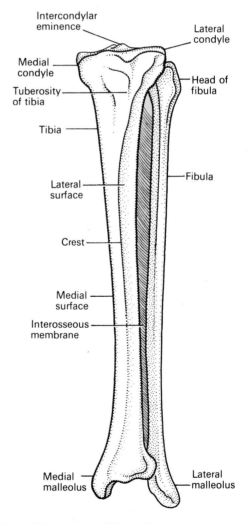

Figure 16:35 The left tibia and fibula with the interosseous membrane.

is the *tuberosity* of the tibia which gives attachment to muscles. The lateral condyle presents an *articular facet* on its inferior surface for articulation with the head of the *fibula* at the *superior tibiofibular joint*.

The shaft of the bone is roughly triangular in shape and is described as having a medial, lateral and posterior surface. The *crest* of the tibia can be felt very close to the surface on the anterior aspect of the leg.

The distal extremity of the tibia is smooth and flat where it forms the *ankle joint* with the *talus*. On the medial aspect there is a downward projecting process called the *medial malleolus*. The lateral surface of this process also takes part in the formation of the ankle joint. The lateral aspect of the distal end of the bone articulates with the fibula at the *inferior tibiofibular* joint.

The fibula (Fig. 16:35)
The fibula is a long slender bone in the leg which is lateral to the tibia. The head or upper extremity articulates with the lateral condyle of the tibia and the lower extremity articulates with the lower extremity of the tibia but projects beyond it to form the *lateral malleolus* and take part in the formation of the ankle joint. The shaft of the bone is ridged for the attachment of muscles. Between the shafts of the tibia and fibula there is an *interosseous membrane*.

The patella or knee cap
This is a *sesamoid* bone associated with the knee joint. It is roughly triangular in shape and lies with the apex pointing downwards. Its anterior surface is in the *patellar tendon*, the tendon of the quadriceps femoris muscle, and its posterior surface articulates with the patellar surface of the femur in the knee joint.

The tarsal or ankle bones
There are seven tarsal bones which form the posterior part of the foot (Fig. 16:36). They are:

    1 talus
    1 calcaneus
    1 navicular
    3 cuneiform
    1 cuboid

The talus articulates with the tibia and fibula at the ankle joint.

The calcaneus or heel bone is markedly roughened for the attachment of muscles which move the ankle joint.

The navicular is situated on the medial aspect of the foot distal to the talus.

The medial, intermediate and *lateral cuneiform bones* and the *cuboid* form a row of bones named from within outwards, which articulate with the other three tarsal bones proximally and with the five metatarsal bones distally.

The metatarsal bones
These are five in number and form the greater part of the dorsum of the foot. At their proximal ends they articulate with the tarsal bones and at their distal ends, with the phalanges. They do not have individual names but are numbered from within outwards.

The phalanges
There are 14 phalanges in each foot which are arranged in a similar manner to those in the fingers, i.e., two in the great toe and three in each of the other toes.

ARCHES OF THE FOOT
The arrangement of the bones of the foot is such that the foot is not a rigid structure. This point is well seen by comparing a normal foot with a 'flat' foot. The bones have a bridge-like arrangement and are supported by muscles and ligaments so that four arches are formed:

    Medial longitudinal arch
    Lateral longitudinal arch
    Two transverse arches

Medial longitudinal arch
This is the highest of the arches and is formed by the calcaneus, navicular, three cuneiform and first three metatarsal bones. Only the calcaneus and the distal end of the metatarsal bones should touch the ground.

Lateral longitudinal arch
The lateral arch is much less marked than its medial counterpart and has only four bony components, the

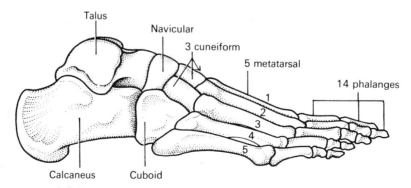

Figure 16:36 The bones of the foot. Lateral view.

calcaneus, cuboid and the two lateral metatarsal bones. Again only the calcaneus and metatarsal bones should touch the ground.

Transverse arches

These arches run across the foot and can be more easily seen by examining the skeleton than the live model. They are most marked at the level of the three cuneiform and cuboid bones.

MUSCLES AND LIGAMENTS WHICH SUPPORT THE ARCHES OF THE FOOT (Fig. 16:37)

As there are movable joints between all the bones of the foot very strong muscles and ligaments are necessary to maintain the strength, resilience and stability of the foot during walking, running and jumping.

Posterior tibialis muscle

This is the most important muscular support of the medial longitudinal arch. It lies on the posterior aspect of the lower leg, originates from the middle third of the tibia and fibula and its tendon passes behind the medial malleolus to be inserted into the navicular, cuneiform, cuboid and metatarsal bones. It can be seen, therefore, that it acts as a sling or 'suspension apparatus' for the arch (Fig. 16:37).

Short muscles of the foot

This group of muscles is mainly concerned with the maintenance of the lateral longitudinal and transverse

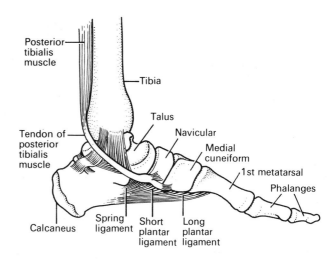

Figure 16:37 The tendons and ligaments supporting the arches of the foot. Medial view.

arches. They make up the fleshy part of the sole of the foot.

Plantar calcaneonavicular ligament or 'spring' ligament

This is a very strong thick ligament stretching from the calcaneus to the navicular bone. It plays an important part in supporting the medial longitudinal arch.

Plantar ligaments and interosseous membranes

These structures support the lateral and transverse arches.

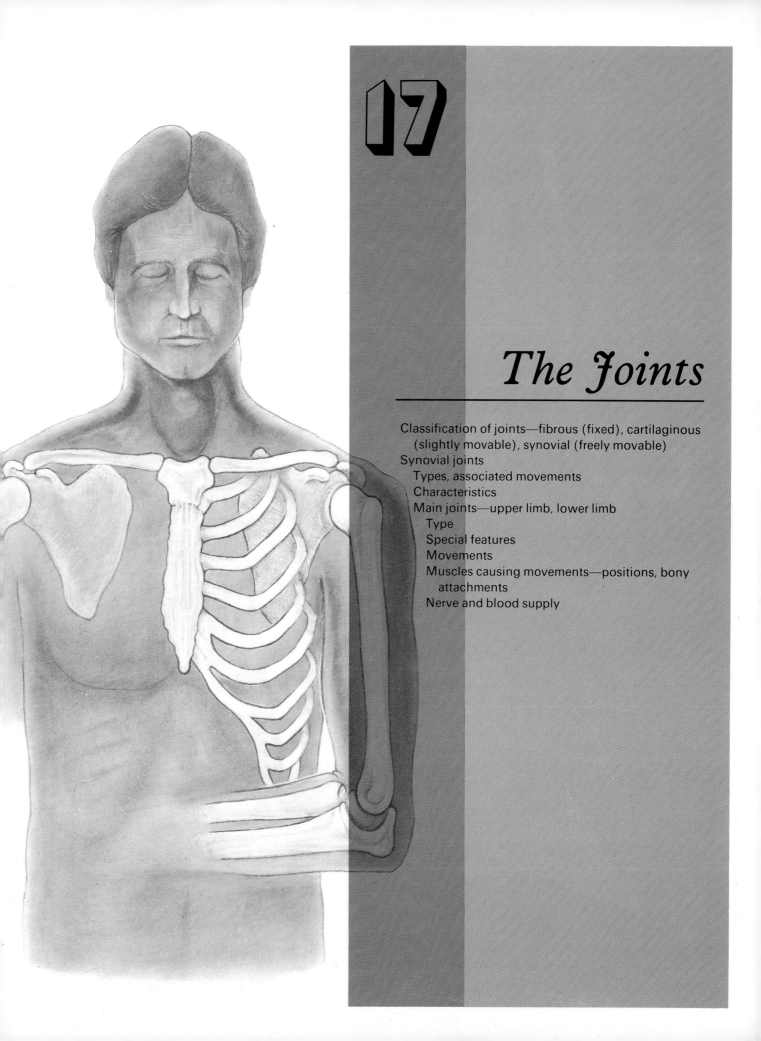

17

The Joints

17. The Joints

A joint is the site at which any two or more bones come together. Some joints have no movement, some only slight movement and some are freely movable. Joints may also be classified as *fibrous, cartilaginous* and *synovial*.

Before going on to describe the main synovial joints of the limbs, the movements which may occur should be considered.

Flexion or bending, usually forward but occasionally backward as in the case of the knee joint.

Extension means straightening or bending backward.

Abduction is movement away from the midline of the body.

Adduction is movement towards the midline of the body.

Rotation is movement round the long axis of a bone.

Pronation means turning the palm of the hand down.

Supination means turning the palm of the hand up.

Circumduction is the combination of flexion, extension, abduction and adduction.

Inversion is the turning of the sole of the foot inwards.

Eversion consists of the opposite movement to inversion, that is, turning the sole of the foot outwards.

Fibrous or Fixed Joints (Fig. 17:1)

In this type of joint there is no movement between the bones concerned. As the name suggests, there is *fibrous tissue* between the ends of the bones. Examples of this type include the joints between the bones of the skull called sutures and the joints between the teeth and the maxilla and mandible.

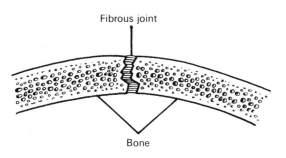

Figure 17:1   A fibrous or fixed joint, e.g., the sutures of the skull.

Cartilaginous or Slightly Movable Joints (Fig. 17:2)

In this case there is a pad of *white fibrocartilage* between the ends of the bones taking part in the joint which allows for very slight movement. Movement is only possible because of compression of the pad of cartilage. Examples of cartilaginous joints include the symphysis pubis and the joints between the bodies of the vertebrae.

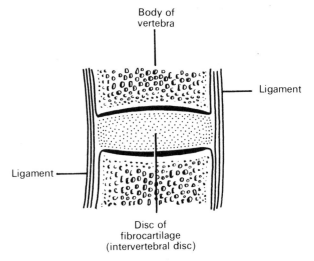

Figure 17:2   A cartilagenous or slightly movable joint, e.g., between the bodies of the vertebrae.

Synovial or Freely Movable Joints

Synovial joints are characterised by the presence of *synovial membrane*.

A considerable amount of movement is possible at all synovial joints and the types are subdivided according to the movements possible. Limitation of movement is due mainly to the shape of the bony surfaces which form the joint.

Ball and socket. These are the most freely movable of all joints. The movements possible are flexion, extension, abduction, adduction, rotation and circumduction. Examples are the shoulder and the hip joints.

Hinge joints. This type permits movement in one plane only. The movements are flexion and extension and examples include the elbow, knee, ankle, the joints between the atlas and the occipital bones and the interphalangeal joints of the fingers and toes.

Gliding joints. In this case the articular surfaces glide over each other. Examples of this type include the sternoclavicular joints, the acromioclavicular joints and the joints between the carpal bones and those between the tarsal bones.

Pivot joints. These joints allow movement round one axis only, that is, a rotatory movement. The classical examples are the proximal and distal radioulnar joints and between the atlas and the odontoid process of the axis.

Condyloid and saddle joints. In these joints the movements take place round two axes thus permitting the movements of flexion, extension, abduction, adduction and circumduction. Examples include the wrist joint, the temporomandibular joint, metacarpophalangeal and metatarsophalangeal joints.

CHARACTERISTICS OF A SYNOVIAL JOINT

All synovial joints have certain characteristics in common and when describing them it is convenient to do so under the following headings.

1. The *type* of synovial joint as defined above.

2. The names of the *parts of the bones* concerned in its formation.

3. *Articular or hyaline cartilage.* The parts of the bones forming the joint are always covered with hyaline cartilage. This tissue provides a smooth articular surface and is strong enough to bear the weight of the body, as is necessary in the ankle joint.

4. *Capsular ligament.* The joint is surrounded and enclosed by a sleeve of fibrous tissue which joins the bones together. It is sufficiently loose to allow for the range of movement of which the joint is capable but strong enough to protect it from injury.

5. *Intracapsular structures.* Some joints have structures within the joint capsule, but outside the synovial membrane, which are necessary to ensure their stability. When these structures do not bear weight they are covered by synovial membrane.

6. *Synovial membrane.* This is composed of secretory epithelial cells which secrete a thick sticky fluid, of the consistency of white of egg called *synovial fluid.* It acts as a lubricant to the joint, provides nutrient materials for the structures within the joint cavity and helps to maintain the stability of the joint. It prevents the ends of the bones from being separated as does a little water between two glass surfaces.

Synovial membrane is found in the joints in well-defined situations:

(*a*) Lining the capsular ligament

(*b*) Covering those parts of the bones within the joint capsule not covered with hyaline cartilage

(*c*) Covering all intracapsular structures which do not bear weight

Little sacs of synovial fluid or *bursae* are found in some joints. Their position is such that they act as cushions to prevent friction between a bone and a ligament or tendon, or the skin where a bone inside a joint capsule is near the surface.

7. *Extracapsular structures.* Most joints have ligaments outside the capsular ligament which strengthen, and lend stability to the joint. They usually blend with the capsular ligament and can only be seen as a thickening of the capsule.

8. *Muscles and movements.* The contraction of muscles is responsible for producing the movements at individual joints and therefore they are best considered together.

9. *Nerve and blood supply.* A useful generalisation is that nerves and blood vessels crossing a joint supply the muscles that move it, and the structures which form the joint.

Main Joints of the Limbs

THE SHOULDER JOINT (Fig. 17:3)

The shoulder joint is a synovial joint of the *ball and socket type.*

The parts of the bones forming the joint are the glenoid cavity of the scapula and the head of the humerus.

Articular cartilage lines the glenoid cavity of the scapula and covers the head of the humerus.

The capsular ligament which surrounds and encloses the joint is very loose inferiorly to allow for the free movement normally possible at this joint.

Intracapsular structures. The glenoid cavity is deepened by a rim of fibrocartilage called the *glenoidal labrum.* By deepening the cavity in this way the joint is made more stable but the range of movement is not reduced. The tendon of the *long head of the biceps muscle*, which lies in the bicipital (intertubercular) groove of the humerus, extends through the joint cavity and is attached to the upper rim of the glenoid cavity. It has an important stabilising effect on the joint.

Synovial membrane lines the capsular ligament and covers that part of the bone within the joint capsule not covered with hyaline cartilage. It forms a sleeve round the part of the tendon of the long head of the biceps muscles which is within the capsular ligament and it covers the glenoidal labrum on both sides. Synovial fluid is secreted by the membrane and lubricates the joint.

Extracapsular structures consist of:

1. *The coracohumeral ligament* extending from the coracoid process of the scapula to the humerus

2. *The transverse humeral ligament* retaining the biceps tendon in the intertubercular groove

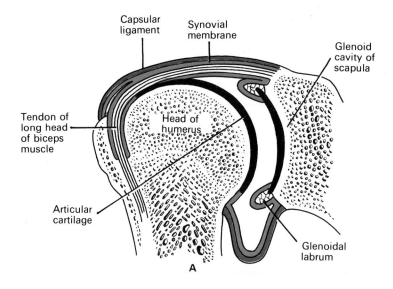

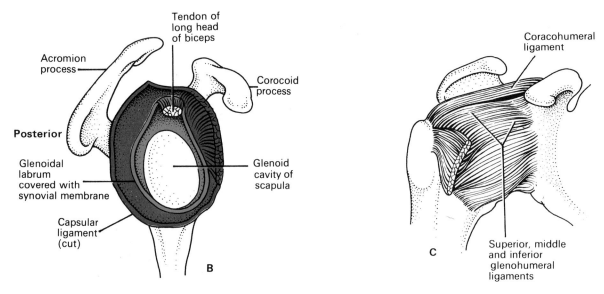

Figure 17:3 The shoulder joint. A. Section viewed from the front. B. The position of glenoidal labrum with the humerus removed, viewed from the side. C. The supporting ligaments viewed from the front.

3. *The tendons* of the muscles which move the joint

MUSCLES AND MOVEMENTS (Fig. 17:6)

Flexion : coracobrachialis, anterior fibres of deltoid and pectoralis major.

Extension : teres major, latissimus dorsi and posterior fibres of deltoid.

Abduction : deltoid.

Adduction : combined action of flexors and extensors.

Medial rotation : pectoralis major, latissimus dorsi, teres major and anterior fibres of deltoid.

Lateral rotation : posterior fibres of deltoid.

Circumduction : flexors, extensors, abductors and adductors acting in series.

Coracobrachialis muscle

This muscle is situated on the upper medial aspect of the arm. It arises from the coracoid process of the scapula, stretches across anterior to the shoulder joint and is inserted into the middle third of the humerus.

Its action is to assist with the movement of flexion of the arm.

Deltoid muscle

The deltoid muscle is triangular in shape and lies directly over the shoulder joint. The fibres originate from the clavicle, the acromion process and spine of the scapula and radiate over the shoulder joint to be inserted into the deltoid tuberosity of the humerus.

This muscle has the ability to act in three separate parts. The anterior fibres assist the pectoralis major in flexing the arm, the middle or main part abducts the arm, and the posterior fibres assist the latissimus dorsi and teres major in extending the arm.

Pectoralis major

This is a broad thick almost fan-shaped muscle which lies on the anterior thoracic wall. The fibres originate from the middle third of the clavicle and from the sternum and are inserted into the lateral lip of the bicipital groove of the humerus.

This muscle draws the arm forward and towards the body, that is, flexes and adducts.

Latissimus dorsi

This is a large triangular-shaped muscle which has its origin in the posterior part of the iliac crest and the spinous processes of the lumbar and lower thoracic vertebrae. It passes obliquely upwards across the back to be inserted by a narrow tendon to the floor of the bicipital groove of the humerus, having passed under the arm. Its action is to adduct the humerus and rotate it medially. It also assists in extension of the arm at the shoulder joint.

Teres major

This muscle originates from the inferior angle of the scapula and is inserted into the humerus just below the shoulder joint. Its action is to draw the humerus backwards and towards the body.

ELBOW JOINT (Fig. 17:4)

The elbow joint is a synovial joint of the *hinge* type.

The parts of the bones concerned are the trochlea and the capitulum of the humerus which articulate with the trochlear notch of the ulna and the head of the radius respectively.

Articular cartilage covers the articular surfaces of the bones.

The capsular ligament acts like a sleeve surrounding and enclosing the joint. It extends from just above the articular surfaces of the humerus to just below the articular surfaces of the radius and ulna.

Intracapsular structures consist of several pads of fat and bursae.

Synovial membrane is found lining the capsular ligament, covering the parts of the bone within the capsule not covered by hyaline cartilage and covering the pads of fat.

Extracapsular structures consist of anterior, posterior, medial and lateral strengthening ligaments.

MUSCLES AND MOVEMENTS (Fig. 17:6)

As the elbow joint is a hinge joint there are only two possible movements.

Flexion : biceps and brachialis.
Extension : triceps.

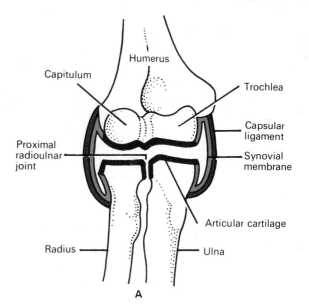

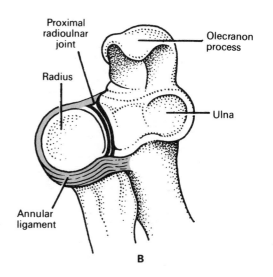

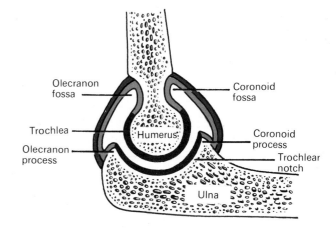

Figure 17:4   The elbow and proximal radioulnar joints. A. Section viewed from the front. B. The proximal radioulnar joint viewed from above. C. Section of the elbow joint, partly flexed, viewed from the side.

Biceps muscle

This is the large muscle lying on the anterior aspect of the upper arm. It is given its name because at its proximal end it is divided and is described as having two heads each of which has its own tendon. The short head rises from the coracoid process of the scapula and passes in front of the shoulder joint down to the arm. The long head originates from the rim of the glenoid cavity and its tendon passes through the joint cavity and the bicipital groove to the arm. The tendon of the long head of the muscle is retained in the bicipital groove of the humerus by a *transverse ligament* which stretches across the groove.

The two parts of the muscle join and are indistinguishable soon after crossing the shoulder joint. The muscle lies on the anterior aspect of the arm and its distal tendon crosses the elbow joint and is inserted into the radial tuberosity. The biceps crosses two joints and acts on both. It helps to stabilise the shoulder joint and at the elbow joint it assists with the movements of flexion and supination.

Brachialis

This muscle lies on the anterior aspect of the upper arm deep to the biceps. It originates from the shaft of the humerus, extends across the elbow joint and is inserted into the ulna just distal to the joint capsule.

It is the main flexor of the elbow joint.

Triceps muscle

This muscle lies on the posterior aspect of the humerus. As the name suggests it arises from three heads, one from the scapula and two from the posterior surface of the humerus. The insertion is by a single tendon to the olecranon process of the ulna.

The most important action of the muscle is to extend the elbow joint.

PROXIMAL AND DISTAL RADIOULNAR JOINTS

Incorporated in the same capsule as the elbow joint is the *proximal radioulnar joint* formed by the rim of the head of the radius rotating in the radial notch of the ulna. The articular surfaces of the bones are covered with hyaline cartilage. The *annular ligament* is a strong extracapsular ligament which encircles the head of the radius and keeps it in contact with the radial notch of the ulna (see Fig. 17:4).

The distal *radioulnar joint* is a synovial pivot joint between the distal end of the radius and the head of the ulna (see Fig. 17:5).

MUSCLES AND MOVEMENTS (Fig. 17:6)

Pronation: pronator teres.
Supination: supinator and biceps.

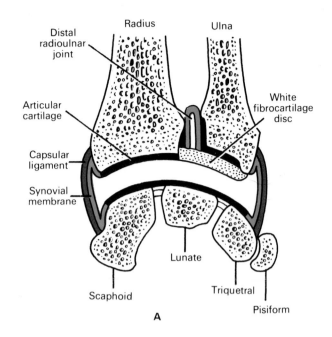

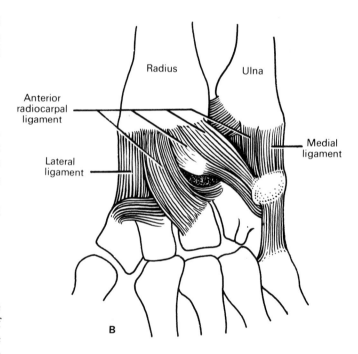

Figure 17:5 The wrist and distal radioulnar joints. Anterior view. A. Section. B. Supporting ligaments.

Pronator teres

This muscle lies obliquely across the front of the forearm. It arises from the medial epicondyle of the humerus and the coronoid process of the ulna, passes obliquely across the forearm to be inserted into the lateral surface of the shaft of the radius.

The action of this muscle is to rotate the radioulnar joints, changing the hand from the anatomical to the writing position, that is, pronating the forearm.

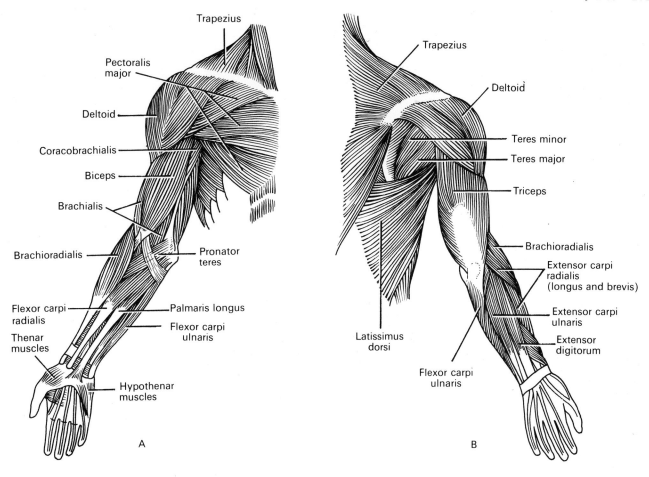

Figure 17:6   The main muscles which move the joints of the upper limb. A. Anterior view. B. Posterior view.

Supinator muscle

This muscle lies obliquely across the posterior and lateral aspects of the forearm. Its fibres arise from the lateral epicondyle of the humerus and the upper part of the ulna and are inserted into the lateral surface of the upper third of the radius.

Its action is to rotate the radioulnar joints, changing the hand from the writing to the anatomical position, i.e. pronation to supination.

WRIST JOINT (Fig. 17:5)

The wrist joint is a synovial joint of the *condyloid type*.

The parts of the bones concerned are the distal end of the radius and the proximal ends of the scaphoid, lunate and triquetral. The distal end of the ulna is separated from the joint cavity by a disc of *white fibrocartilage*, the under surface of which articulates with the carpal bones of the joint. This disc of fibrocartilage also separates the inferior radioulnar joint from the wrist joint.

Articular cartilage covers the articular surfaces of the bones.

Capsular ligament surrounds and encloses the joint.

Intracapsular structures. There are none of significance in this joint.

Synovial membrane is found lining the capsular ligament and covering the parts of the bones within the capsule not covered by hyaline cartilage.

Extracapsular structures consist of medial and lateral ligaments and anterior and posterior radiocarpal ligaments.

MUSCLES AND MOVEMENTS (Fig. 17:6)

The wrist joint is capable of a wide range of movement:

Flexion : flexor carpi radialis and the flexor carpi ulnaris.

Extension : extensors carpi radialis (longus and brevis) and the extensor carpi ulnaris.

Abduction : flexor and extensors carpi radialis.

Adduction : flexor and extensor carpi ulnaris.

Flexor carpi radialis

This muscle originates from the medial epicondyle of the humerus, lies on the anterior surface of the forearm and is inserted into the second and third metacarpal bones.

It is one of the flexor muscles of wrist joint and when it acts with the extensor carpi radialis it abducts the joint.

Flexor carpi ulnaris

This muscle lies on the medial aspect of the forearm. It originates from the medial epicondyle of the humerus and the upper parts of the ulna and is inserted into the pisiform, the hamate and the fifth metacarpal bones.

When contracted with the flexor carpi radialis it flexes the wrist, and when it acts with the extensor carpi ulnaris it adducts the joint.

Extensor carpi radialis longus and brevis

These muscles lie on the posterior aspect of the forearm. The fibres originate from the lateral epicondyle of the humerus and are inserted by a long tendon into the second and third metacarpal bones.

They are associated with extension and abduction of the wrist joint.

Extensor carpi ulnaris

This muscle, like the other extensor muscle, lies on the posterior surface of the forearm. It originates from the lateral epicondyle of the humerus and is inserted into the fifth metacarpal bone.

It is associated with the movements of extension and adduction of the wrist.

JOINTS OF THE HANDS AND FINGERS

There are synovial joints between the carpal bones, between the carpal and metacarpal bones, between the metacarpal bones and proximal phalanges and between the phalanges. The powerful movements which take place at these joints are produced by muscles in the forearm which have tendons extending into the hand. These tendons are encased in a sleeve of synovial membrane and they are held close to the wrist bones by strong ligaments. The tendons move smoothly within the synovial sheath as the joints move. Many of the finer movements of the fingers are produced by numerous small muscles in the hand.

HIP JOINT (Fig. 17:7)

The hip joint is a synovial joint of the *ball and socket type*.

The parts of the bones concerned are the cup-shaped acetabulum of the innominate bone and the almost spherical head of the femur.

Articular cartilage lines the acetabulum and covers the head of the femur.

The capsular ligament is a thick strong sleeve of fibrous tissue enclosing the joint. It extends from just above the rim of the acetabulum to about half-way along the neck of the femur.

Intracapsular structures. The acetabular labrum, like the glenoidal labrum of the shoulder joint, is a ring of fibrocartilage attached to the rim of the acetabulum

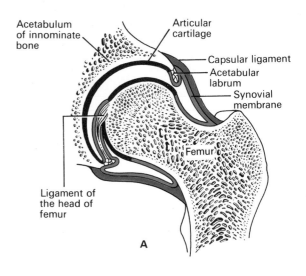

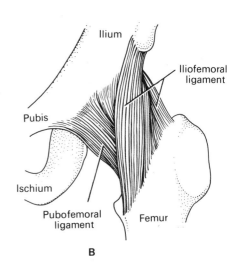

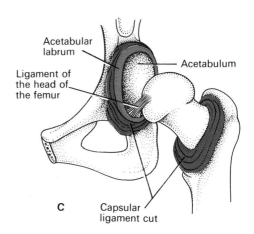

Figure 17:7 The hip joint. Anterior view. A. Section. B. Supporting ligaments. C. Head of femur and acetabulum separated to show acetabular labrum and ligament of head of femur.

which deepens the cavity, lends stability to the joint but does not limit its range of movement. The *ligament of the head of the femur* extends from the shallow depression in the middle of the head of the femur to the lower edge of the acetabulum. It conveys a blood vessel to the head of the femur.

Synovial membrane is found lining the capsular ligament and covering the parts of the bone within the joint cavity which are not covered with articular cartilage. It covers both sides of the acetabular labrum and forms a sleeve around the ligament of the head of the femur.

Extracapsular structures. There are three important ligaments surrounding and giving strength to the joint:
1. The iliofemoral ligament is an inverted Y-shaped ligament lying anteriorly.
2. The ischiofemoral ligament lies posteriorly.
3. The pubofemoral ligament lies inferiorly.

MUSCLES AND MOVEMENTS (Figs. 17:8 and 17:9)

Flexion: psoas, illiacus, rectus femoris and sartorius.
Extension: gluteus maximus and the hamstrings.
Abduction: gluteus medius and minimus, sartorius and others.
Adduction: adductor group.
Lateral rotation: mainly gluteal muscles and adductor group.
Medial rotation: gluteus medius and minimus and others.

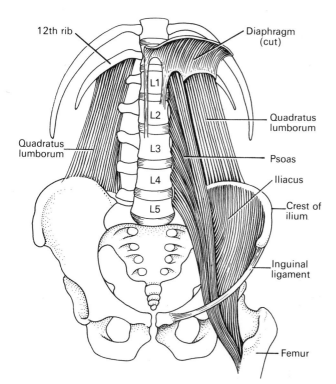

Figure 17:8   The muscles of the posterior abdominal wall and pelvis which flex the hip joint.

Psoas muscle

This muscle arises from the transverse processes and the bodies of the lumbar vertebrae. It passes across the bones of the greater pelvis and behind the inguinal ligament to be inserted into the lesser trochanter of the femur.

Its action together with the iliacus is to cause flexion at the hip joint.

Iliacus muscle

This is a triangular-shaped muscle which lies in the iliac fossa of the innominate bone. It originates from the inner aspect of the iliac crest, passes over the iliac fossa and joins the tendon of the psoas muscle to be inserted into the femur.

The combined action of iliacus and psoas causes flexion at the hip joint.

Quadriceps femoris

This is a group of four muscles which lie on the front of the thigh. They are the *rectus femoris* and *three vasti muscles*. The rectus femoris originates from the ilium and the three vasti from the upper end of the femur. Together they pass over the front of the knee joint to be inserted into the tuberosity of the tibia by the patellar tendon.

Only the rectus femoris plays a part in flexing the hip joint. Together the group acts as a very strong extensor of the knee joint.

Gluteal muscles

This group consists of the gluteus maximus, medius and minimus.

Gluteus maximus forms the fleshy part of the buttock. It originates from the outer surface of the iliac bone and the sacrum and is inserted into the upper end of the femur. It is the muscle chiefly concerned with the maintenance of the erect position by extension of the hip joint.

Gluteus medius lies deep to gluteus maximus. It originates from the outer surface of the ilium and is inserted into the greater trochanter of the femur. It is associated with abduction and medial rotation of the thigh.

Gluteus minimus is the most deeply situated of the three muscles with its origin and insertion similar to those of gluteus medius. Its action is to abduct and medially rotate the thigh.

Sartorius

This is the longest individual muscle in the body. It originates from the anterior superior iliac spine and passes obliquely across the thigh to be inserted into the medial surface of the upper part of the tibia.

It is associated with the actions of flexion and abduction at the hip joint and flexion at the knee.

Adductor group

This group lies on the medial aspect of the thigh. The

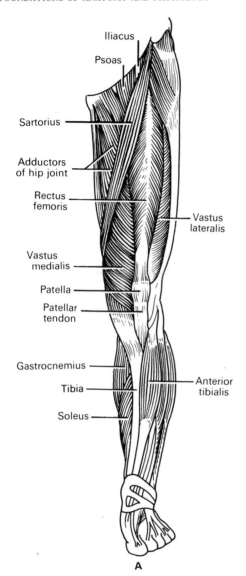

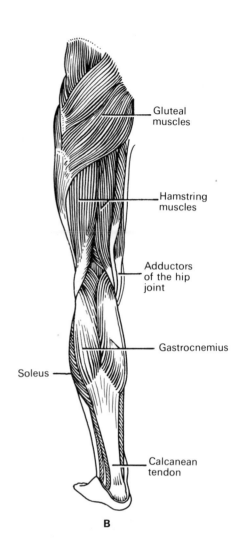

Figure 17:9 The main muscles of the lower limb. A. Anterior view. B. Posterior view.

muscles making up the group all originate from the pubic bone and are inserted into the linea aspera of the femur.

As the name suggests this group adducts the thigh.

KNEE JOINT (Fig. 17:10)

The knee joint is a synovial joint of the *hinge type*.

The parts of the bones concerned are the condyles of the femur, the condyles of the tibia and the posterior surface of the patella.

Articular cartilage covers the articular surfaces of the three bones.

The capsular ligament encloses the joint posteriorly, medially and laterally. The anterior part consists of the tendon of the quadriceps femoris muscle which at the same time supports the patella.

Intracapsular structures. These include:

1. *Cruciate ligaments.* There are two cruciate ligaments which cross each other extending from the inter-condylar notch of the femur to the intercondylar eminence of the tibia. They have an important stabilising effect on the joint.

2. *Semilunar cartilages or menisci.* These are incomplete discs of white fibrocartilage which lie on top of the articular condyles of the tibia. They are wedge-shaped, being thicker at their outer edges. They have a stabilising effect on the joint by preventing lateral displacement of the bones.

3. *Bursae and pads of fat.* Numerous bursae and pads of fat are to be found within the capsule of this joint. They are situated so that they prevent friction between a bone and a ligament or tendon and between the skin and the patella.

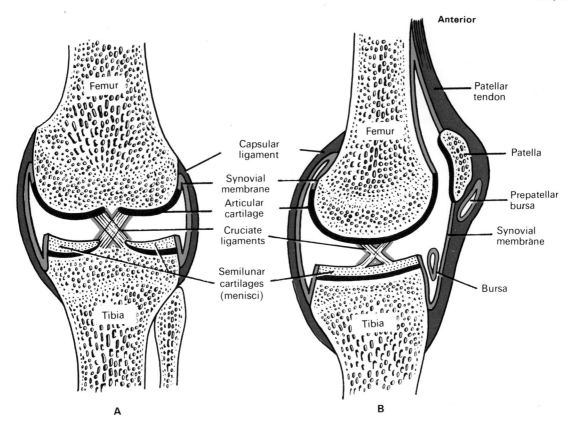

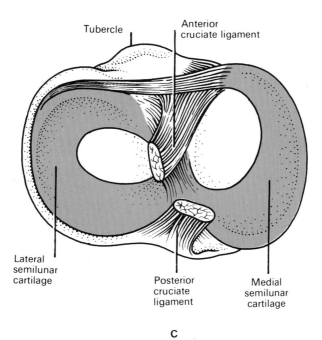

Figure 17:10 The knee joint. A. Section viewed from the front. B. Section viewed from the side. C. The superior surface of the tibia showing the semilunar cartilages and the cruciate ligaments.

Synovial membrane. This membrane lines the capsular ligament and the inner surface of the patellar tendon and covers the parts of the bone within the capsule which are not covered with hyaline cartilage. It covers the cruciate ligaments and the pads of fat. Only the outer edges of the menisci are covered with synovial membrane. It is not found on the weight-bearing surface.

Extracapsular structures. The two most important strengthening ligaments of the knee joint are the medial and lateral ligaments.

MUSCLES AND MOVEMENTS (Fig. 17:9)

The main movements are flexion and extension. There is a slight rotatory movement which '*locks*' the joint when it is fully extended. This is important in relation to the maintenance of balance because when the joint is locked balance is maintained with less muscular effort than when it is slightly flexed.

Flexion (bending backwards): gastrocnemius and hamstrings.

Extension (straightening): quadriceps femoris muscle.

Hamstring muscles

This group of muscles lies on the posterior aspect of the thigh. All the muscles making up the group originate from the ischium and are inserted into the upper end of the tibia. They are the *biceps femoris, semimembranosus* and *semitendinosus muscles.*

Their chief action is to flex the knee joint.

Gastrocnemius

This is the muscle which forms the bulk of the calf of the leg. It arises by two heads, one from each condyle of the femur, and passes down behind the tibia to be inserted into the calcaneus by the *calcanean tendon* (*Achilles* tendon).

This muscle crosses the knee joint and the ankle joint. It flexes the knee and is a very powerful *plantarflexor* of the ankle. It is used extensively in walking, running and jumping.

Quadriceps femoris

This group has already been described in relation to the hip joint. It is the only muscle involved in extension of the knee joint.

ANKLE JOINT (Fig. 17:11)

The ankle joint is a synovial joint of the *hinge type*.

The parts of the bones concerned are the distal end of the tibia and its malleolus (medial malleolus), the distal end of the fibula (lateral malleolus) and the talus.

Articular cartilage covers the articular surfaces of the bones.

The capsular ligament surrounds and encloses the joint.

Intracapsular structures. There are no structures of significance in this joint.

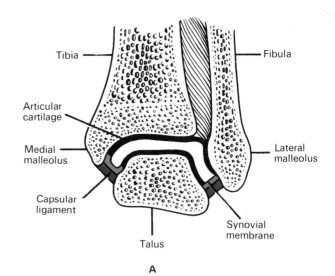

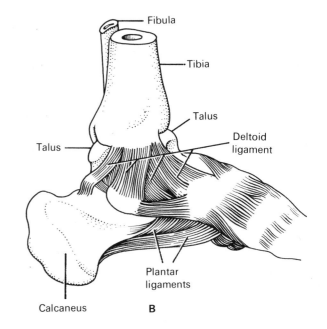

Figure 17:11 The left ankle joint. A. Section viewed from the front. B. Supporting ligaments. Medial view.

Synovial membrane is found lining the capsular ligament and covering the parts of the bone within the joint capsule not covered with articular cartilage.

Extracapsular structures. There are four important ligaments strengthening this joint. They are the anterior, posterior and lateral ligaments and a very strong medial ligament called the deltoid ligament.

MUSCLES AND MOVEMENTS (Fig. 17:9)

Special names are given to the movements of this joint:

Flexion or dorsiflexion: anterior tibialis assisted by the muscles which extend the toes.

Extension or plantarflexion: gastrocnemius and soleus assisted by the muscles which flex the toes.

The movements of *inversion* and *eversion* occur between the tarsal bones and not at the ankle joint.

Anterior tibialis muscle

This muscle originates from the upper end of the tibia and passes down the anterior surface of the leg to be inserted into the middle cuneiform bone by a long tendon.

It is associated with dorsiflexion of the foot.

Soleus

This is one of the main muscles of the calf of the leg lying immediately deep to the gastrocnemius. It originates from the head and upper part of the fibula and the tibia. Its tendon joins that of the gastrocnemius so that they have a common insertion into the calcaneus by the calcanean tendon.

This muscle is chiefly concerned with plantarflexion at the ankle joint and helps to stabilise the joint when the individual is standing up.

Gastrocnemius

This muscle was described previously in relation to the movements of the knee joint. It is a powerful plantarflexor.

JOINTS OF THE FOOT AND TOES

There is a number of synovial joints between the tarsal bones, between the tarsal and metatarsal bones, between the metatarsals and proximal phalanges and between the phalanges. The movements at these joints are produced by muscles in the leg with long tendons which cross the ankle joint and by muscles of the foot. The tendons which cross the ankle joint are encased in synovial sheaths and are held close to the bones by strong ligaments. They move smoothly within their sheaths as the joints move.

Movement at joints in the foot occur during walking, running and jumping. They also have important functions in the maintenance of balance and supporting the arches of the foot.

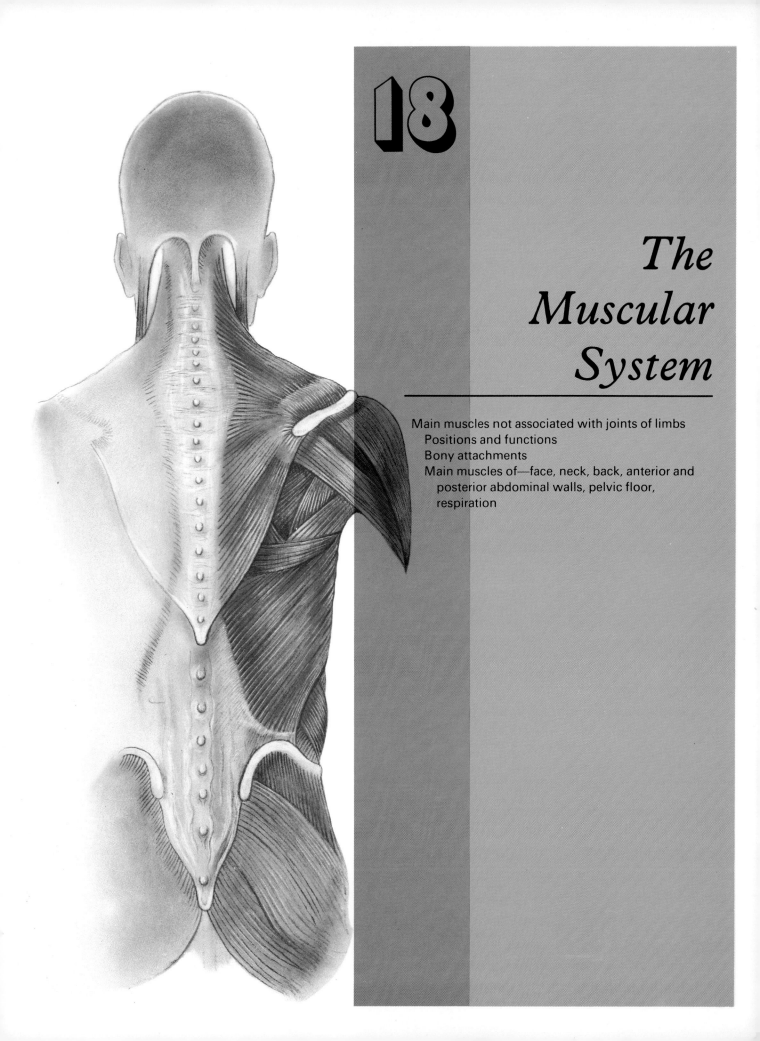

The Muscular System

Main muscles not associated with joints of limbs
  Positions and functions
  Bony attachments
  Main muscles of—face, neck, back, anterior and
    posterior abdominal walls, pelvic floor,
    respiration

18. The Muscular System

The skeletal muscles described in this chapter are those which are not directly involved in the movements of the joints of the limbs.

Muscles of the Face and the Neck (Fig. 18:1)

There is a large number of muscles concerned with alteration of facial expression and with the movements of the lower jaw during chewing and speaking. Only the main muscles concerned will be considered here. Except where indicated the muscles described in this section are present in pairs, one on each side.

Occipitofrontalis (unpaired)
This consists of a posterior muscular part which lies over the occipital bone, an anterior part over the frontal bone and an extensive flat tendon or *aponeurosis* which stretches over the dome of the skull and joins the two muscular parts.

The action of this muscle is to raise the eyebrows.

Levator palpebrae superioris
This muscle extends from the posterior part of the orbital cavity to the upper eyelid.

The action of this muscle is to raise the eyelid.

Orbicularis oculi
This muscle surrounds the eye, the eyelid and the area immediately round the orbital cavity; it closes the eye and when strongly contracted 'screws up' the eyes.

Buccinator
This is the flat muscle of the cheek. Its action is to draw the cheeks in towards the teeth in chewing and to expel

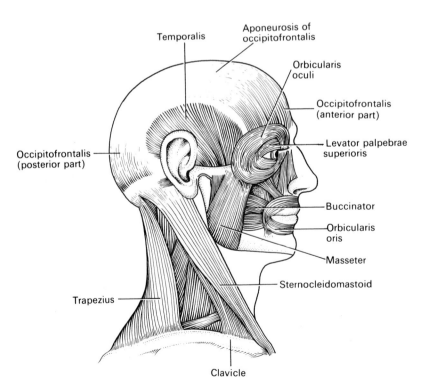

Figure 18:1   The main muscles on the right side of the face, head and neck.

280

air from the mouth. It is sometimes described as 'the trumpeter's muscle'.

Orbicularis oris (unpaired)
This is a circular muscle which surrounds the mouth and blends with the muscles of the cheeks.

Its action is to close the lips and, when strongly contracted, shapes the mouth for whistling.

Masseter
This is a broad muscle which extends from the zygomatic arch to the angle of the jaw. In chewing it draws the mandible up to the maxilla and exerts considerable pressure on the food.

Temporalis
This muscle is shaped like a fan, covering the squamous portion of the temporal bone. It passes behind the zygomatic arch to be inserted into the coronoid process of the mandible. Its action is to assist in closing the mouth.

Pterygoid muscle
This muscle extends from the greater wing of the sphenoid bone to the mandible. Its action is to close the mouth and pull the lower jaw forward.

There are a great many muscles situated in the neck but only two of the large ones are considered here:
2 sternocleidomastoid muscles
2 trapezius muscles

Sternocleidomastoid muscle
This muscle arises from the manubrium of the sternum and the clavicle and extends upwards to the mastoid process of the temporal bone. It assists in turning the head from side to side.

When the muscle on one side contracts it draws the head towards the shoulder.

When both muscles contract together they:

1. Flex the cervical vertebrae.
2. Draw the sternum and clavicle upwards when the head is maintained in a fixed position. This is a less common movement which occurs mainly in forced respiration.

Trapezius muscle
This muscle covers the shoulder and the upper part of the back of the neck. The upper attachment is to the occipital protuberance, the medial attachment is to the transverse processes of the cervical and thoracic vertebrae and the lateral attachment is to the lateral part of the clavicle and to the spinous and acromion processes of the scapula.

The actions of this muscle include:

1. Pulling the head backwards

2. Squaring the shoulders and controlling the movements of the scapula when the shoulder joint is in use

Muscles of the Back (Fig. 18:2)

There are six pairs of large muscles in the back in addition to those which form the posterior part of the abdominal wall. The arrangement of these muscles is the same on each side of the vertebral column.
2 trapezius
2 teres major
2 psoas
2 sacrospinalis
2 latissimus dorsi
2 quadratus lumborum

Teres major and latissimus dorsi
These muscles were described previously in relation to the movements of the shoulder joint which they extend.

Psoas muscle
This muscle was described in relation to the movement at the hip joint. It is one of the main flexors.

Quadratus lumborum
This muscle originates from the posterior part of the crest of the ilium then it passes upwards, parallel and close to the vertebral column and is inserted into the 12th rib.

Together the two muscles have the action of fixing the lower rib during respiration and causing backward flexion or extension of the vertebral column. If one muscle contracts it will cause lateral flexion of the lumbar region of the vertebral column.

Sacrospinalis muscle
This is the name given to a group of muscles which lie between the spinous and transverse processes of the vertebrae. They originate from the sacrum and are finally inserted into the occipital bone.

The main action of this group is to produce backward flexion or extension of the vertebral column.

Muscles of the Anterior Abdominal Wall (Figs. 18:3 and 18:4)

There are four layers of muscle which make up the anterior abdominal wall:
2 rectus abdominis
2 external oblique
2 internal oblique
2 transversus abdominis

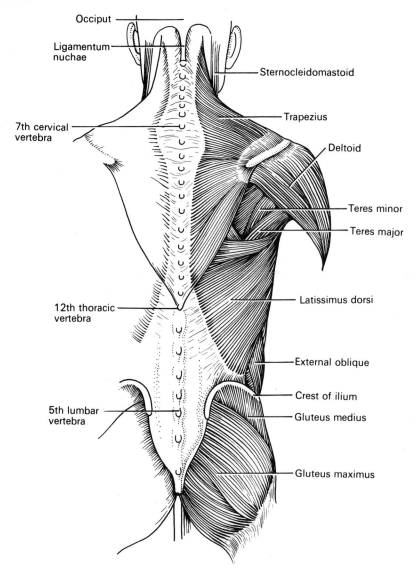

Occiput

Ligamentum
nuchae

Sternocleidomastoid

Trapezius

7th cervical
vertebra

Deltoid

Teres minor

Teres major

Latissimus dorsi

12th thoracic
vertebra

External oblique

Crest of ilium

5th lumbar
vertebra

Gluteus medius

Gluteus maximus

Figure 18:2   The main muscles of the back.

The abdominal wall is divided longitudinally into two equal parts by a very strong tendinous cord called the *linea alba* which extends from the xiphoid process of the sternum to the symphysis pubis. The structure of the abdominal wall on each side of the linea alba is identical.

Rectus abdominis

This is the most superficial of the four muscles. It is a broad flat muscle originating from the transverse part of the pubic bone then passing upwards to be inserted into the lower ribs and the xiphoid process of the sternum. Medially the two muscles are attached to and separated by the linea alba.

External oblique

This muscle extends from the lower ribs *downwards and*

forward to be inserted into the iliac crest and, by an aponeurosis, to the linea alba.

Internal oblique

This muscle lies deep to the external oblique muscle. Its fibres arise from the crest of the ilium and by a broad band of fascia from the spinous processes of the lumbar vertebrae. The fibres pass *upwards towards the midline* to be inserted into the lower ribs and, by an aponeurosis, to the linea alba. The fibres of this muscle are at right angles to those of the external oblique.

Transversus abdominis

This is the deepest layer of muscle tissue of the abdominal wall. The fibres arise from the iliac crest and the lumbar vertebrae and pass across the abdominal wall to be

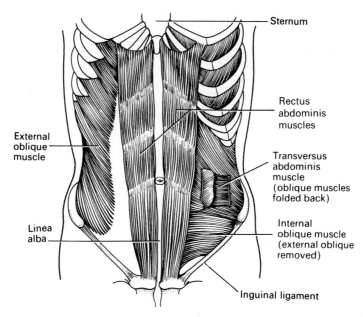

Figure 18:3 The muscles of the anterior abdominal wall.

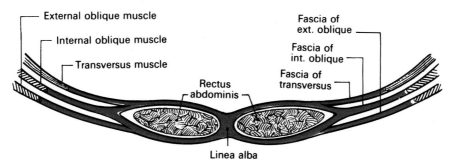

Figure 18:4 Diagram of the arrangement of the fascia of the muscles of the anterior abdominal wall. Cross section.

inserted into the linea alba by an aponeurosis. The fibres of this muscle are at right angles to those of the rectus abdominis.

FUNCTIONS

The main function of the four muscles is to form a strong muscular anterior wall to the abdominal cavity. When the muscles contract together they have the effect of:

1. Compressing the abdominal organs
2. Flexing the vertebral column in the lumbar region

Contraction of the muscles on one side only, bends the trunk towards that side. Contraction of the oblique muscles on one side, rotates the trunk.

THE INGUINAL CANAL

This is a canal 2·5 to 4 cm long which passes obliquely through the abdominal wall. It runs parallel to and is immediately in front of the transversalis fascia and part of the inguinal ligament. In the male it contains the *spermatic cord* and in the female, the *round ligament*. This constitutes a weak point in the otherwise strong abdominal wall, but the fact that the canal runs obliquely through the wall does, to some extent, reduce the weakness.

Muscles of the Posterior Abdominal Wall

The muscles of the posterior part of the abdominal wall have been described already (see Figs. 18:5 and 18:6). They are:

Quadratus lumborum
Psoas
Internal oblique
Transversus abdominis

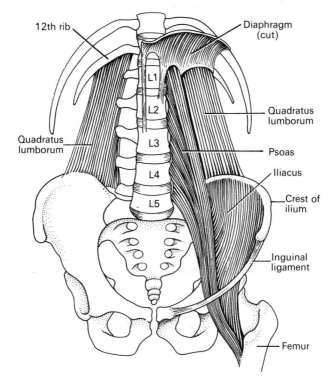

Figure 18:5   The deep muscles of the posterior abdominal wall.

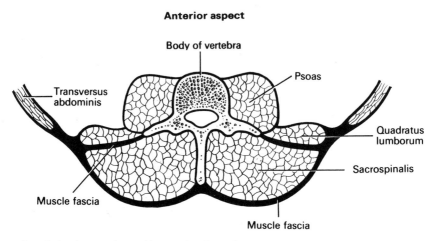

Figure 18:6   Transverse section of a lumbar vertebra and its associated muscles.

Muscles of the Pelvic Floor (Fig. 18:7)

The pelvic floor is divided into two identical parts at the midline. Each half is made up of muscles and fascia. The muscles are:

    Levator ani

    Coccygeus

Levator anus

This is a broad flat muscle which forms the anterior part of the pelvic floor. It originates from the inner surface of the lesser pelvis and unites in the midline with the muscle from the opposite side. Together they form a sling which supports the organs of the pelvic cavity.

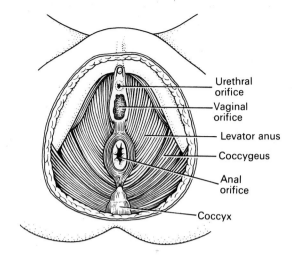

Figure 18:7   The muscles of the pelvic floor.

Labels on figure:
- Urethral orifice
- Vaginal orifice
- Levator anus
- Coccygeus
- Anal orifice
- Coccyx

Coccygeus

This muscle is triangular in shape and is situated behind the levator ani. It originates from the medial surface of the ischium and is inserted into the sacrum and coccyx.

These two muscles complete the formation of the pelvic floor, which is perforated in the male, by the urethra and the anus and in the female, by the urethra, the vagina and the anus.

Muscles of Respiration

The muscles of respiration have been described previously (Ch. 7). They are:

External intercostal	11 pairs
Internal intercostal	11 pairs
Diaphragm	1

Index

Each page number listed in the index represents a separate mention of an item in the text. Where an item continues on the following page(s) only the first page number is given.